Biomolecules in Organic Solvents

Editor

Armando Gómez-Puyou, M.D., Ph.D.
Institute of Cellular Physiology
National Autonomous University of México
México, D.F., México

Associate Editors

Alberto Darszon, Ph.D.
Center for Research on Genetic Engineering and Biotechnology
National Autonomous University of México
Cuernavaca, Morelos, México

Marietta Tuena de Gómez-Puyou, M.D., Ph.D.
Institute of Cellular Physiology
National Autonomous University of México
México, D.F., México

CRC Press
Boca Raton   Ann Arbor   London   Tokyo

Library of Congress Cataloging-in-Publication Data

Biomolecules in organic solvents / editor, Armando Gómez-Puyou ;
   associate editors, Alberto Darszon, M. Tuena de Gómez-Puyou.
      p. cm.
    Includes bibliographical references and index.
    ISBN 0-8493-4823-4
     1. Enzymes—Biotechnology. 2. Organic solvents—Biotechnology.
   3. Reversed micelles. 4. Enzymes—Solubility. I. Gómez-Puyou,
Armando. II. Darszon, Alberto. III. Tuena de Gómez-Puyou, M.
   TP248.65.E59B58   1992
   661′.807—dc20   91-27051
                                                     CIP

Direct all inquiries to CRC Press, Inc., 2000 Corporate Blvd., N.W., Boca Raton, Florida 33431.

International Standard Book Number 0-8493-4823-4

Library of Congress Card Number 91-27051

Printed in the United States of America 1 2 3 4 5 6 7 8 9 0

FOREWORD

The observation that enzymes could work in organic solvents was an exciting and surprising finding. The discovery led to novel perspectives in biotechnology and gave rise to basic research that attempted to establish how enzymes could function under such conditions. These two intimately related challenges attracted scientists from different disciplines and, as a result, the last years have witnessed a remarkable increase in the number of reports that deal with the behavior of enzymes in such unnatural conditions. Nevertheless, there are still many questions yet to be answered; these questions deal with the structure of biomolecules, enzyme kinetics, and the forces that operate in systems that are predominantly formed with organic solvents. Also, there is the question as to level of biological complexity that can be functionally operative in these nonconventional systems.

All these questions are of importance simply because biological function has been observed in organic solvents. However, there is perhaps a more fundamental question: When experimentalists detect enzyme function in systems principally formed with organic solvents, are they observing the action of an enzyme in a microenvironment distinct from water? The available data indicate that in all the described nonconventional systems, enzymes must be in contact with water in order to express catalytic activity. Indeed, water seems to be an indispensable ingredient of enzyme function. Thus, these observations give rise to a paradox, i.e., in order to take full advantage of the recently acquired organic solvent systems, one must learn more of water and its interactions with biomolecules.

In the mind of many researchers, there is also another question: How far will the field of biomolecules in organic solvents go? Some believe that there are (or were) too many expectations. They may be right, but it is a fact that enzyme function in organic solvents can take place with monomeric and polymeric enzymes, with soluble and integral membrane proteins, with macromolecular complexes, and with whole cells. Magnificent constructions stated with a stone; from a brief review of what is available, it would seem that at the moment there is more than a stone.

THE EDITOR

Armando Gómez-Puyou, M.D., Ph.D. is a researcher at the Institute of Cellular Physiology of the National Autonomous University of México. Dr. Gómez-Puyou obtained his M.D. in 1961 and his Ph.D. in Biochemistry in 1976, both from the University of México.

Dr. Gómez-Puyou is a member of the Mexican Society of Biochemistry, the American Society of Biological Chemists, the Third World Academy of Sciences, and is founding Member of the Latin American Academy of Sciences. He was a Guggenheim Fellow at the Medical School of Johns Hopkins University and spent a sabbatical year at the Arrhenius Laboratory of the University of Stockholm. He has presented over 20 invited lectures in national and international meetings and has published close to 100 papers. He has received grants from Consejo Nacional de Ciencia y Tecnología (México) and from the Organization of American States. His current interests include the mechanisms of energy transduction in biological membranes and the behavior of enzymes in low water systems.

ASSOCIATE EDITORS

Dr. Alberto Darszon, Ph.D., is a researcher at the Center for Research on Genetic Engineering and Biotechnology of the National Autonomous University of México in Cuernavaca, México. Dr. Darszon received a Ph.D. in Biochemistry in 1977 from the Center for Research and Advanced Studies in México City. He did postdoctoral work in biophysics at the University of California in San Diego and joined the Faculty of the Center for Research and Advanced Studies in 1979. In 1991, he moved to his current position. In 1985, Dr. Darszon received the National Science Award from the Mexican Academy for Scientific Research, and in 1989 he received the Award in Biomedicine from the Miguel Aleman Foundation. Other honors include Fellowships from the John Simon Guggenheim Foundation and the World Health Organization. He has published over 50 papers and reviews; his main interests are the mechanism of ion channels in biological membranes and the properties of biolomolecules in reverse micelle systems.

Dr. Marietta Tuena de Gómez-Puyou, M.D., Ph.D., is a full-time Researcher at the Institute of Cellular Physiology of the National Autonomous University of México. She is a member of the Mexican Society of Biochemistry and of the National Research System of México; she is also a member of the Board of the International Union of Women Bioscientists. Dr. Tuena de Gómez-Puyou received the award in Natural Sciences from the National University of México in 1989. She was Professor of Biochemistry at the Faculty of Medicine of the University of México from 1963 to 1982; at that time she moved to the Institute of Cellular Physiology. She has been in the Organizing Committee of two international advanced courses in biomembranes, which were sponsored by UNESCO and the World Health Organization. She has spent sabbatical leaves at the Medical School of Johns Hopkins University and the Arrhenius Laboratory in Stockholm. Her research has been supported by grants from the Consejo Nacional de Ciencia y Tecnología. She has published more than 90 papers, mostly on bioenergetics; in the last years she has also been studying the behavior of biological systems in low-water media.

CONTRIBUTORS

Guadalupe Ayala, Ph.D.
Department of Biochemistry
Faculty of Medicine
National Autonomous University of
  México
México, D.F., México

Roque Bru, Ph.D.
Department of Biochemistry
Faculty of Biology
University of Murcia
Murcia, Spain

Sudipta Chatterjee
Department of Chemical Engineering
University of Pittsburgh
Pittsburgh, Pennsylvania

Martha Lucinda Contreras-Zentella
Department of Microbiology
Institute of Cellular Physiology
National Autonomous University of
  México
México, D.F., México

Alberto Darszon, Ph.D.
Department of Biochemistry
Center for Research on Genetic
  Engineering and Biotechnology
National Autonomous University of
  México
Cuernavaca, Morelos, México
and Department of Biochemistry
Center for Research and Advanced
  Studies
National Polytechnic Institute
México, D.F., México

J. Edgardo Escamilla, Ph.D.
Department of Microbiology
Institute of Cellular Physiology
National Autonomous University of
  México
México, D.F., México

Maria Famiglietti
Institute for Polymers
ETH-Zentrum
Zurich, Switzerland

Janos H. Fendler, Ph.D., D.Sc.
Department of Chemistry
Syracuse University
Syracuse, New York

D. Alejandro Fernandez-Velasco
Department of Bioenergetics
Institute of Cellular Physiology
National Autonomous University of
  México
México, D.F., México

Sergio T. Ferreira, Ph.D.
Department of Biochemistry
Institute of Biomedical Sciences
Federal University of Rio de Janeiro
Rio de Janeiro, Brazil

John L. Finney, Ph.D.
ISIS Facility
Rutherford Appleton Laboratory
Oxon, England

Francisco García-Carmona, Ph.D.
Department of Biochemistry
Faculty of Biology
University of Murcia
Murcia, Spain

Georgina Garza-Ramos, M.Sc.
Department of Bioenergetics
Institute of Cellular Physiology
National Autonomous University of
  México
México, D.F., México

Armando Gómez-Puyou, M.D., Ph.D.
Institute of Cellular Physiology
National Autonomous University of
  México
México, D.F., México

**M. Tuena de Gómez-Puyou, M.D.,
  Ph.D.**
Department of Bioenergetics
Institute of Cellular Physiology
National Autonomous University of
  México
México, D.F., México

James G. Goodwin, Jr.
Department of Chemical Engineering
University of Pittsburgh
Pittsburgh, Pennsylvania

Enrico Gratton
Laboratory of Fluorescence Dynamics
Department of Physics
University of Illinois
Urbana, Illinois

Alejandro Hochköppler, Ph.D.
Institute for Polymers
ETH-Zentrum
Zürich, Switzerland

Riet Hilhorst, Ph.D.
Department of Biochemistry
Agricultural University
Wageningen, The Netherlands

Pier Luigi Luisi
Institute for Polymers
ETH-Zentrum
Zürich, Switzerland

Iván Ortega-Blake, Ph.D.
Cuernavaca Laboratory
Institute of Physics
National Autonomous University of
  México
Cuernavaca, Morelos, México

Laura Escobar Pérez, Ph.D.
Unidad de Investigacion Biomedica
Centro Médico Nacional
Instituto Mexicano del Seguro Social
México, D.F., México

Nestor Pfammatter
Institute of Polymers
ETH-Zentrum
Zürich, Switzerland

Igor Rapanovich
Department of Chemical Engineering
University of Pittsburgh
Pittsburgh, Pennsylvania

Alan J. Russell, Ph.D.
Department of Chemical Engineering
University of Pittsburgh
Pittsburgh, Pennsylvania

Humberto Saint-Martin
Cuernavaca Laboratory
Institute of Physics
National Autonomous University of
  México
Cuernavaca, Morelos, México

Alvaro Sánchez-Ferrer, Ph.D.
Department of Biochemistry
Faculty of Biology
University of Murcia
Murcia, Spain

Hugh F. J. Savage, Ph.D.
Chemistry Department
University of York
York, England

Liora Shoshani, M.Sc.
Department of Biochemistry
Center for Research and Advanced
  Studies
National Polytechnic Institute
México, D.F., México

Cees Veeger, Ph.D.
Department of Biochemistry
Agricultural University
Wageningen, The Netherlands

Raymond M. D. Verhaert, Ph.D.
Department of Biochemistry
Agricultural University
Wageningen, The Netherlands

Antonie J. W. G. Visser, Ph.D.
Department of Biochemistry
Agricultural University
Wageningen, The Netherlands

TABLE OF CONTENTS

Chapter 1

NEW APPROACHES IN STUDYING BIOMOLECULE-WATER INTERACTIONS

Hugh F. J. Savage and John L. Finney

TABLE OF CONTENTS

I. INTRODUCTION

Water is one of the main solvent environments in which proteins are located and function; others are lipids, high salts, or carbohydrate. However, neither the stabilization of water-protein systems nor their interactions, e.g., with ligands such as other protein molecules, coenzymes, or substrates, are well understood despite a great deal of research over the last 30 to 40 years. The free energy balance for most proteins — i.e., the difference between the free energies of the folded and unfolded states — is marginal ($\Delta G \sim 10$ to 20 kcal/mol) and the contributions of the protein and water interactions are difficult to assess with accuracy. A major reason for this difficulty is that such systems are multicomponent, and the inherent complexity prevents one from isolating a small part of the system without perturbing other parts. Hence, no single experiment is able to given an unambiguous interpretation. In addition, many of the techniques that have been used to study water-protein interactions are indirect; interpretation of such experiments requires the use of models, the nature of which may strongly influence the conclusions drawn.

The role of water in biological systems including proteins has been reviewed many times, and the interested reader is referred to these reviews for comprehensive coverage of the field.[1-5] In recent years, however, the development of experimental techniques have allowed the use of direct methods — those which can give information without involving possibly erroneous interpretive models — to tackle various aspects of this complex but highly relevant area. Therefore, we have chosen here to be very selective and to concentrate on recent work using direct techniques which, in our view, has both led to significant progress in our understanding of water-biomolecule interactions and promises further advances in the future. The techniques we stress are neutron and X-ray diffraction, where the advent of intense radiation sources coupled with advances in instrumentation have allowed questions to be answered that previously could not be sensibly addressed. This work relates both to "model systems" in which particular aspects are isolated (e.g., hydrogen bonding, water structure at interfaces, and the hydrophobic interaction) and the macromolecular systems themselves.

In the following section, we set the scene with a discussion of the role of molecular-level solvation effects in typical biomolecular processes such as protein folding. This illustrates the complexity of the many effects that contribute to the small free energy difference between the native and unfolded states. We also stress the importance of understanding the intermolecular forces involving hydration and the relevance of structural and dynamic perturbations of solvent molecules close to relevant (polar, charged, and apolar) chemical groups.

Section III describes the use of diffraction methods to obtain direct information on solvation in single crystals. Emphasis is placed on some of the problems of analyzing water-water interactions and deriving nonbonded geometrical restraints between water molecules that can be used to examine the water structure around proteins where high resolution data on the solvent cannot be obtained experimentally.

In Section IV, the investigation of aqueous solutions of small molecules of biological relevance using neutron diffraction techniques is discussed. We emphasize the use of isotope substitution which now allows us for the first time to obtain detailed structural information on not only biomolecule hydration, but also on the perturbation of the solvent close to chemical groups of importance in biomolecule interactions (e.g., nonpolar groups).

In Section V, some recent work, still at a qualitative level, on some theoretical aspects of protein stability is discussed. Elements of the stabilizing and destabilizing forces are estimated from geometrical calculations made possibly by the earlier diffraction work. Two such interactions, which are well known to involve the surrounding water environment, have been examined — and speculated upon — extensively over the last 20 years. These are the association of buried surface areas with a quantitative estimation of the hydrophobic effect[6,7] and H-bonding in the water-protein system.[8,9] These interactions appear to be correlated in all the proteins analyzed, and some interesting results are presented pertaining to the overall stability of whole proteins, local structures such as secondary structures, and possible structures in the early stages of the folding process.

Finally, a summary is offered which brings together the work presented and looks forward to areas of future work we believe are particularly promising.

II. THERMODYNAMIC AND STRUCTURAL CONSIDERATIONS

Just how globular protein molecules form their native structures and remain stable has been a popular area of research over many years. The need to understand protein stability becomes increasingly urgent now that many thousands of amino acid sequences of proteins are continually being processed and deposited in banks, while their three-dimensional (3-D) structure determination is a relatively long task (only $\sim$500 unique structures to date have been solved by X-ray crystallography and 2-D NMR). An understanding of the processes involved in protein stability and folding would be very useful in developing a protein folding model which is urgently needed in current biotechnology. It is, for example, pointless modifying a protein for a particular use if the modification destabilizes the protein in its working environment.

Thermodynamically, it is known experimentally that the free energy change on folding a typical protein is of the order of 10 to 20 kcal/mol. This is equivalent to the potential energy of only two to five hydrogen bonds. Bearing in mind there may be several hundred hydrogen bonds in the protein, failure to make only one or two of these — e.g., through a mutation — could dramatically affect the protein stability. The marginal nature of protein stability can also be seen in comparison with a typical thermal fluctuation at room temperature of kT $\sim$0.5 kcal/mol: the protein stability is only 20 to 40 kT. That most proteins are thus only marginally stable is itself of biological significance. Free energy changes of similar magnitude drive typical enzyme-substrate, hormone-receptor, and protein-protein associations which are other important processes in which solvent interactions are strongly implicated.

Understanding how the solvent is involved in the folding of a protein requires a knowledge of the changes in the solvent-protein interactions during folding, i.e., we need to know the nature and extent of these interactions in both the initial and final states. With respect to the former, this "denatured" state appears to be quite flexible and is generally thought to be of a random coil nature; it is thus impossible to crystallize and analyze its structure by diffraction (although 2-D NMR may bring some success in characterizing the denatured state in future). For the latter "folded" state, there are a substantial number ($>$500) of crystal structures determined by X-ray crystallography. From these data many of the characteristics of the chemical interactions involved in protein stabilization can be analyzed within the protein itself and with the solvent.

From this body of work, we can draw conclusions which include the following. First, there is a very high degree of H-bonding by the polar groups of the protein: more than 95% of the main-chain amide nitrogens form H-bonds either to other main-chain polar atoms (e.g., in helical and beta-sheet secondary structure) or to water molecules. In addition to H-bonds, covalent disulfide bridges between sulfur-containing amino acids occur and topologically link together distant parts of the chain. Salt bridges between oppositely charged side chains help to stabilize the native structure. The majority of charged and polar side chains lie on the surfaces of globular proteins. Very few neutral polar and almost no charged side chains are buried in the interior — unless they can form H-bonds to other polar atoms.

The still often-stated picture of the protein surface as being largely polar with nearly all the apolar groups tucked away inside the protein out of contact with the solvent is somewhat incorrect as has been amply demonstrated by characterizations of the surface accessibility[7,10-12] of various groups that make up a protein. Whereas this "polar-in, polar-out" generalization seems to be the case in lipid systems such as bilayer membranes, for globular proteins the polar/apolar segregation is not clear. This is not surprising when we recall that amino acids have chemical and structural characteristics that are different from those of lipid molecules. First, amino acids are much smaller and their side chains may contain polar or apolar groups only, or both polar and apolar groups may be present in the same side chain, thus making it impossible to designate a side chain as either polar or apolar. Second, unlike bilayers in which each lipid molecule is free to move and orientate itself, the amino acids are covalently linked through their peptide groups and their relative movements are very restricted. Thus, overall, it is relatively more difficult to separate out the polar and apolar groups in proteins.

Despite the above restrictions, it has been well established since the first globular proteins were solved that, in the majority of cases, the protein interiors are essentially composed of apolar cores.[4] These cores contain very few polar side chains, although apolar groups are readily located on the surface. Detailed studies of the interior geometry show that the internal packing of the mainly apolar groups within the protein is extremely efficient[11,13,14] so that protein interiors have even been likened to a "glassy" structure of closely-packed spheres.

The folded native protein can thus be perceived of as a molecule which is densely packed, but in a way that makes good use of possible hydrogen bonding (between both main-chain and side-chain groups), with very few internal hydrogen bonds remaining unmade. This efficiency of interactions of different kinds in the same system is quite remarkable and probably requires the bonding versatility of the surrounding water molecules to make it possible. The protein presents charged side-chain, polar side-chain, and polar main-chain groups to the surface, all of which are potentially capable of making specific interactions with the solvent. In addition, there is a significant area of nonpolar surface with which the water interactions are less specific.

We stress that the above picture of a protein is a time averaged one, the reality being of a dynamic nature, with a constant sampling through the thermal fluctuations of the protein of many connected configurations of approximately equal energy.

These regularities apparent in folded globular proteins — in essence relating to a **simultaneous** satisfaction of several chemically distinguishable kinds of interaction (e.g., hydrogen bonding, disulfide bonds, and efficiency of packing) — are major constraints that allow further more detailed analysis with a view to understanding the folded structures of globular proteins in general. This is discussed in some detail in Section V.

Being unable to crystallize it, the denatured state is much more difficult to characterize; indeed, it seems likely from a variety of studies[15] that many different unfolded configurations of the same molecule can be obtained depending on the method used to denature a native protein. A reasonable working picture is probably one in which the structure is intermediate between a completely random coil configuration — in which the polypeptide is fully extended in space (Figure 1a) — and the folded state, perhaps retaining some elements of alpha-helical secondary structure (Figure 1b). In the following brief discussion, which tries to relate the free energy change on folding to the changes in the local intermolecular interactions with the solvent, we assume the denatured state is the extended random coil conformation fully exposed to solvent (Figure 1a). Although quite sufficient for this qualitative discussion, any attempt to be quantitative may consequently overestimate the contributions of the various energetic and entropic terms to the total free energy of stability ΔG.

A. POLAR GROUPS

Consider the marked polar groups P,P' in Figure 1, perhaps a peptide NH and CO, respectively. In the ideal unfolded state (Figure 1a), both groups will be capable of interaction with solvent water molecules as indicated schematically (though these water molecules will

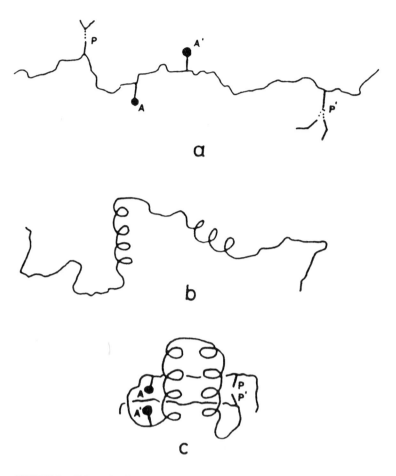

FIGURE 1. Schematic diagram of a protein molecule (a) in extended random coil conformation, (b) (partially) denatured, with remnants of secondary structure, and (c) fully folded to its native state. P,P′ represent polar groups which hydrogen bond to water in (a) and to each other in (c), while A,A′ represent apolar groups that become buried in the folded conformation.

not be "bound" in the accepted sense of the word, but will exchange rapidly with other solvent molecules). In the folded conformation (Figure 1c), these particular polar groups — together with a large number of other pairs, typically a hundred or two for a protein of the size of hen egg white lysozyme — have been "switched" from hydrogen bonding with solvent to hydrogen bonding with each other. This will have several thermodynamic consequences which will be included in the total free energy difference ΔG between the folded and unfolded states. First, there may be a change in potential energy in that the hydrogen bond interactions (including the solvent) are now of different kinds. This is represented schematically:

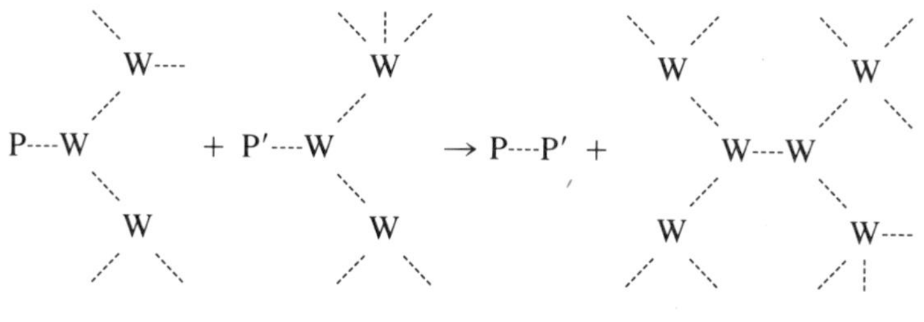

$$
P\text{----}W \;+\; P'\text{----}W \;\rightarrow\; P\text{---}P' \;+\; W\text{----}W
$$

with surrounding W groups

SCHEME 1

where the W's represent water molecules. The potential energy change from this switching process will depend upon the relative "strengths" of the polar group-water interactions (which will likely be different for each type of polar group) and the water-water interactions. Indications from quantum mechanical calculations are that generally the hydrogen bond strengths decrease in the order P–P > P–W > W–W.[2]

In addition to hydrogen bonding rearrangements between protein and water, there may also be a decrease in the number of hydrogen bonds made by a protein on folding. In the fully extended polypeptide chain, each polar atom could make its full "ideal complement" of hydrogen bonds to local water molecules, e.g., two for a carbonyl C=O, etc. On the other hand, in the folded state a buried carbonyl oxygen often only makes one H-bond (e.g., to an NH one turn along a helix). Thus, a not insignificant number of potential hydrogen bonds is not realized in the folded native structure producing a destabilizing effect with respect to the extended state. Details of the assessment of potential hydrogen bonding within a large number of known protein structures are discussed in Section V.

There is also an entropic contribution to consider. From the water's point of view, the expulsion of a molecule from a relatively restricted region of configurational space (interacting with P or P' in Figure 1a) to the bulk solvent will result in an entropic gain to the system on folding the protein — a gain of a magnitude that is uncertain and depends crucially on understanding solvent interactions both with the polar groups on the protein and in the bulk. Offsetting this entropic gain is the entropic loss as far as the polar groups P,P' are concerned, having been removed from a relatively free to a highly restricted environment. Again, the magnitude of this effect, usually considered in terms of a loss in configuration entropy of the polypeptide chain on folding, is highly uncertain.

B. APOLAR GROUPS

In a similar way, we can focus on apolar groups A and A' as their interactions and environment change as the protein folds. Again, we can consider the process in terms of energetic and entropic effects, of which the latter are usually referred to as the hydrophobic interaction and are conventionally considered the more important. In principle, however, the effect is qualitatively similar to the entropy change in release of water from interaction with polar groups P,P' to the bulk solvent. In the apolar case, however, the effect of A,A' when fully exposed to solvent (Figure 1a) is to restrict, in a much less well characterized way, the surrounding solvent, probably both dynamically and configurationally, resulting in a loss of entropy of this water when compared to the surrounding bulk. On folding, these restricted waters are expelled to the solvent with a consequent entropic gain, but again, as in the polar case, are compensated by a loss of chain entropy of uncertain magnitude.

Energetically, it seems likely that there will be a clear gain for the apolar groups on entering the interior of the protein, where the density is greater by about a factor of two, with a consequent increase in their van der Waals' interactions. This latter effect is usually overlooked, though it is being increasingly recognized.[2,5,16-18] A rough calculation of the magnitude of the reduction in potential energy on folding suggests the van der Waals' contribution may be quite significant. In other treatments, however, this energy component may be already counted in the estimated hydrophobic free energy change.

The above discussion of possible contributions to the free energy stabilization of protein folding is far from complete, but it highlights the point that many of the different forces concerned involve solvent interactions. A fuller discussion shows that although the changes at the level of the individual (polar, apolar, or charged) group are probably quite subtle perturbations, and therefore quantitatively small (perhaps less than or of the order of kT), the large number of such groups (perhaps a few hundred in a reasonably large protein) can result in the totals for the various contributions being large, possibly an order of magnitude larger than the free energy difference on folding.

Thus, protein stability seems to depend upon the existence of several strong effects of opposite sign which partly compensate each other. The degree to which this compensation is incomplete is the overall small free-energy difference in favor of the folded configuration. In one attempt to draw up a "balance sheet" of these effects,[2] individual contributions of up to ± 100 to 200 kcal mol^{-1} were estimated. Thus, an uncertainty of only 5 to 10% in each of these estimated contributions is of the same order as the overall free energy of stability. Thus, considering the quantitative uncertainties there are in the individual molecular level terms, we may begin to appreciate the problems in discussing the stability of proteins quantitatively, and we may be aware of the danger of asserting that one effect in particular dominates the process.

In fact, although since the landmark paper of Kauzmann in 1959,[19] conventional wisdom has tended to assert that the hydrophobic interaction dominates both protein folding and interaction processes; this assertion is not borne out by attempts at drawing up free energy balance sheets for these processes.[2] Increasingly, experimental evidence is also being brought forward which questions the dominance of the hydrophobic effect.[16,20]

Some recent work which suggests that two of these contributions may well effectively compensate each other — namely, potentially unmade ("lost") hydrogen bonds and buried surface area — is presented below in Section V.

To gain a better understanding of protein-water interactions involved in protein stability and other protein functions, we clearly need much better information on both energy- and entropy-related aspects. The "state" of water close to the various polar and apolar groups needs to be fully characterized energetically and entropically with respect to changes that take place in moving from the initial unfolded state to the final folded protein. The assessment of the perturbations of the solvent at the biomolecular interfaces (partly addressed in Sections IV and V) will probably provide an essential key to understanding these systems.

In trying to make progress in the area of biomolecular solvation, we need to work not only on the larger, complex protein-water systems, but also in parallel on carefully selected smaller model systems. The latter systems can be well characterized and, from appropriate measurements of them, reliable structural and energetic information on hydrogen bonding, charged group-water and apolar group-water interactions can be built up. This knowledge may then be applied both to develop and test our understanding of these interactions and in the interpretation of the behavior of the larger biomolecular systems. We now proceed to discuss some of the recent approaches used in trying to improve our understanding of protein-water interactions which we find particularly promising.

III. DIFFRACTION FROM SINGLE CRYSTALS

As mentioned in the introduction, diffraction methods provide a reliable direct method of obtaining structural information about protein-water interactions; the results depend negligibly on interpretive methods and thus give us relatively unbiased data on solvation. The structural results can be used to understand the geometry of solvation of various groups on biomolecules. Geometrical characteristics from small, well-defined systems[21,22] can be used as stereochemical restraints in the analysis of solvent in larger protein systems and also to test potential functions used to model the interface interactions, e.g., through computer simulation.

In this section, we consider diffraction methods applied to single crystals of biomolecules. Structural data from protein crystals are of the prime interest here. However, such data are generally much less accurate than those obtained from smaller hydrates, especially in the location of water hydrogen atoms for which there is usually no information from protein systems. The hydrogens are a very important feature of water structure and their relative positions need to be located. As noted above, any structural characteristics that can be observed in the smaller systems are very useful in the interpretation of solvent in protein crystals (cf., the use of covalent restraints used in fitting the protein molecule to electron density in protein structure analyses).

Below, we give a brief outline of some relevant points relating to diffraction methods and summarize the analysis of water in protein crystals. Then follows a more detailed discussion of water structure in more accurately determined small crystal hydrates and ice structures analyzed by neutron diffraction which allows the all-important hydrogen atoms to be located. These results from small-molecule hydrate studies are then related to the structure of water around a protein with the example of insulin. We conclude this section with a brief comment on apolar group hydration structures.

A. SOME TECHNICAL CONSIDERATIONS

The X-ray or neutron diffraction pattern of a crystal is essentially the Fourier transform of the distribution of scattering centers (electrons or nuclei, respectively) in the target crystal. In principle, a complete description of the crystal structure is obtained by back-transforming the scattered amplitude distribution. In practice, only intensities can be measured and no phase information can be obtained; special techniques are therefore used to obtain an initial set of phases before the back-transform can be performed. Details of these procedures can be found in standard crystallographic texts.[23-26]

Once an electron density map (in the X-ray case) has been obtained, a model of the molecule is fitted to it using computer graphics in the case of macromolecules. The final stage involves the overall refinement of the structure to obtain the best structural model consistent with the data. For small molecules, such as amino acid hydrates, a larger number of "reflections" (essentially Fourier components of the structure) will normally be measurable; as the number of positional and thermal parameters to be refined will be relatively small, the Q value — the ratio of the number of observed data to the number of parameters refined — will be high. Thus, these structures will tend to be overdetermined, so the refinement procedure will normally give structures that are generally well-determined with accurate distance and angle data on the solvent molecules as well as on the solute.

For a protein, Q will generally be much lower, for, in addition to the larger number of atoms — and therefore parameters to be refined — few macromolecules diffract strongly enough at high angles to allow us to determine their structures to atomic resolution. In most cases, therefore, the structure refinement is assisted by restraining a number of structural characteristics known from small molecules studies, such as bond lengths and angles, torsion angles, planarity, etc. For small structures, restraints are not normally needed unless disorder is a problem. Thus, an **unbiased** assessment of the solvent structure can be made for these systems.

For X-rays, the scattering power of hydrogen is low and, except in the high-resolution structures (mainly limited to small molecules), only the nonhydrogen atoms are seen. Thus, information on the important hydrogen bonding orientations will be missing. This is not true of neutron diffraction in which the hydrogen is a relatively strong scatterer. However, as hydrogen is also a very strong incoherent scatterer, neutron scattering patterns from hydrogenous samples tend to have a much higher background. To reduce this, neutron diffraction measurements are often made from samples in which the solvent and easily exchangeable protons on the protein have been replaced by deuterium by crystallizing from a D_2O-based solvent.

Although neutron diffraction might appear ideal for investigating solvent organization, additional problems arise. For example, recent experience has shown[27,28] it is often not clear whether a given solvent peak should be assigned as a deuteron or an oxygen, as their scattering powers are similar. Such confusions can lead to problems in interpreting partly ordered solvent regions. In such cases, the X-ray data can be invaluable in identifying the oxygen positions and the two techniques should therefore be thought of as complementary.

a neutron

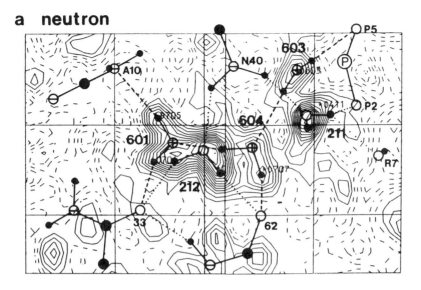

b X-ray2

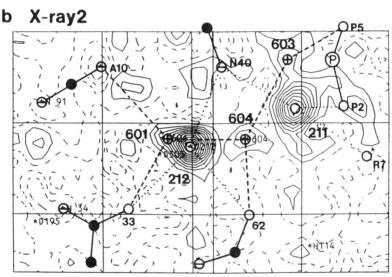

FIGURE 2.   Solvent density around a disordered side chain (N40) of the coenzyme B_{12} molecule. (a) Neutron; (b) X-ray Fo-Fc difference Fourier maps (the alternative position for the sidechain [N640] is not shown, but lies ~1.8 Å behind N40). Two partially occupied solvent networks are present: NETWORK A: waters 211 and 212, occupancies — 0.9 (X-ray), 0.6 (neutron); and NETWORK B: waters 601, 603, and 604, occupancies — 0.1 (X-ray), 0.4 (neutron).

Besides the problem of chemical identity, there may also be difficulties in interpreting and modeling disordered regions of density. The handling of static disorder is fairly straight-forward, but where alternative sites for a particular atom or group of atoms tend to overlap (especially in the case of neutron maps), care must be taken when interpreting the local structure. An example of this is shown in Figure 2 for one of the solvent regions in the medium-sized crystal hydrate of coenzyme B_{12}.[29,30] The X-ray oxygen positions (b) can be used as guidelines in the interpretation of the neutron diffraction map (a).

Modeling dynamic disorder is more of a problem, especially when the solvent density does not have a relatively simple shape that can be represented by spheres (isotropic) or

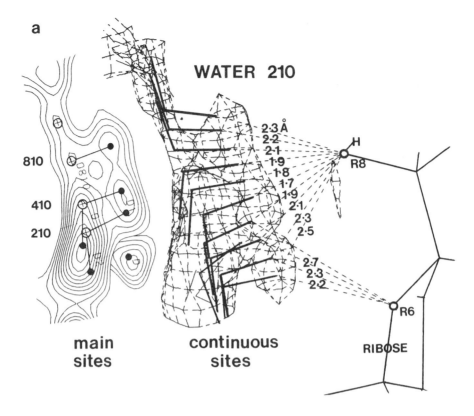

FIGURE 3.   Interpretation of solvent regions in the coenzyme B_{12} structural analysis (a) over the 210 solvent region; (b) over the 211 and 214 solvent regions. "Main" sites initially assigned to well-defined solvent density. "Continuous" sites assigned to the elongated and diffuse regions of solvent density.

ellipsoids (anisotropic). One way to interpret the solvent is as follows (see References 27 and 28 for details). First, assign main sites to the more ordered regions of solvent density and assign "continuous" sites to the more diffuse and elongated regions between the main sites (an extension of this is to assign a grid of solvent sites). An example of this is shown in Figure 3. From the main sites, alternative water networks can be formulated using the standard criteria of H-bonding (O . . . O = 2.5 to 3.2 Å, H . . . O = 1.5 to 2.4 Å) and van der Waals' radii. Figure 4 shows the main solvent networks in the main solvent regions of the coenzyme B_{12} hydrate,[30] a system which is well characterized and from which much has been learned about solvent structure.

B. DIFFRACTION FROM PROTEIN CRYSTALS

Protein crystals may contain between 25 and 90% water by volume, generally located in the interstices between the protein molecules. The nature of the water structure at the protein interface usually appears to be significantly ordered, but becomes more disordered and bulk-liquid-like further away from the surface — usually outside the first hydration shell.

Over 500 protein structures have been solved using X-ray diffraction. However, the data for most of these structures lie at resolutions of 1.5 to 3.0 Å which are substantially worse than atomic resolutions of ~1.0 Å. Only a small proportion of proteins have been studied to better than 1.5 Å resolution, mainly because the crystals involved do not diffract well at higher resolutions. Examples of high resolution analyses in which the solvent regions have been well characterized are crambin,[31] rubredoxin,[32] and insulin.[33] Crambin has been studied to ~0.9 Å resolution with ~80% of the solvent observed to be well ordered: 73

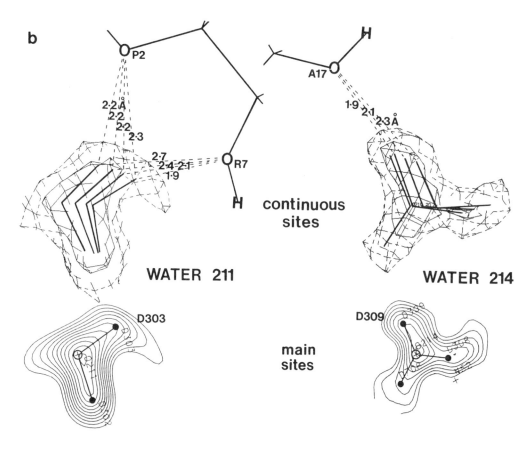

FIGURE 3 (continued).

solvent sites were assigned, 4 of them as ethanol molecules. Around some of the apolar groups, several distorted pentagonal water rings were present,[34] forming hydrogen-bonded bridge structures between the polar groups of the protein. In 2Zn insulin crystals, ~70% of the solvent content was modeled used ~340 solvent sites, while in rubredoxin, 127 sites were assigned accounting for 50% of the solvent.

From X-ray analyses of most proteins, only the water oxygen positions are obtained with any reliability and the description of the geometries of the hydration of solvent-exposed polar and apolar protein groups is thus restricted to the nonhydrogen atoms (unless postulated calculated positions are included, generally making further assumptions about hydrogen bonding geometries which may not be justified). Within this limitation, the hydrogen-bond geometries agree quite well overall with the more accurate geometries obtained from small-molecule data. The O(W) . . . O(P) distances vary from 2.2 to 3.5 Å with a main peak around 2.9 Å; O(W) . . . N(P) distances range from 2.4 to 3.5 Å with a peak around 3.0 Å.[35] Distances below ~2.5 Å are mainly due to alternative partially occupied solvent sites. The Y . . . O(W) . . . Y angles (where Y = O or N) range between ~70 and ~150° with the majority around 100 to 120°. To obtain more detailed and accurate geometries, especially involving the hydrogen positions, neutron diffraction studies are required, most of which have been made on small hydrate systems. Nevertheless, some neutron analyses have been carried out on small proteins at resolutions of 1.2 to 1.8 Å: crambin at 1.2 Å (46 amino acids[36]), bovine pancreatic trypsin inhibitor at 1.8 Å (BPTI; 58 amino acids[37]), and hen egg white lysozyme at ~1.4 Å (129 amino acids[38]). The details of these analyses are much less precise than their equivalent high-resolution X-ray studies. They have as yet to be fully reported and thus will not be discussed further here.

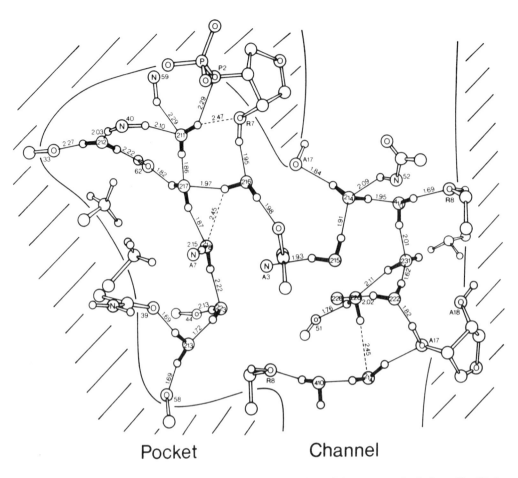

FIGURE 4. Main solvent networks in the pocket and channel regions of the coenzyme B_{12} hydrate. The filled bonds represent the O–D bonds of the water molecules.

C. DETAILS OF WATER STRUCTURE IN SMALL HYDRATES AND ICES

Many of these structures have been studied at very high resolutions (better than 1.0 Å) using neutron diffraction, and on average the accuracy of the bond lengths are of the order of 0.01 Å (cf., average of 0.1 Å for protein crystals). Thus, the details of the geometries involved in the interactions can be examined with a great deal more confidence.[21,22,28,39] The hydrogen-bonded geometries involving waters within small hydrates and ices display very wide ranges: O . . . O distance between 2.5 and 3.2 Å and O–H . . . O angles between 120 and 180°. The O . . . O . . . O angles range from 70 to 150°; as this range corresponds to 109 ± 40°, large deviations from the expected tetrahedrality occur.

Although hydrogen bonds are directional in character, this known directionality does not appear to fully explain why such a wide range of geometries occurs for water molecules. However, a better understanding is gained when the local short-range contacts are considered, information which has come to light only recently, and for which hydrogen location using neutrons was essential.[39] In this work, details of the O . . . O, H . . . O, and H . . . H atomic geometries were examined in ice and small hydrate structures[28,39] and some interesting "repulsive regularities" (RRs) in terms of minimum contact distances were revealed. Four different short-range repulsive regularities were identified, listed as follows:

RR1. **O . . . O repulsion of the H-bonds** — Standard plots of the O . . . O and H . . . O H-bond distances vs. the O–H . . . O H-bond angle involving water molecules are shown in Figure 5. Usually, a regression line is drawn through the points on the plots representing an overall decrease in the bond angle as the hydrogen-bond distance increases. However, the scatter of the geometries is way outside the accuracy of these data points (~0.01 Å for distance) for this correlation to be quantitatively significant. A more useful way of looking at this plot is to accept that the H-bonds can bend when required, but that **there is a limit to the amount of bending possible**. This limit is represented by the full lines in Figure 5, which are in effect H-bond bending limit curves. The actual angles and distances a water H-bond may have appear to depend on the local packing arrangements determined by the surrounding short-range contact of the following regularities labeled RR2, RR3, and RR4.

RR2. **O . . . O nonbonded contacts** — This regularity related to the significant anisotropy found experimentally in the water oxygen-oxygen contact which range from 3.1 to 3.6 Å. Figure 6a shows the minimum O . . . O nonbonded contacts in ices and hydrates with respect to the two orientational angles defined in the inset. Three main regions are apparent and can be assigned the following van der Waals' radii: region A ~ 1.8 Å over the lone pair region, region B ~ 1.7 Å between the hydrogens, and region C ~ 1.6 Å between the hydrogens and the lone pair region. In conjunction with RR4, more detailed assignments are discussed below.

RR3. **H . . . O interactions** — In the configuration O1–H1 . . . O2–H2, there are two main types of interactions: H1 . . . O2, the normal H-bonds with distances of 1.5 to 2.4 Å, and H2 . . . O1, the longer remote nonbonded interactions, which are seen to have a minimum contact distance of around 3.0 Å in the ice and hydrate structures at ambient pressure. Such H2 . . . O1 contacts for water in coenzyme B_{12} crystals are shown in Figure 7. Figure 8a shows a plot of H2–O2 . . . O1 angle of the O1–H1 . . . O2–H2 configuration against the O1 . . . O2 H-bond distance. A limiting (full) line can again be drawn representing the limits for H-bond geometries. As in the case of RR1, a correlation line can be replaced by a line delineating an **excluded region**, a very powerful constraint that can then be used in understanding water orientations in any hydrated structure. As the O1 . . . O2 H-bond distance decreases, the H2–O2 . . . O1 angle increases in order to maintain the minimum RR3 H2 . . . O1 contact distance of ~3.0 Å. This repulsive restraint appears to control the orientational structures of hydrogen bonded systems and hence is very powerful in helping us understand solvent structure at protein — or other — interfaces.

RR4. **H . . . H interactions** — These refer to the H . . . H repulsions involving water hydrogens. Contacts down to 2.1 Å occur, with a spread of 2.1 to 2.6 Å. Figure 8b shows a plot of the H . . . H contacts vs. the sum of the angles (χ) subtended at the hydrogens (see inset). As the angle χ increases, the H . . . H contacts tend to decrease and again a limiting line can be drawn.

Although at first sight these repulsive regularities may seem a little abstruse, this replacement of a "correlation line" to which hydrogen-bond geometries are assumed to tend toward — by an "excluded region" in which hydrogen bond geometries are just not allowed — is a major change in approach in rationalizing hydrogen-bonded structures. In considering protein hydration, it allows us to replace a "soft restraint" (bond angles and distances tending towards ideal values) by a hard excluded volume constraint, which actually **prohibits** the occurrence of a wide range of geometries. We believe this use of excluded volumes of configurational space for water geometries is a major step forward in understanding hydration structures in general and hence promises to be of value in improving our understanding of macromolecule-water interactions in particular. The above merely sketches the overall picture concerning repulsive restraints; more detailed discussions are given elsewhere.[28,30,39,40]

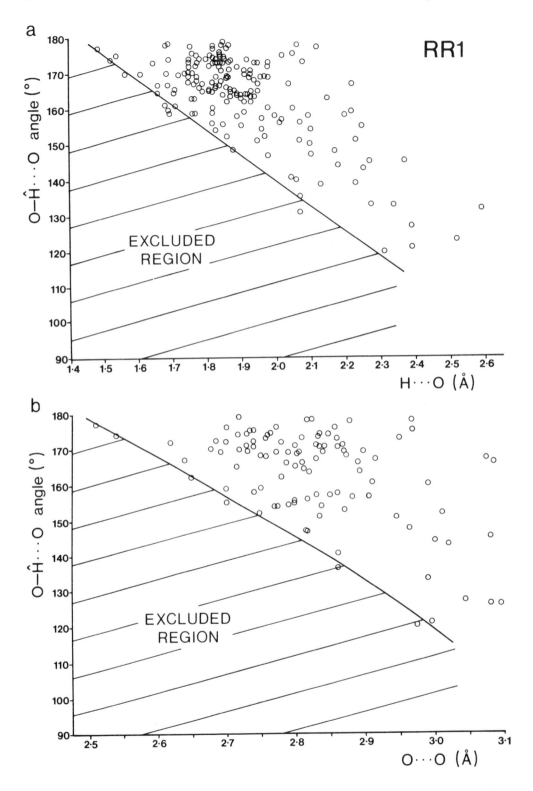

FIGURE 5.   Plot of O–H . . . O angles vs. (a) H . . . O distances and (b) O . . . O distances for water H-bonds.
The solid curves approximately represent minimum allowed values.

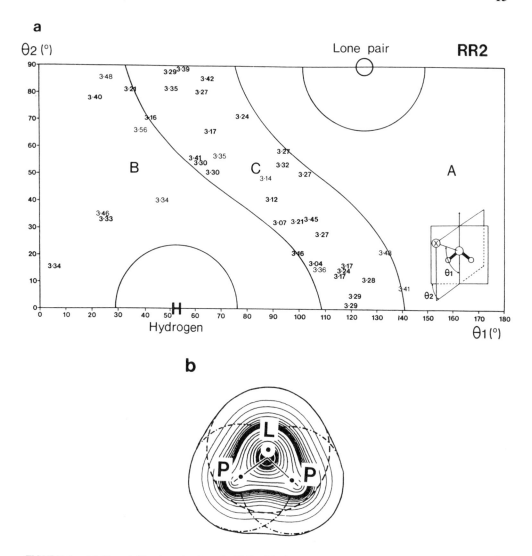

FIGURE 6. (a) θ1 and θ2 orientational angle (defined in inset) plot for nonbonded O . . . O,C contacts (Å) around four-coordinated water molecules. Bold numbers are O(W) . . . O contacts and lighter numbers are O(W) . . . C contacts. (b) Displaced van der Waals' model for the RR nonbonded short-range interactions around a water molecule. Contours represent the water electron density (from quantum mechanical calculations) and dashed circles represent displaced van der Waals' spheres: centered at L of the oxygen (radius = 1.5 Å) and P for hydrogen (radius = 1.3 Å). The L and P centers are displaced by 0.2 Å from the nuclear positions of O and H. Note: Two L sites are present in approximately the tetrahedral lone pair positions. Adapted from Reference 41.

To a first approximation, the variable nature of the RR2 and RR4 minimum short-range contacts seen in studies of hydrate structures so far examined can be rationalized further in terms of a significantly nonspherical model of the water molecule in which the center of four van der Waals' spheres are slightly displaced from each other in a way which is consistent with the known asphericity of the electron density over a water molecule. Figure 6b shows the total molecular electron density for a water molecule obtained from recent quantum mechanical calculations.[41] The surface of the van der Waals' sphere for an atom is usually considered to correspond to the outer electron density levels of the atom. On this basis, four spheres can be used to account specifically for the anisotropic character of the RR2 and RR4 contacts. The centers of the spheres reside at locations that are displaced as follows:

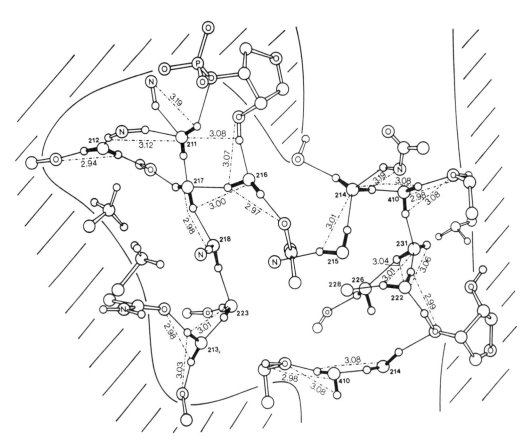

FIGURE 7. Close H2 . . . O1 contacts (dashed-dotted lines) involving the main water positions in the coenzyme B$_{12}$ hydrate. Distances in angstroms.

Oxygen — 0.2 Å from the oxygen nucleus along the ideal lone pair direction (centers at L), roughly corresponding to the conventional lone pair positions
Hydrogen — 0.2 Å from the hydrogen nucleus towards the oxygen nucleus (centers at P), corresponding to the hydrogen positions in high resolution X-ray studies

The displaced spheres are shown as dashed circles in Figure 6b. When geometries have been recalculated in the known ice structures with L and P centers replacing the O and H atom positions, the results indicate that the anisotropic minimum contacts for RR2 and RR4 have been taken into account: the minimum L . . . L and P . . . P distances become isotropic with consistent minimum values of 3.0 Å ± 0.02 Å and 2.6 Å ± 0.02 Å, respectively. Thus, van der Waals' radii of 1.5 Å for the oxygen (centered on L) and 1.3 Å for hydrogen (centered on P) can be assigned.

The inclusion of the RR restraints in addition to "normal" H-bonding considerations allows many of the specific characteristics of water structure to be explained in both ices and hydrates. The short-range contacts appear to largely control the final orientations of individual water molecules. For example, the basic tetrahedral characteristics of water structure can be related to a minimization of the RR repulsive interactions (especially RR3: remote H2 . . . O1 contacts) and a maximization of the number (four) and strengths (2.7 to 2.8 Å) of H-bonds formed. Shorter H-bonds (<2.7 Å) require the acceptor water to adopt a more trigonal planar conformation to reduce the strain of the H2 . . . O1 contacts, while longer ones of 2.8 to 3.2 Å allow significant angular distortions from tetrahedrality which are often required to maximize H-bonding around a water.

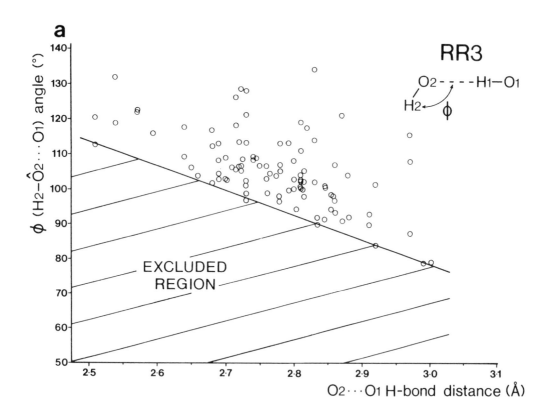

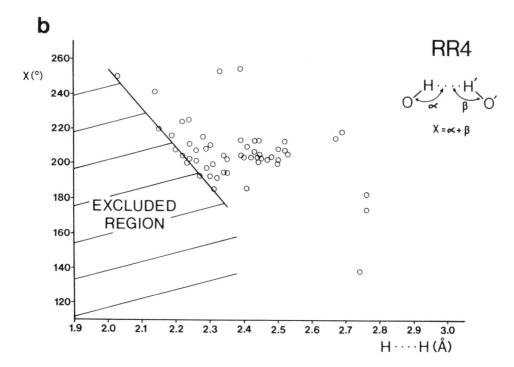

FIGURE 8.   (a) Plot of H2–O2 . . . O1 angles (φ) vs. O2 . . . O1 H-bond distances for waters involved in H-bonds. The solid line represents the minimum allowed values. (b) Plots of water-water H . . . H contacts vs. χ, the sum of the angles α and β subtended at the hydrogen atoms (see inset).

The H-bonded coordinations around waters in small hydrates range from two to six, with values of three or four being the most frequent. In the ice polymorph structures, four is the minimum number. Generally, there is a distinct lack of directionality of the H-bonds over the lone pair regions — an observation well established in small-molecule crystals.[42] The local H-bonded coordination values depend on the number and presence of local polar groups and on the maintaining of the minimum RR values. In regions where a significant number of apolar groups are present, waters that are less than three coordinated may be present.

The RR restraints segregate the geometries of a water molecule into an excluded (area below the limit line in the RR plots of Figures 5 and 8) and an allowed accessible space (above each curve). The RR restraints can be applied to larger proteins systems in which only the water oxygen positions are known, with no hydrogen sites having been specifically located experimentally. We can use the RR restraints along with standard H-bond criteria (e.g., O . . . O H-bond distance = 2.5 to 3.2 Å, etc.) to assign with unprecedented confidence the missing water hydrogens and so formulate water networks.[28]

As an example of the use of these restraints obtained in small-molecule hydrate studies, we now return to a protein system. As mentioned above, the solvent regions in 2Zn insulin crystals[33] have been extensively studied and have had oxygens assigned. Figure 9a shows the electron density over one of the solvent regions along with water sites. In Figure 9b, one of several possible water networks formulated using the above criteria is shown. Several alternative sites separated by less than 2.5 Å are present and alternative H-bonded networks can also be formulated. This is an early example of the use of the repulsive restraints in evaluating hydration structures in proteins; further work is expected which will lead to improved understanding of the detailed nature of the protein-water interface and ultimately to its relevance in protein stability and interactions.

D. CLATHRATE WATER STRUCTURES

The possible existence of clathrate-like water structures around apolar groups and their possible relevance to the hydrophobic effect in biomolecular systems has been a subject of intense interest over the last 40 years[43,44] since the seminal paper of Kauzmann.[19] From crystallographic studies of water structures of molecules with apolar groups,[45] the water molecules in the cage structures identified make H-bonds only with other water molecules or polar atoms nearby, and avoid "H-bonded type" interactions in which the water proton points towards an apolar group, thus preventing the loss of H-bonds. Clathrate-like cages of water molecules are found to be formed around relatively small apolar molecules in certain crystals and, because of the small size of these hydrate systems, the water in these cage structures is well ordered and can be well characterized.

This does not appear to be the case for protein crystals. The solvent close to nonpolar groups appears experimentally to be relatively disordered in most cases, and from the protein crystal data itself it is not possible to clearly state that clathrate cages are present around apolar groups. In the case of crambin, several distorted five-membered rings are found (a geometry commonly expected in water structures), but there are no extensive partial cages. It appears that in most crystal hydrates with water-apolar contacts present, the waters tend to maximize the number of H-bonds they make — which does not necessarily mean they form regular water cages, but instead make the best of the environment they find themselves in. This underlines the versatility and flexibility of water molecules, attributes which appear central to the role of water in biomolecular stability and interactions.

IV. SCATTERING FROM AQUEOUS SOLUTIONS

Until recently, high-quality structural information on hydrate systems has been limited to crystal studies such as those summarized above. From this kind of work, much has been

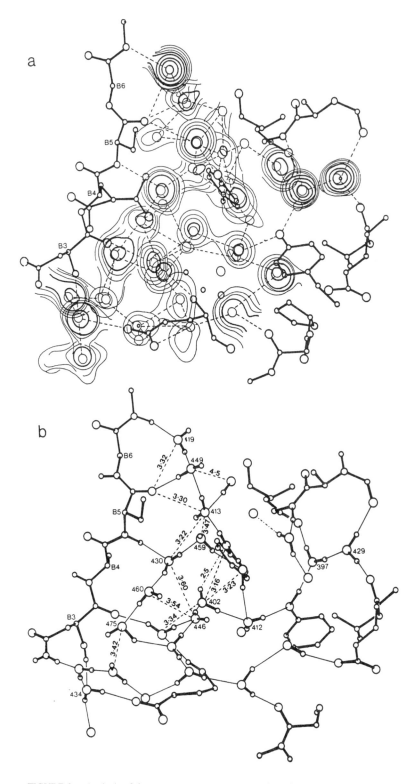

FIGURE 9. Analysis of the water structure over one region of solvent density adjacent to residues B3-B7 in porcine 2Zn insulin crystals. (a) Solvent density at 1.5 Å resolution from X-ray refinement; (b) one of several possible water networks constructed in which all the nonbonded minimum constraints are satisfied (broken lines represent nonbonded O . . . O contacts).

learned about the details of solvent interactions — both within a macromolecular solute (charged, polar, and nonpolar groups) and with other water molecules themselves. However, the crystal is an artificial environment for a protein: although the solvent may be structurally disordered with respect to ice, additional order is likely to be imposed on the solvent by, for example, the presence of neighboring protein molecules.

Biomolecular interactions, however, take place in solution, and it is here that recent developments in neutron scattering promise to throw new light on hydration structures in solutions, initially of small biologically relevant molecules containing chemical groups of interest. Moreover, capitalizing on the advantages of pulsed neutron sources such as ISIS at the Rutherford Appleton Laboratory in the U.K., we can now begin to look at how the presence of a solute **perturbs** the surrounding solvent. As the hydrophobic interaction is traditionally thought to operate through the "ordering" of solvent close to an apolar group — thus reducing the system entropy which is then increased when previously "ordered" waters are expelled to the bulk as two apolar groups come together — the ability to probe possible solvent perturbations in complex systems is potentially of major significance with respect to biomolecular interactions.

Structural information obtainable from liquids is by definition less complete than that obtainable from crystals, but nevertheless a Fourier transform of the liquid diffraction pattern yields structural information on a function called a pair correlation function, or radial distribution function g(r). This is a one-dimensional average function which gives the probability of finding an atom at a distance r from any other atom. For a multicomponent system such as water, to describe the structure, we need to specify three partial correlation functions, $g_{OO}(r)$, $g_{OH}(r)$, and $g_{HH}(r)$, which again give the probabilities of finding an atom of type j (e.g., H) at a distance r from an atom of type i (e.g, O). For example, $g_{OH}(r)$ gives a probability of finding a hydrogen atom at a distance r from the central oxygen. For the solution of a molecule in water, several more such correlation functions are required relating to both solute-solute and solute-solvent distances in order to obtain a full description of the structure.

In a standard scattering experiment, our Fourier transform yields a weighted sum of these partial correlation functions, the weighting factors depending on the relative concentrations and scattering powers of each species. Using neutrons, however, we can begin to separate out subsets of these partials which then allow us to obtain much more detailed information on solution structure. This is done by taking advantage of the fact that, for certain elements, different isotopes may scatter neutrons differently: the neutron scattering factors of different isotopes are sometimes different. Thus, by performing experiments on solutions which are chemically the same, though having different isotopes of certain elements, they will be seen differently by neutrons; we can in effect turn ourselves into a Maxwell demon and effectively "sit on" a selected atom (on the solute or solvent) and look at our surrounding neighborhood from the vantage point of the atom we have isotopically substituted.

The technique was first developed by Enderby in application to aqueous electrolyte solutions and this work has revolutionized our understanding of ionic solvation.[46] More recently, it has been extended to solutions of polar, charged, and nonpolar molecules, and it has begun to throw new light on the hydration of polar molecules such as amides[47,48] (including protein denaturants such as urea), bifunctional molecules (e.g., the cryoprotectant DMSO,[49] and methanol,[50,52] and molecular ions such as the tetramethylammonium ions.[51,52] The interested reader is referred to the quoted references both for the details of the technique and specific applications to date.

As an example, we restrict ourselves here to stating the tentative conclusions of one ongoing study, namely, on the tetramethylammonium (TMA) chloride system.[51] In this extensive study, the hydration of the TMA ion itself has been probed through substitution of the central nitrogen atom. The results suggest that TMA — which although being charged presents significant apolar surface to the solvent — hydrates not as an ion, but as a nonpolar molecule. Around 20 water molecules hydrate the TMA moiety, forming a structure resembling a somewhat disordered cage reminiscent of a clathrate. This hydration structure does not seem to significantly change with concentration, though at high concentration the chloride anion probably accommodates itself into the cage structure (again illustrating the flexibility of the water molecule in accommodating itself to distinctly different environments). Using H/D substitution on the water molecules, we can also probe any changes in water structure caused by the solute:[53] in these experiments, perhaps surprisingly, no evidence was found for any significant structural perturbation of the water from its bulk structure. This result could have significant implications for our understanding of the hydrophobic interaction. Finally, the effects on the hydration structure of a protein denaturant urea were examined. Conventional wisdom has asserted that such denaturants may act through perturbing hydration structures around nonpolar groups exposed on protein surfaces and so any experimental detection of such ''structure breaking'' action would be significant support for this hypothesis. However, contrary to expectation, no significant perturbation of the TMA hydration ''shell'' was observed. There was no evidence for urea acting as a water structure breaker. On the contrary, the urea molecule appears to fit well into the water network, a result which is consistent with other neutron-scattering work on urea solutions.

Such neutron-scattering isotope substitution studies on hydration of model biologically relevant molecules is at an early stage and the above conclusions remain tentative. However, the approach is already allowing us to obtain very new information on the hydration of important functional groups in molecules of intense interest in the study of biomolecular interactions. As neutron sources and instrumentation continue to progress, the further applications of the technique promise to throw new light on such problems which could not previously be tackled with direct methods such as diffraction.

V. A THEORETICAL PERSPECTIVE

Much work has been undertaken over the past 20 years in the development of potential functions for water-protein interactions. These developments were largely stimulated by the advent of computer simulation calculations to relatively complex systems, made possible by the increasing power of computers over the past decade or so. As noted in Section II on thermodynamics, the details of these interactions appear to be very complex and known only within uncomfortably wide margins of error. With this in mind, more qualitative approaches may be developed which are consistent with the principle of Occam's razor: trying to keep the model within reasonable bounds of simplicity. This approach is not dissimilar in philosophy to the early protein structure work of Linus Pauling in deriving the alpha-helical and beta-sheet structures from a polypeptide model[54,55] which included two main parameters, namely, planar peptide bonds and H-bonds formed between main-chain polar atoms. What we discuss below leans in this direction.

The initial aim of this approach is to find consistent patterns or regularities (i.e., >95% consistency; cf., RR restraints in Section III) within water-protein systems which relate to the overall stability of a protein system. Many different forces are involved in their stabilization (Section II) and several of the important ones are discussed here in more detail without using complicated and imperfectly known potential functions to estimate energies. Two of these ''forces'' that have been examined extensively over the last 30 years, both at the geometrical level and using potential functions, are H-bonding in proteins[8] and the hydrophobic effect and its apparent empirical correlation with buried apolar groups.[6,7,10,11]

As stated in Section II, from our knowledge of protein structure derived from crystallographic studies, over 90% of the polar atoms in proteins have been estimated[12] to form

H-bonds either to other protein atoms or to water molecules, and this near-maximization of H-bonds appears to be a very consistent regularity in proteins — especially of the amide nitrogens (>95%). In addition, very few polar side-chains appear to be buried in the protein interior, and almost none without forming H-bonds. The cores of the majority of globular proteins are mainly composed of buried apolar groups; however, apolar groups are also found on the surfaces of these proteins, sometimes very extensively.[7]

It thus appears that there is a strong restraint on proteins to form as many H-bonds as possible, similar to the situation for water. This is apparently a stronger restraint than burying the apolar groups (few polar side chains are buried whereas many apolar groups may be surface exposed). However, not all possible H-bonds appear to be made; the number of unmade ones has therefore been assessed. From this assessment, we can conclude that there is a strong correlation between the number of unmade H-bonds and the buried surface area of the carbon and sulfur atoms within known protein structures.

The unmade H-bonds can be defined as potentially "lost" H-bonds (LHBs) as follows: the denatured protein is assumed to have all its polar atoms fully exposed to the surrounding solvent with each polar atom making its full complement of standard H-bonds to waters. For example, a carbonyl C=O oxygen may form two H-bonds, a hydroxyl O–H oxygen may form three H-bonds (two as a proton acceptor and one as proton donor), main-chain N–H nitrogen may form one H-bond, and so on. In the folded state, on the other hand, buried carbonyl oxygen may make only one H-bond and thus it loses one H-bond. The number of LHBs can be summed over all the polar atoms. Three contributions to the LHBs were included as follows: (1) LHBs of the protein atoms, (2) LHBs of any internal waters present, and (3) LHBs of surface waters. These contributions were based on geometrical calculations outlined in Table 1.

The LHBs are essentially destabilizing interactions while other interactions can be considered as stabilizing contributions. Three such stabilizing interactions were also calculated: (1) HP: "hydrophobic energy" associated with buried carbons and sulfurs, (2) IP: salt bridges, and (3) SS: disulfide bridges. The basis of these calculations is again given in Table 1.

The above six contributions (referred to collectively as the LHB model) were calculated using the protein structure coordinates in the Brookhaven protein databank (1990 version).[56] Structures were included if they satisfied the following restrictions: resolution was better than 2.0 Å and restraints were used in refinement. About 120 crystallographically independent protein molecules were included for which there were approximately 75 uniquely different protein structures.

Figure 10a shows a plot of the total LHB contribution vs. the three stabilizing factors (HP + SS + IP). The line drawn is the least squares' line; the correlation coefficient of 0.993 indicates a very strong correlation between the LHB component and the sum of the three stabilizing factors, namely, the HP + SS + IP components. This strongly suggests that the three stabilizing factors considered could compensate for the destabilizing LHB factors.

Using this assumption, the LHB contribution can be scaled by a factor of 0.714 such that the stabilizing factors just out-balance the destabilizing LHB factors (Figure 10b: solid line is of unit slope). Thus, where a point lies below the solid line, the LHB model assumes the protein has a relatively stable conformation; above the line, the protein is unstable. The dashed line shows a window of ~30 kcal/mol in which the majority of the protein structures lie. The LHB model can thus be used as a structural check for the correctness of noncovalent geometries within protein structures — particularly for those known only to lower resolution (2.0 Å to 3.5 Å).

Figure 11a shows a plot for the LHB model involving individual alpha-helices isolated from ~30 native structures. Most of the points lie above the unit slope line indicating that

TABLE 1
Stabilizing and Destabilizing Forces Included in Lost Hydrogen Bond (LHB) Model

Destabilizing Contributions — LHBs

Three contributions were included:

1. LHBs for protein atoms: In the extended state, each of the polar atoms of the protein is assumed to make its full complement of H-bonds to the surrounding water: e.g., C=O carbonyls make two, O–H hydroxyls make three, etc. But, in the folded state, some of the polar atoms do not make all their potential-bonds on folding. The H-bonds were calculated geometrically from the coordinates of the native structures in the PDB data-bank.[56] The criteria of standard H-bond cutoff distances, H-bond bending angle contribution, and accessible areas of exposed polar atoms were used to estimate the number of H-bonds made by each polar atom.
2. LHBs of internally buried waters: A water is assumed to form a maximum of four H-bonds. Thus, if it only forms two H-bonds, then two H-bonds are lost. H-bonds were estimated geometrically as in 1 above.
3. LHBs of surface waters: These are estimated from the surface points assigned to protein atoms. Surface points were generated using a subroutine of the surface/volume analysis program radial.[62] For each surface point (S) all surrounding triplets of surface points (within 2.0 to 3.5 Å) were taken in turn and a least-squares plane calculated. If point S lies more than 0.75 Å below the plane (i.e., on the protein side toward the atom of the surface point S), then an H-bond is assumed to be lost. If S lies on the other side (above), no H-bonds are lost.

Although the surface water model is quite crude, leaving this contribution out of the LHB summation did not significantly affect the LHB plots: the correlation coefficient only dropped from 0.993 to 0.989.

Stabilizing Contributions — HP, IP, SS

1. Hydrophobic energy model, HP: This is calculated in the standard way from the correlation between buried surface area and experimental transfer free energy values of the amino acids from organic solvent to water.[6] Only carbons and sulfurs were taken into account using values of 16 and 21 cal/Å,[2] respectively, as derived from a fitting by Eisenberg and McLachlan[63] of five atom types to the experimental free energies of transfer from octanol to water.[64]
2. Ion-pair model, IP: Oppositely charged side chains are considered to form salt bridges if they are less than 4 Å apart.[65] Stabilization energies of 1 kcal/mol are given for surface salt bridges[66] and 3 kcal/mol for internal salt bridges.[67]
3. Disulfide model, SS: Contributions of covalent disulfide bonds to the stability of proteins are calculated using the Flory formula.[68]

in isolation alpha-helices are quite unstable with respect to the model. This is in agreement with most solubility studies which indicate that the majority of helical structures are only soluble in organic solvents which are fairly apolar. A few of the helical points lie below the line and appear to be marginally stable: one such point is the C-peptide of ribonuclease A (i.e., first 13 residues of this protein), which has indeed been reported by Baldwin et al.[57] to be stable experimentally (mainly through ion pair interactions). In addition to this, when individual helices from a known protein structure are built up — i.e., calculations are performed for one helix, then two helices, then three, and so on — the points usually start above the line and for three or more helices in contact, the points lie mainly below the line. Examples of these results were seen for myoglobin (six helices) and insulin (three helices). In the latter, one and two helix points lie above the line, indicating they are unstable, while the three helix points lie below the line, indicating stability.

Figure 11b shows a plot for isolated beta sheets of 2 to 6 strands from ~30 protein structures. Here, most of the points lie below the line indicating that such structures are fairly stable. Most of the beta-sheet points that lie above the line (on the left side of the plot) are from two strand fragments isolated from larger sheets. This is because some of the apolar groups which are well buried in, for instance, a six stranded beta-barrel, may become exposed in the isolated two strands decreasing the HP contribution. The larger sheets lie to

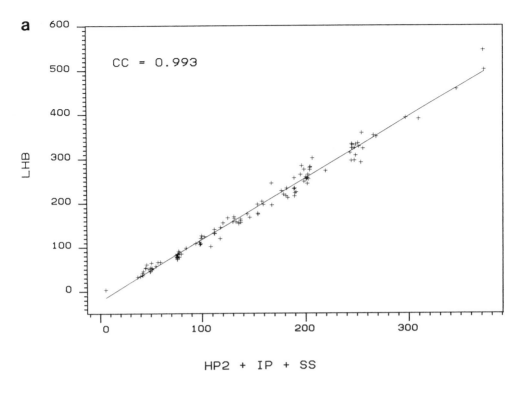

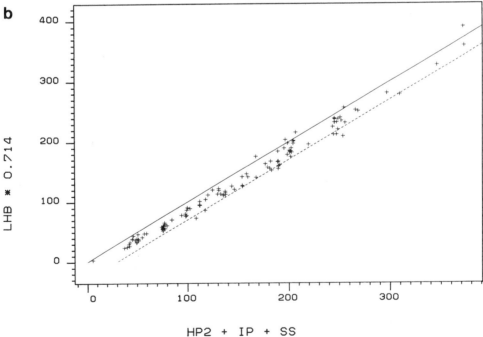

FIGURE 10.   (a) Plot of LHB (sum of three lost hydrogen-bond contributions) vs. sums of HP (hydrophobic energy), IP (ion pairs), and SS (disulfide bonds); (b) same as Figure 10a, with LHB scaled by 0.714.

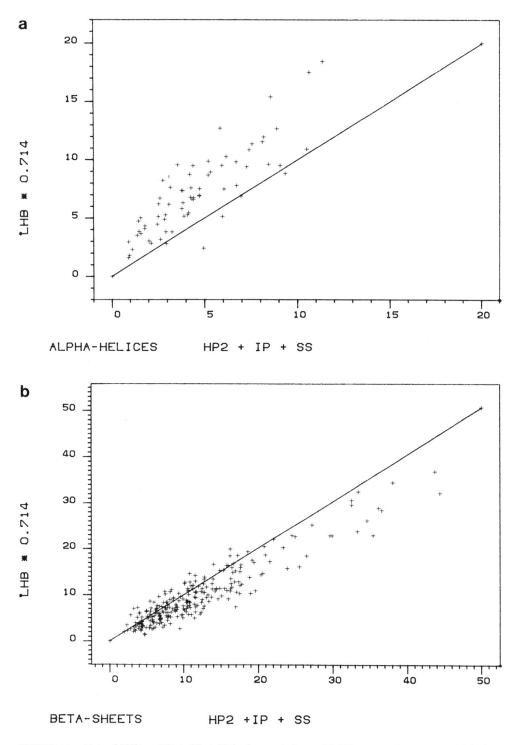

FIGURE 11.   Plots of LHB vs. HP + SS + IP for fragments from ~30 different protein structures. (a) Alpha-helices; (b) beta-sheets.

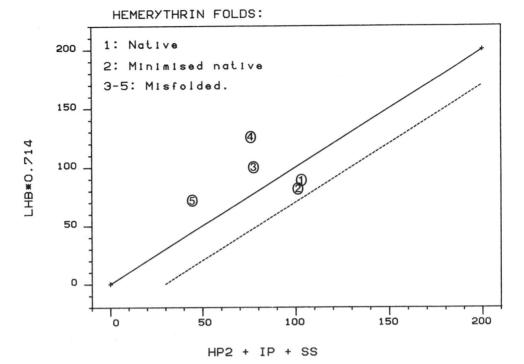

FIGURE 12. Plot of LHB vs. HP + SS + IP for misfolded hemerythrin structures (113 amino acid residues). 1: Native structure; 2: energy-minimized native structure; 3 and 4: misfolded structures; 5: hemerythrin sequence placed on RHE immunoglobulin beta-sheet backbone. The energies obtained from the energy minimizations were as follows: model 2 (native) = +1110 kcal/mol, model 3 = +1213 kcal/mol, and model 4 = +1341 kcal/mol.

the right of the plot. Calculations were also undertaken on isolated loop-type structures. The points for these fragments were well distributed above and below the line, showing no clear tendency for isolated loops.

This LHB model — based on relatively simple assumptions concerning the interactions both within the protein and to the solvent — provides an effective way of discriminating between non-native (unstable) and native or near-native (stable) protein structures. The alpha-helical hemerythrin structure[58] was used as an example to further test its usefulness, with some misfolded structures being generated by Freeman at the Royal Institution.[59] The 113-residue structure was randomly folded from an initial extended sheet by putting in a few turns and placing distance constraints between some hydrogens (turns and hydrogens were chosen at random). The structure was then energy-minimized *in vacuo*. Another misfolded protein was produced by superimposing the all-alpha-helical hemerythrin sequence onto the equal length all beta-sheet backbone of RHE immunoglobulin[60] and then minimizing the energy (similar to the work of Novotny et al.[61]).

Calculations using the LHB model were then undertaken on the misfolded structures. Figure 12 shows the resulting plot. The native structure (point 1) along with an energy-minimized native structure (point 2) lie below the line and both are stable. The two misfolded structures (points 3, 4, and 5) lie above the line indicating that they are unstable. The energies obtained from the minimization are given in the legend of Figure 12. Compared with the 10% differences in minimized energies, the LHB model appears to offer a more effective way of filtering out the probable non-native structures.

The application of the LHB model to polyamino acid structures produced some very surprising and useful results. Although several types of polyamino acids (including two residue types, e.g., leu-gln) were included, only the polyleucine and polyvaline results are discussed here. Ten residue strands were built up into four different model structures (Figure 13a):

1. B model: standard beta-sheet structure
2. R model: radial structure
3. P model: planar bilayer structure
4. A model: combination of structures R and P

Figures 13b and 13c show plots of the number of strands in the model structures vs. the energy calculated from the LHB model (i.e., energy = LHB − HP − IP − SS) for polyleucine and polyvaline. Several interesting features are apparent. First, the R model is far more stable (e.g., for the case of two strands) than is the B model. This arises from the fact that in the R model (for a low number of strands) all the polar atoms are very solvent accessible and are able to form their full complement of H-bonds to surrounding waters (Figure 13a). However, in the beta-sheet case B, the inner sheet carbonyl oxygens are quite inaccessible to solvent, thus losing a large number of H-bonds (Figure 13a, top). This suggests that radial-type structures may have implications for the early stages of protein folding.

Another finding was that for more than four strands in the case of polyvaline, and more than six strands in the polyleucine cases, the A structure is more stable than the other structures. This structure is very much like that of a lipid bilayer. It would appear that for a low number of strands (up to four to six depending on the size of the side chain) a radial structure is preferred, but as the number of strands increases, a hole develops in the R structure making it very unstable (the surface waters in the hole lose many H-bonds and the accessibility of the polar atoms becomes much reduced). The R-type structure can easily collapse into the more stable A bilayer structure resembling a mini-micelle as the strand number increases.

Sequential two-stranded radial structures (two main chains are not directly H-bonded to one another) are found at or near the center of many beta-sheet proteins and probably form small prefolding units early on in the folding process. Figure 14 shows such strands highlighted in the structures of several beta-sheet proteins (immunoglobulins, prealbumin, and superoxide dismutase). Another example of these non-H-bonded radial structures are the alpha-beta units which occur readily in the alpha-beta family of protein structures such as triose phosphate isomerase (TIM) (Figure 14).

In summary, the above qualitative calculations of the LHB model show a strong compensatory correlation in protein-water systems between the destabilizing factor of potentially lost H-bonds (LHBs) and known stabilizing factors relating to the assumed hydrophobic interaction (associated with buried surface areas), ion pairs, and disulfide bridges. A maximization and change of the H-bonding capacity appear to play important roles in the different structures protein-water systems adopt whether the proteins are fully unfolded or partially or fully folded.

VI. SUMMARY

Although solvent effects play a significant role in biomolecular processes, such as protein folding, enzyme-substrate interactions, and protein association, our understanding of the detailed mechanisms involved remains sketchy. This is partly because of the complexity of such systems together with the apparent subtlety of many of the solvent-related effects thought to be important.

Thermodynamic and structural considerations suggest that protein stability — which is

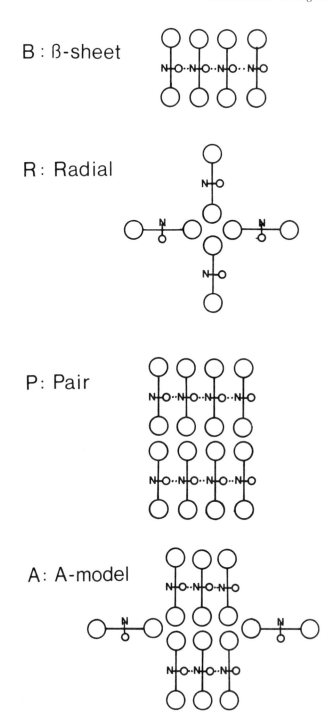

FIGURE 13.   (a) Schematic of four model polyamino acid structures built from beta-strands containing ten residues each: B, standard beta-sheet structure; R, radial structure; P, planar bilayer structure; and A, combination of R and P structures. The beta-strands are viewed looking down the chain with the circles representing the side chains. Shown in (b) and (c) are plots of the number of strands in the model structures vs. the energy calculated from the LHB model, that is, E = LHB − HP − SS − IP, for polyleucine (b) and polyvaline (c).

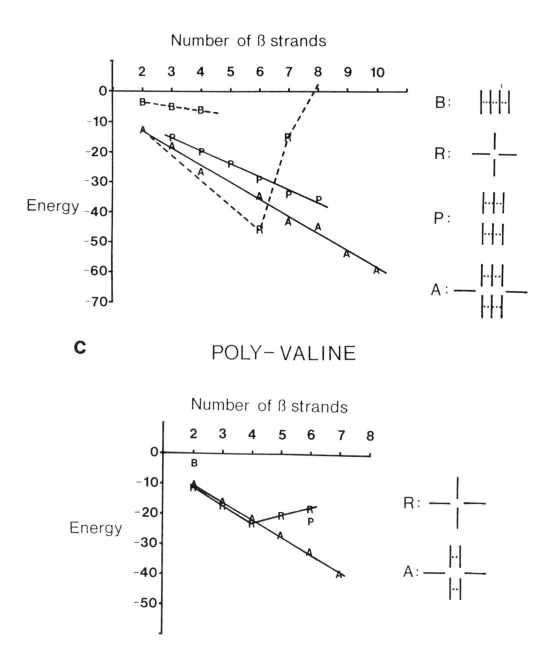

FIGURE 13 (continued)

marginal — relates to the small difference between several much larger competing driving forces. Although conventional wisdom has for a long time asserted that the so-called hydrophobic interaction dominates such processes, we find theoretical evidence for this assertion unconvincing and experimental evidence weak. In fact, there is increasing evidence suggesting that hydrophobic dominance is an erroneous assumption. That the free energy of stability is so small in relation to the quality of our knowledge of the elementary

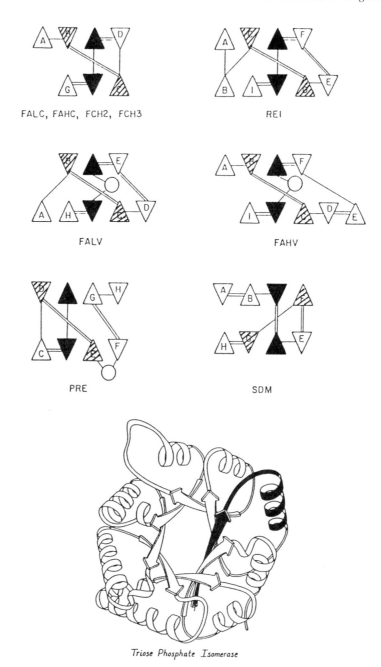

FALC, FAHC, FCH2, FCH3

REI

FALV

FAHV

PRE

SDM

Triose Phosphate Isomerase

FIGURE 14.   Known protein structures in which consecutive two-strand segments forming radial-type structures occur. The beta-sheet structures of some immunoglobulins (FALC, REI, FALV, FAHV), prealbumin (PRE), and superoxide dismutase (SDM) are shown. Triangles represent beta-strands and circles represent alpha-helices. There are no mainchain-mainchain H-bonds between the highlighted sequential strands, but buried apolar groups are included between the strands. In these structures, the two-stranded radial-type arrangements are shaded — either stripped or filled — and usually occur near the center of each protein. An example of an alpha/beta protein is shown at the bottom — triose phosphate isomerase (TIM) — with one of the alpha/beta units highlighted (filled in). The arrows represent beta-sheet strands and the ribbon helices represent alpha-helices.

intermolecular interactions concerned suggests that following this route of improving potential functions to obtain a better estimate of this small difference of several large numbers may not be the most profitable way forward.

Structural regularities both in respect to folded protein structures and neighboring solvent molecules have been identified from accurate diffraction studies of crystals of both proteins and small molecules. Particularly useful in the past few years have been high-resolution neutron studies capable of locating hydrogen atoms, work which has uncovered relatively simple "repulsive regularities" between water molecules which allow us for the first time to reliably rationalize the orientational structures of hydrogen-bonded systems including water. In addition to helping us understand water itself, these regularities allow us to build up more reliable hydration structures around proteins, where the resolution available from crystallographic studies is usually insufficient to allow direct determination of the details of the solvent structure.

Using these regularities found both in protein and in solvent interactions, a relatively simple "lost hydrogen bond" model can be developed which suggests the stability of proteins — and their component secondary structures as well as other related assemblies — can be qualitatively assessed to a high degree of reliability. The lost hydrogen bonds in a structure appear to be compensated effectively by three stabilizing components which can be quantitatively estimated using semi-empirical rules. Despite its relative simplicity, this theoretical device appears to be much more promising than trying to improve our quantitative knowledge of the details of the many intermolecular potential functions which ultimately must control the stability of protein-water systems.

In leading to the regularities used in this theoretical approach to protein and water systems, studies on smaller model systems have been crucial and new techniques presently being exploited in their early stages show further promise. One example discussed relates to structures in solution where neutron scattering can now give not only high quality information on the hydration of small (and increasingly medium-sized) molecules, but also on the perturbation of hydration by, for example, protein protectants and denaturants. Other areas of neutron scattering showing promise but not discussed include inelastic scattering studies of both model molecules and real biological molecules themselves. An example of the latter is work on the dynamics of proteins such as bovine pancreatic trypsin inhibitor[69] and myoglobin[70] as well as initial studies on nucleic acids. Probing changes in protein dynamics as a function of hydration can potentially throw more light on the role of the solvent in the behavior of such biologically important systems.

REFERENCES

1. **Cooke R. and Kuntz, I. D.,** The properties of water in biological systems, *Annu. Rev. Biophys. Bioeng.,* 3, 95, 1974.
2. **Finney, J. L., Gellatly, B. J., Golton, I. C., and Goodfellow, J.,** Solvent effects and polar interactions in the structural stability and dynamics of globular proteins, *Biophys. J.,* 32, 17, 1980.
3. **Finney, J. L., Goodfellow, J. M., and Golton, I. C.,** The structure and dynamics of water in globular proteins, in *Structural Molecular Biology,* Davies, B. D., Danyluk, S., and Saenger, W., Eds., Plenum Press, New York, 1982, 387.
4. **Edsall, J. T. and Mackenzie, H. A.,** Water and proteins. I and II, *Adv. Biophys.,* 10, 137, 1978; 16, 53, 1983.
5. **Finney, J. L.,** The role of water perturbations in biological processes, in *Water and Aqueous Solutions,* Neilson, G. W. and Enderby, J. E., Eds., Adam Hilger, Bristol, 1986, 227.
6. **Chothia, C.,** Hydrophobic bonding and accessible surface area in proteins, *Nature,* 248, 338, 1974.
7. **Lee, B. and Richards, F. M.,** The interpretation of protein structures: estimation of static accessibility, *J. Mol. Biol.,* 55, 379, 1971.

8. **Finney, J. L.,** The organisation and function of water in protein crystals, in *Water, A Comprehensive Treatise,* Vol. 6, Franks, F., Ed., Plenum Press, New York, 47, 1979.

9. **Baker, E. N. and Hubbard, R. E.,** Hydrogen bonding in globular proteins, *Prog. Biophys. Mol. Biol.,* 44, 179, 1984.

10. **Shrake, A. and Rupley, J. A.,** Environment and exposure to solvent of protein atoms lysozyme and insulin, *J. Mol. Biol.,* 79, 351, 1973.

11. **Finney, J. L.,** Volume occupation, environment and accessibility in proteins: the problem of the protein surface, *J. Mol. Biol.,* 96, 721, 1975.

12. **Finney, J. L.,** Volume occupation, environment and accessibility in proteins. Environment and molecular area of RNase-S, *J. Mol. Biol.,* 119, 415, 1978.

13. **Richards, F. M.,** Areas, volumes, packing and protein structure, *Annu. Rev. Biophys. Bioeng.,* 6, 151, 1978.

14. **Richards, F. M.,** The interpretation of protein structures: total volume, group volume and distributions, and packing density, *J. Mol. Biol.,* 82, 1, 1974.

15. **Pain, R. H.,** Molecular hydration and biological function, in *Biophysics of Water,* Franks, F. and Mathias, S., Eds., John Wiley & Sons, Chichester, 1982, 3.

16. **Ross, P. D. and Subramanian, S.,** Thermodynamics of protein association reactions: forces contributing to stability, *Biochem.,* 20, 3096, 1981.

17. **Ben-Naim, A.,** Solvent-induced interactions: hydrophilic and hydrophobic phenomena, *J. Chem. Phys.,* 90, 7412, 1989.

18. **Dill, K. A., Privalov, P. L., Gill, S. J., and Murphy, K. P.,** The meaning of hydrophobicity, *Science,* 250, 297, 1990.

19. **Kauzmann, W.,** Some factors in the interpretation of protein denaturation, *Adv. Prot. Chem.,* 14, 1, 1959.

20. **Privalov, P. L. and Gill, S. J.,** The stability of protein structure and hydrophobic interaction, *Adv. Prot. Chem.,* 39, 191, 1988.

21. **Ferraris, G. and Franchini-Angela, M.,** Survey of the geometry and environment of water molecules in crystalline hydrates studied by neutron diffraction, *Acta Crystallogr. Sect. B,* 28, 3572, 1972.

22. **Chiari, G. and Ferraris, G.,** The water molecule in crystalline hydrates studied by neutron diffraction, *Acta Crystallogr. Sect. B,* 38, 2331, 1982.

23. **Blundell, T. L. and Johnson, L. N.,** *Protein Crystallograhy,* Academic Press, New York, 1976.

24. **Jeffrey, J. W.,** *Methods in X-Ray Crystallography,* Academic Press, London, 1971.

25. **Stout, G. H. and Jensen, L. H.,** *X-Ray Structure Determination,* Macmillan, New York, 1968.

26. **Ladd, M. F. C. and Palmer, R. A.,** *Structure Determination by X-Ray Crystallography,* Plenum Press, New York, 1977.

27. **Savage, H. F. J. and Wlodawer, A.,** Determination of water structure around biomolecules using X-ray and neutron diffraction methods, *Methods Enzymol.,* 127, 162, 1986.

28. **Savage, H. F. J.,** Water structure in crystalline solids: ices to proteins, *Water Sci. Rev.,* 2, 67, 1986.

29. **Savage, H. F. J.,** A study of the water structure in crystals of the vitamin B_{12} coenzyme, Ph.D. thesis, University of London, 1983.

30. **Savage, H. F. J.,** Water structure in vitamin B_{12} coenzyme crystals, *Biophys. J.,* 50, 947, 1986.

31. **Hendrickson, W. A. and Teeter, M. M.,** Structure of the hydrophobic protein crambin determined directly from the anomalous scattering of sulphur, *Nature,* 290, 107, 1981.

32. **Watenpaugh, K. D., Maregulis, T. N., Sieker, L. C., and Jensen, L. H.,** Water structure in a protein crystal: rubredoxin at 1.2 Å resolution, *J. Mol. Biol.,* 122, 175, 1978.

33. **Baker, E. N., Blundell, T. L., Cutfield, J. F., Cutfield, S. M., Dodson, E. J., Dodson, G. G., Crowfoot Hodgkin, D. M., Hubbard, R. E., Isaacs, N. W., Reynolds, C. D., Sakabe, K., Sakebe, N., and Vijayan, N. M.,** The structure of 2Zn pig insulin crystals at 1.5 Å resolution, *Philos. Trans. R. Soc. London,* 319, 369, 1988.

34. **Teeter, M. M.,** Water structure of a hydrophobic protein at atomic resolution: pentagonal rings of water molecules in crystals of crambin, *Proc. Natl. Acad. Sci.,* 81, 6014, 1984.

35. **Thanki, N., Thornton, J. M., and Goodfellow, J. M.,** Distributions of water around amino acid residues in proteins, *J. Mol. Biol.,* 202, 637, 1988.

36. **Teeter, M. M. and Kossiakoff, A. A.,** The neutron structure of the hydrophobic plant protein crambin, in *Neutrons in Biology,* Schoenborn, B. P., Ed., Plenum Press, New York, 1984, 335.

37. **Wlodawer, A., Walter, J., Huber, R., and Sjolin, L.,** Structure of bovine pancreatic trypsin inhibitor, *J. Mol. Biol.,* 108, 301, 1984.

38. **Mason, S. A., Bentley, G. A., and McIntyre, G. J.,** Deuterium exchange in lysozyme at 1.4 Å resolution, in *Neutrons in Biology,* Schoenborn, B. P., Ed., Plenum Press, New York, 1984, 323.

39. **Savage, H. F. J. and Finney, J. L.,** Repulsive regularities of water structure in ices and crystalline hydrates, *Nature,* 322, 717, 1986.

40. **Finney, J. L. and Savage, H. F. J.,** Impenetrability revisited: new light on hydrogen bonding from neutron studies on biomolecule crystal hydrates, *J. Mol. Struct.,* 177, 23, 1988.

41. **Hermansson, K.,** The Electron Distribution in the Bound Water Molecule, Ph.D. thesis, *Acta Univ. Ups.,* 9, 1984.

42. **Olovsson, I. and Jönsson, P. G.,** X-ray and neutron diffraction studies of hydrogen bonded systems, in *The Hydrogen Bond, Recent Developments in Theory and Experiments,* Schuster, P., Zundel, G., and Sandorfy, C., Eds., North-Holland, Amsterdam, 1976, 393.

43. **Franks, F.,** The hydrophobic interaction, in *Water: A Comprehensive Treatise,* Franks, F., Ed., Plenum Press, New York, 1975, 1.

44. **Tanford, C.,** *The Hydrophobic Effect,* John Wiley & Sons, New York, 1980.

45. **Jeffrey, G. A.,** Water structure in organic hydrates, *Acc. Chem. Res.,* 2, 344, 1969.

46. **Enderby, J. E. and Neilson, G. W.,** X-ray and neutron scattering by electrolytes, in *Water: A Comprehensive Treatise,* Franks, F., Ed., Plenum Press, New York, 6, 1, 1979.

47. **Finney, J. L. and Turner, J.,** Neutron scattering studies of molecular hydration in solution, *Electrochim. Acta,* 33, 1183, 1988.

48. **Finney, J. L. and Turner, J.,** Neutron diffraction studies of aqueous solutions of biological importance: an approach to liquid state structural chemistry, *Ann. N.Y. Acad. Sci.,* 482, 127, 1986.

49. **Luzar, A. and Soper, A. K.,** unpublished, 1990.

50. **Soper, A. K. and Finney, J. L.,** unpublished, 1990.

51. **Turner, J., Soper, A. K., and Finney, J. L.,** A neutron diffraction study of tetramethylammonium chloride in aqueous solution, *Mol. Phys.,* 70, 769, 1990.

52. **Turner, J., Finney, J. L., and Soper, A. K.,** Neutron diffraction studies of structure in aqueous solutions of urea and tetramethylammionium chloride and in methanol, *Z. Naturforsch.,* 46a, 73, 1990.

53. **Soper, A. K. and Silver, R. N.,** Hydrogen-hydrogen pair correlation functions in liquid water, *Phys. Rev. Lett.,* 49, 471, 1982.

54. **Pauling, L. and Corey, R. B.,** Atomic coordinates and structure factors for two helical configurations of polypeptide chains, *Proc. Natl. Acad. Sci.,* 37, 235, 1951.

55. **Pauling, L. and Corey, R. B.,** The pleated sheet: a new layer configuration of polypeptide chains, *Proc. Natl. Acad. Sci.,* 37, 51, 1951.

56. **Bernstein, F. C., Koetzle, T. F., Williams, J. B., Meyer, E. F., Brice, M. D., Rodgers, J. R., Kennard, O., Shimanouchi, T., and Tasumi, M.,** The protein databank, a computer based archival file for macromolecular structures, *Eur. J. Biochem.,* 80, 319, 1977.

57. **Bierzynski, A., Kim, P. S., and Baldwin, R. L.,** A salt bridge stabilises the helix formed by isolated C-peptide of RNase A, *Proc. Natl. Acad. Sci.,* 79, 2470, 1982.

58. **Klotz, I. M., Klippenstein, G. L., and Hendrickson, W. A.,** Hemerythrin: alternative oxygen carrier, *Science,* 192, 335, 1976.

59. **Freeman, C.,** Energy Calculations on Misfolded Proteins, unpublished, 1990.

60. **Furey, W., Wang, B. C., Yoo, C. S., and Sax, M.,** Structure of a novel Bence-Jones protein (rhe) fragment at 1.6 Å resolution, *J. Mol. Biol.,* 167, 661, 1983.

61. **Novotny, J., Rashin, A. A., and Bruccoleri, R. E.,** Criteria that discriminate between native proteins and incorrectly folded model, *Protein Struct. Function Genet.,* 4, 19, 1988.

62. **Gellatly, B. J. and Finney, J. L.,** Calculation of protein volumes: an alternative to the Voronoi procedure, *J. Mol. Biol.,* 161, 305, 1982.

63. **Eisenberg, D. and McLachlan, A. D.,** Solvation energy in protein folding and binding, *Nature,* 319, 199, 1986.

64. **Fauchere, J.-L. and Pliska, V.,** *Eur. J. Med. Chem. Chim. Ther.,* 18, 369, 1983.

65. **Barlow, D. J. and Thornton, J. M.,** Ion pairs in proteins, *J. Mol. Biol.,* 168, 867, 1983.

66. **Perutz, M. F.,** Electrostatic effects in proteins, *Science,* 201, 1187, 1978.

67. **Fersht, A. R.,** Conformational equilibria in alpha and delta-chymotrypsin, the energies and importance of the salt bridge, *J. Mol. Biol.,* 64, 497, 1972.

68. **Flory, P. J.,** Theory of elastic mechanisms in fibrous proteins, *J. Am. Chem. Soc.,* 78, 5222, 1956.

69. **Cusack, S., Smith, J., Finney, J. L., Tidor, B., and Karplus, M.,** Inelastic neutron scattering analysis of picosecond internal protein dynamics: comparisons of harmonic theory with experiment, *J. Mol. Biol.,* 202, 903, 1988.

70. **Doster, W., Cusack, S., and Petry, W.,** Dynamic transition of myoglobin revealed by inelastic neutron scattering, *Nature,* 337, 754, 1989.

Chapter 2

ENZYMES IN REVERSE MICELLES CONTAINING PHOSPHOLIPIDS

Alberto Darszon and Liora Shoshani

TABLE OF CONTENTS

I. INTRODUCTION

A considerable effort is now being devoted to the study of enzymes in media that are predominantly composed of organic solvents.[1-3] There are multiple reasons for this, among which some of the main ones are

1. (Many interesting and industrially important reactions must take place in hydrophobic environments with apolar substrates. Thus, resuspending or trapping functional enzymes, which are extremely competent catalysts, in hydrophobic environments has become very attractive. In addition, these systems also allow the use of soluble substrates)
2. In such apolar environments, under certain conditions, enzymes behave quantitatively and/or qualitatively different than in aqueous media. Some gain enormously in thermal stability, and others are capable of performing reactions that do not occur to a significant degree in water, i.e., reversing hydrolytic reactions.
3. Water participates directly or indirectly in all processes leading to protein thermal denaturation.[4-6] Thus, it is clear that systems which allow the experimenter to control the amount of water to which the enzyme is exposed can be advantageously used to study the water requirements of enzymatic catalysis and the participation of this solvent in thermostability and protein structure.

Basically two approaches have been used to study the properties of enzymes in organic media: (1) the direct suspension of a protein at various levels of hydration in organic solvents[7,8] and (2)(the transfer of proteins to organic solvents via "reverse micelles" that are formed with natural or synthetic surfactants with or without a cosurfactant.[9-17] In apolar organic solvents, under certain conditions which depend on the nature and concentration of the surfactant, reverse micellar aggregates form.[18,19] These aggregates, called reverse micelles, can be schematically described by a small water pool (20 to 200 Å diameter) enclosed by a surfactant monolayer. The hydrophobic chains of the surfactant molecules orient themselves toward the organic phase, while the polar heads define the surfactant-water interphase in the interior of the micelle) In this system, more than 40 enzymes have been shown to be active,[1,2] and it is thought that (soluble enzymes arrange themselves in the internal water space of the micelle)[20-23]

This chapter will focus on the second approach and particularly on enzymes in reverse micelles formed with phospholipids.

II. WHY USE PHOSPHOLIPIDS IN REVERSE MICELLE FORMATION?

(Phospholipids are the natural surfactants that constitute the backbone of biological membranes.) Membrane proteins immersed in the phospholipid bilayer are modulated by the physicochemical properties of the lipid ensemble and so are many soluble proteins which interact with the membrane interphase. Thus, phospholipids have a special interest as building blocks of model systems. Also, as will be discussed below, there are indications that reverse micelles may be present under certain conditions in biological membranes and that they may play an important role.[24]

In the study of membrane function, it has been difficult to obtain separate information about the physicochemical behavior of the internal and external compartments. This is due to the fact that various measurements record a weighted average of the intra- and extravesicular regions because of the exchange between them.[25] In this regard, and in others to be discussed later, reverse micelles could be a reasonable model for the membrane interior since the exchange with the outside is significantly reduced or slowed.[26]

In addition, cytoplasmic proteins seem to achieve extraordinarily high concentrations (150 to 250 mg/ml) which are difficult to obtain in a test tube.[27] Considering this, it is worthwhile to ask how much free water is present inside the cell and how the metabolites are transferred from one protein site to another. It has been argued that, (in many metabolic pathways, the enzymes interact closely forming multienzyme systems where substrates and products are transferred from one protein to the other without ever being in the bulk solution.[28,29] Reverse micelles offer the possibility of attaining such high protein concentrations in their reduced water space and, in principle, could allow two or more different proteins to be present in the same micelle. In this sense, reverse micelles, and particularly those formed from phospholipids, may offer an interesting model for what may occur inside the cell.)

A. THE BIOLOGICAL MEMBRANE

Biological membranes are essential elements of all living organisms. They are highly selective permeability barriers which compartmentalize cells and organelles creating specialized environments. They play a key role in the processing of information, excitability, secretion, endocytosis, and the interconversion of energy.

The main components of biological membranes are lipids, proteins, carbohydrates, ions, and water. The lipid-protein proportion varies greatly within different membranes, from ~20% protein in the myelin sheath to 75% in the inner mitochondrial membrane. The carbohydrates usually range from 1 to 10% of the total membrane dry weight.

Even though basically all biological membranes share the same components, each type contains special proteins and a particular mixture of lipids which endows it with functional specificity. In addition, the distribution of these membrane components is asymmetric. The proteins are vectorially oriented in the bilayer and each monolayer contains a different amount of individual lipid species. As a result of the asymmetric organization of the membrane, its functions are vectorial too. Different processes may take place at each side of the bilayer.

The current picture of the biological membrane is shown in Figure 1.[30] In the closely packed lipid bilayer, the hydrophylic head groups are oriented to the adjacent aqueous phase, while their long hydrophobic chains are secluded in the membrane interior where those of each monolayer meet and interdigitate. The membrane has a thickness of about 50 to 75 Å. Mainly, three types of proteins are present in membranes:

1. Proteins embedded in the bilayer, usually called intrinsic or integral — These proteins cross the bilayer via hydrophobic protein sequences and usually have exposed intracellular and extracellular domains.[31] They can transverse the bilayer once, as is commonly found for growth factor receptors, cell surface antigens, and adhesion molecules, or several times, as is commonly found in transport and energy transduction proteins. Integral membrane proteins are more or less free to move laterally depending on their interaction with other membrane proteins or those present in the cytoskeleton.[31]
2. Proteins which interact with the membrane interphase, either with membrane proteins or with the polar phospholipid heads — These proteins are sometimes called extrinsic or peripheral.
3. Proteins attached to the membrane via a fatty acid or a glycosyl-phosphatidylinositol anchor attached at the COOH-terminus through ethanolamine like alkaline phosphatase, acetylcholinesterase, and 5'-nucleotidase.[32]

B. LIPIDS

The lipids present in biological membranes can be divided into polar lipids (phospholipids and glycolipids) and nonpolar lipids (mono-, di-, and triglycerols, sterols, and steryl esters of long-chain fatty acids). Cholesterol is the most frequent sterol found, particularly in plasma membranes.

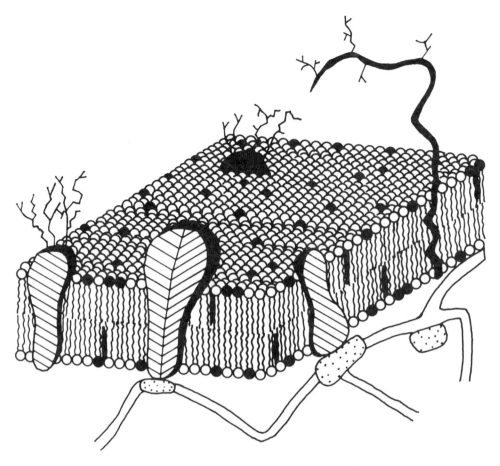

FIGURE 1.   Model of the structure of biological membranes.

Phospholipids are the most prominent components of biological membranes. They contain phosphoric acid as a mono- or diester. The two main members of the phospholipid family are the glycerophospholipids and sphingophospholipids.[33]

Probably the most important class of membrane lipids is the glycerophospholipids, which are derived from glycerol (Figure 2). These are the lipids that have been used to form reverse micelles and, because of this, special attention will be given to them. Glycerophospholipids contain two fatty acids, usually of 14, 16, or 18 carbons, esterified to the first two hydroxy groups of glycerol. The third primary hydroxy group is esterified to a phosphate which carries the polar head group. Frequently, glycerophospholipids from biological membranes contain an insaturated fatty acid in the middle position although both fatty acids may be insaturated.[34]

The most abundant phospholipids in higher plants and animals are phosphatidylcholine (PC, 40 to 50%) and phosphatidylethanolamine (PE, 30%). Phosphatidylserine (PS) and phosphotidylinositol (PI) are minor components of biological membranes (1 to 10%). Phosphatidylglycerol (PG) is generally present at a low concentration in animal membranes; however, it can amount to 10 to 20% in plant membranes, 50 to 60% in chloroplasts, and up to 70% of the membrane phospholipid in Gram-positive bacteria.

Sphingolipids are the second major class of lipids. They are derived from sphingosine or a similar long-chain amino alcohol (Figure 3). A hydrophylic head group (R) is esterified to the phosphate, and an amide bond binds a fatty acid (R^1) to the second carbon of sphingosine. The simplest sphingolipids are the ceramides, an example of which is sphingomyelin (SM), where R = choline ($-CH_2CH_2N^+(CH_3)_3$).

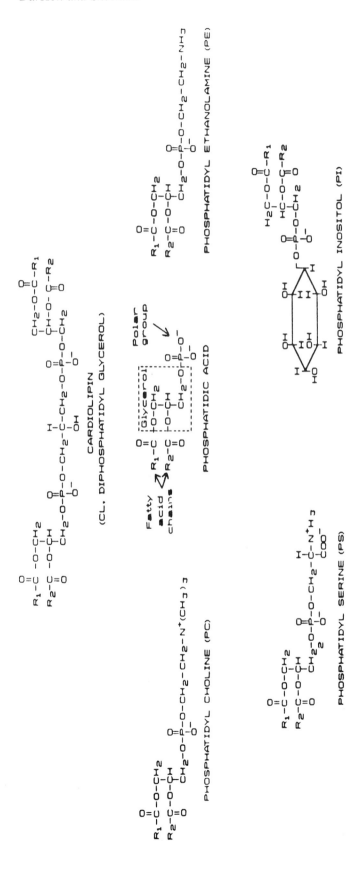

FIGURE 2. Structural representation of the main glycerophospholipids.

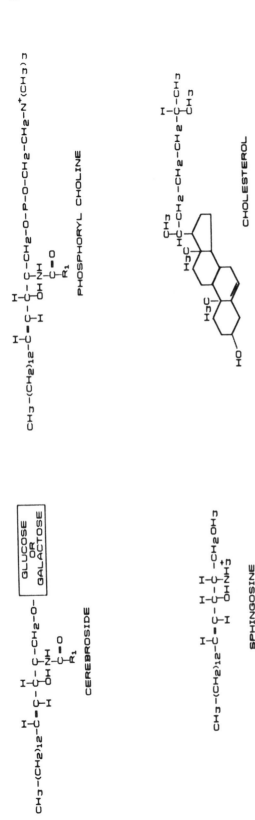

FIGURE 3.   Structural representation of other important biological lipids.

Phospholipids are the most polar lipids. At pH 7.0, the phosphate group has a negative charge since its pK_a is 1 to 2. Choline and ethanolamine are positively charged at neutral pH since the pK_a values of their amino groups are 13 and 10, respectively. Therefore, PC, PE, and SM are dipolar zwitterions with no net charge. PS, PI, and PG have a net negative charge at neutral pH.

Thus, it can be seen that there is a considerable range of polar head groups among phospholipids differing in charge, polarity, and size. In addition, there are multiple variations in length and degree of insaturation of the fatty acid chains of phospholipids. It is to be expected that the properties of the lipid bilayer are influenced by these variations which, together with the asymmetric distribution of these lipids between and within the two monolayers, must somehow modulate the functional status of the proteins found interacting with the membrane.

In spite of the enormous variety of phospholipids, they all have a fundamental common characteristic: they are amphipathic molecules. Two distinct domains are present in amphipaths: a polar hydrophylic region (the head) and the hydrophobic chains (tails). The consequence of the simultaneous presence of these two domains in phospholipids governs their schizophrenic behavior toward water. This behavior, referred to as lyotropic and termotropic mesomorphism, is responsible for their capacity to associate in various hydrated phases depending on the temperature and the water content. Under physiological conditions, biological phospholipids spontaneously form hydrophobic boundaries.

III. PHOSPHOLIPID-WATER AND PHOSPHOLIPID-PHOSPHOLIPID INTERACTIONS

A. THEORETICAL CONSIDERATIONS

Theoretical studies have contributed valuable information towards the understanding of the forces that underlie lipid-lipid interactions in water leading to the formation of micelles and bilayers.[35-36] These studies will be summarized since it is thought that their conclusions illuminate some of the properties and possible interactions of phospholipids in reverse micelles. It is worth emphasizing that all the interactions that will be discussed are interdependent.

It has been argued that the reluctance to break the hydrogen bonds between water molecules is the main reason it is energetically favorable for lipids to aggregate.[35] Tanford has called this the "hydrophobic effect" and believes that it is the main driving force of lipid organization rather than the attraction between the hydrophobic alkyl chains of phospholipids.

London-van der Waals' dispersion forces are also thought to participate in the interactions between nonpolar residues. They depend on steric factors and on the sixth power of distance between the residues. The attraction energy for two CH_2 groups 5 Å apart is about 100 cal mol^{-1}. If hydrocarbon chains are closely packed, the total interaction between them can reach 1 to 20 kcal mol^{-1}.[37] Increasing the chain length of the acyl chains increases the van der Waals attractive forces between them, while increasing the number of *cis*-insaturations decreases these forces.

Coulombic forces are also important in the interactions between the phospholipid polar heads and with the polar residues of proteins. These interactions occur mainly at the bilayer-water interface. Such forces depend on the first power of the distance between the charges and would result in a repulsive force of around 4 kcal mol^{-1} for two unit charges found close to each other.[37]

Since phospholipids are closely packed in the lipid aggregates, steric repulsion forces must also be considered. These interactions only depend on parameters of size and shape and can be modeled by particles such as cylinders, cones, and spheres.[38]

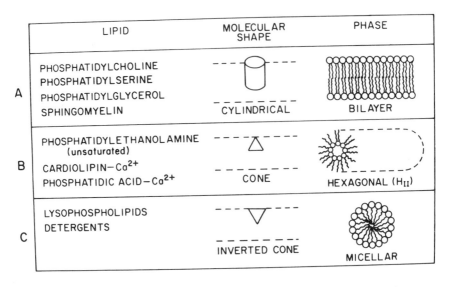

LIPID	MOLECULAR SHAPE	PHASE
A	PHOSPHATIDYLCHOLINE PHOSPHATIDYLSERINE PHOSPHATIDYLGLYCEROL SPHINGOMYELIN	CYLINDRICAL	BILAYER
B	PHOSPHATIDYLETHANOLAMINE (unsaturated) CARDIOLIPIN—Ca²⁺ PHOSPHATIDIC ACID—Ca²⁺	CONE	HEXAGONAL (H_II)
C	LYSOPHOSPHOLIPIDS DETERGENTS	INVERTED CONE	MICELLAR

FIGURE 4.   Organization of lipids (and detergents) in water and their molecular shape.

The changes in the energy of association of lipids determines which aggregates are thermodynamically favorable and what type of structures will be generated. For example, in the transfer of an alkyl chain from water into bulk liquid hydrocarbon, about 820 cal mol^{-1} are gained per CH_2 and 2100 cal mol^{-1} for the terminal CH_3 groups. It has been demonstrated that the density of molecules (X) incorporated into aggregates of number N can be expressed in the following equation:

$$X_N = N\{X_1 exp[(\mu_1^0 - \mu_N^0)kT]\}^N \tag{1}$$

where N is the aggregation number, μ_N^0 is the mean free energy per molecule in the aggregate, k is Boltzmann's constant, and T is the temperature. As N increases, μ_N^0 must decrease for large stable aggregates to form. The variation of μ_N^0 with N is probably complex and different types of structures may coexist. On the other hand, the size and shape of the aggregates will depend on how the molecules can pack geometrically and the intermolecular forces between them. The interrelationship of these parameters has been shown to determine whether bilayers, spherical micelles, or reverse micelles will form.[39]

B. EXPERIMENTAL RESULTS

The most common phases adopted by phospholipids in water have been reviewed recently.[40] These structures have been determined mainly by X-ray diffraction,[41,42] nuclear magnetic resonance (NMR),[34] and electron microscopy.[43]

Figure 4 schematically illustrates the three most common lipid phases and how they relate to the critical packing shape of these molecules. Lysophospholipids have a larger volume in the head than in the acyl chain and thus have an inverted cone shape which favors the formation of micelles (Figure 4A).

In lipids such as PC, PS, and PG, the hydrocarbon volume of the two acyl chains is similar to that of the polar head, favoring the formation of bilayers (Figure 4B). In the bilayer, the polar heads of phospholipids lie parallel to the plane of the bilayer.[34,44] The phospholipid acyl chains extend into the membrane bilayer forming its core. However, the first two atoms of the sn-2 chain lie parallel to the bilayer surface. The consequence of this is that, for a phospholipid with the same fatty acyl groups, the sn-2 chain extends less into the bilayer than the sn-1.[34,44]

When isolated, unsaturated PE and phosphatidic acid (PA) and cardiolipin in the presence of Ca^{2+} are among the lipids which can form a hexagonal 7II phase (H_{II})[24] (Figure 4C). The formation of such phases, where the lipids are packed as 20 Å-diameter aqueous cylinders, is related to the form of the molecule. Lipids with head areas smaller than their tails are cone shaped and form inverted micelles of the H_{II} types.[24,45] This type of nonbilayer lipid phase will be further discussed later.

Irregularities in experimental parameters have been detected in the behavior of pure lipid-water systems, with respect to temperature and water content, by X-ray diffraction, dilatometry, NMR, and electron spin resonance (ESR).[46,47] These irregularities or discontinuities in the experimentally observable parameters are called "phase changes" or "phase transitions" and are characterized by a transition temperature (T_c) at which they occur.[34]

At T_c, model membranes undergo a phase transition in which the phospholipid acyl chains change from an ordered (gel) to a more disordered (liquid crystalline) state.[24,48] In the gel state, the acyl chains are in the extended all-*trans* configuration, relatively rigid, and packed in a hexagonal lattice, but generally tilted relative to the plane of the bilayer. Above T_c, in the liquid crystalline phase, the acyl chains lie perpendicular to the surface. Both the acyl chains and the polar head become more mobile. A bilayer in the liquid crystalline phase ($\sim$45 Å)[44] is $\sim$15% thinner than in the gel phase.[49] As a result of this, the acyl chains of the two monolayers must either interdigitate in the fluid state or the methyl ends of the chains fold and curl within each monolayer.[50,51]

When mixtures of lipids are present, as is the case in biological membranes, the physical properties of each lipid species will influence their arrangement. Ideal mixing in both the liquid crystalline and gel phase is expected to occur for lipids that only differ by one or two carbon atoms in their acyl chains; in contrast, dissimilar lipids having a heterogeneous gel phase will tend to undergo phase separations into domains corresponding to the pure components (Figure 5). Lipid domains have been documented under physiological conditions.[52-55]

The melting of the acyl chains, the transitions from one mesophase to another, and the phase separations into lateral domains involve changes in the heat content of the system.[34] Valuable information has been obtained using differential scanning calorimetry to measure T_c and the excess specific heat of the transition enthalpy (ΔH).[56,57] Among the main conclusions reached are

1. An increase in the chain length of the acyl chains increases ΔH, T_c, and the transition temperature ΔS.
2. The addition of *cis* and *trans* double bonds lowers T_c. *Cis* double bonds have a more profound effect than *trans* double bonds. Two double bonds on the sn-2 position have the maximum effect on T_c for the case of PC; further increases in insaturation had no effects.[58]
3. T_c depends on the polar head. For lipids with the same saturated acyl chains the order is T_c (choline) $=$ T_c (glycerol) $<$ T_c (serine) $<$ T_c (ethanolamine). The polar head groups have little effect on the transition enthalpies and entropies at a given chain length.

What is the reason for the extraordinary lipid diversity of biological membranes? — Approximately 240 different species of phospholipids have been estimated to be present in the human erythrocyte membrane.[59] There are more than 23 naturally occurring fatty acids and a variety of polar heads in phospholipids. Why is such a diversity necessary? One possible reason would be to generate microdomains to accommodate and subtly regulate membrane protein function. Cullis and De Kruijff[60] have indicated that this diversity may allow the simultaneous presence of lipids that have the ability to form nonbilayer phases and others that stabilize or form bilayer phases. It has already been mentioned that certain

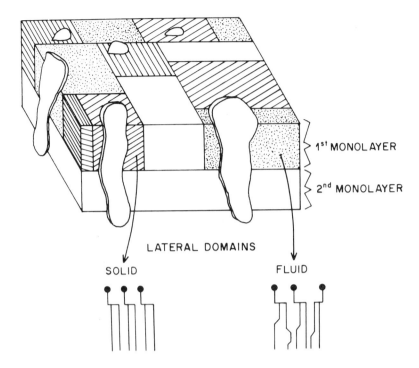

FIGURE 5. Schematic representation of lipid phase separations in biological membranes. Immiscibility between regions of different fluidity (fluid-less fluid, fluid-solid, etc.) may give rise to distinct domains. Only the different domains of the 1st monolayer are shown; however, the domains of the 2nd monolayer would be different. (Modified from Ref. 52.)

lipids (i.e., unsaturated PE) can adopt hexagonal phases.[61,62] Freeze fracture electron microscopy on mixed lipid model membranes indicates the likelihood of intrabilayer reverse micelles in biological membranes.[63,64] These nonbilayer phases could play an important role in two processes, membrane fusion and transbilayer exchange of lipids (flip-flop). The participation of reverse micelles in the process of fusion has been supported by studies with different vesicular model systems.[60,65,66] Apparently, reverse micellar structures appear at the regions where fusion occurs. On the other hand, Noordam et al.,[67] found that changes from a bilayer to a nonbilayer phase in a lipid dispersion containing 50% PE were accompanied by a threefold increase in the flip-flop of PC. However, nonbilayer lipid phases have not been clearly demonstrated in biological membranes.

IV. GENERAL CHARACTERISTICS OF REVERSE MICELLES CONTAINING PHOSPHOLIPIDS

Reverse micelles are spheroidal aggregates formed by certain amphiphilic molecules in a polar media. In this chapter, we will refer to reverse micelles containing phospholipids as those reverse micelles in which the phospholipids are the only amphiphilic molecules or those in which reverse micelles are formed of phospholipids and another cosurfactant.

The studies carried out with phospholipid-containing reverse micelles in organic solvent deal mainly with two parameters. The first includes the dimensions of reverse micelles in different organic solvents and the arrangement of the phospholipids in them; the second includes the amount of intramicellar water and its interaction with phospholipids. Unfortunately, very little information is available with respect to the first parameter. This is, in part, due to the fact that the strategies and models usually used to study the dynamics of micellar formation in water[39] are not suitable for reverse micelles in organic solvents.

A. PHOSPHOLIPID AGGREGATION IN ORGANIC SOLVENTS WITH LOW WATER CONTENT

One of the first attempts to determine the structure of phospholipid aggregates in organic solvents involved studying the osmotic pressure and the diffusion coefficient of lecithin (PC) in benzene.[68,69] The authors of this work found two types of reverse micelles in their system: small micelles formed by approximately four lecithin molecules with a molecular weight of about 3180 Da, and larger micelles that exist above a critical concentration of 730 μg/ml, formed by about 70 PC molecules with a molecular weight of 55,000 Da. According to their theoretical calculations, they suggested a laminar arrangement of monomers in the reverse micelles of PC in benzene. The same authors[70] have used light scattering to study PC reverse micelles formed in aliphatic alcohols. Their results indicated that, as the polarity of the solvent decreased, the micellar size increased and proposed a rod-like arrangement of PC molecules in the reverse micelles. Later, a few attempts were made to calculate the critical micellar concentration (CMC) for PC reverse micelles using NMR[71,72] (^{2}H, ^{13}C, and ^{31}P). A CMC of 11.1 mg/ml at 24°C and an aggregation number of about 3 were reported[71] for reverse micelles in CHCl$_3$. On the other hand, Boicelli and co-workers[72] have calculated that each PC reverse micelle in benzene contains 700 to 800 PC molecules and a molecular weight of 570,000 Da. The average micellar diameter they report for these reverse micelles is about 130 ± 20 Å (their system contains 30 μl of water per milliliter).

The above-mentioned results and the data from the literature indicate that phospholipid reverse micelles have the following general characteristics: (1) egg PC and DLPC (dilinoleyl PC) can aggregate to form small micelles of a few molecules in solvents such as chloroform, carbon tetrachloride, methanol, or ethanol. Larger micelles of about 20 to 100 molecules are formed in solvents such as benzene, chlorobenzene, hexane, and toluene. Thus, the molecular weight of reverse micelles of phosphatidylcholine can range from 3000 to 80,000; (2) usually, the aggregation number of phospholipid increases with Wo.

B. HYDRATION OF PHOSPHOLIPIDS IN REVERSE MICELLES

As mentioned earlier, the other parameter that has been studied extensively with reverse micelles containing phospholipids is the interaction between water and phospholipids. Most reports about the hydration of phospholipids (especially PC) in various systems of reverse micelles deal with two or three different phospholipid hydration states. Nevertheless, they do not all agree on the number of molecules required for each state.

Fung and McAdams[73] studied the interaction of water with phospholipids in reverse micelles by means of NMR. They showed that there is one water molecule tightly bound to the polar head in PC reverse micelles in CCl$_4$ which rapidly exchanges with the rest of the water molecules. In addition, Davenport and Fisher[74] have shown that this first water molecule is tightly bound to the phosphate group of PC and, as more water is added, the water molecules are less strongly bound, but are still much more motionally restricted than in free water.

Several water populations with distinct motional properties have been reported based on various methods. Poon and Wells[75] proposed the existence of three distinct hydration states for PC in an ethereal solution. The first involves about seven water molecules, and the second and the third hydration states require another 25 to 30 water molecules each.

Three different groups have studied the hydration of PC reverse micelles in benzene.[72,76,77] Walter and Hayes[76] determined two to three water molecules for the first hydration shell and four to six molecules for the second. The data of Gotthard and Stelzner[77] support the existence of one molecule of water tightly bound per phospholipid molecule plus another molecule that forms the first region of water. The second region is formed by five molecules of water per phospholipid and, when the water concentration is further increased, a third region is formed probably of water that aggregates in the micellar core. Their findings are

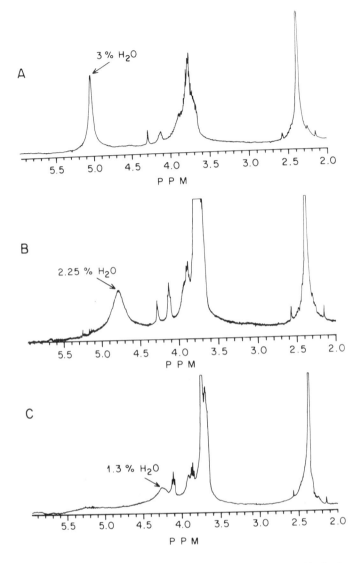

FIGURE 6.   300-MHz ^{1}H-NMR spectra of the TPT system (Asolectin 8.5 mg/ml in toluene/Triton X-100 85:15 %v/v) with H_2O. (A) 3%, (B) 2.25%, and (C) 1.3%. Notice the chemical shift and signal-broadening changes of the water peak. The toluene methyl signal was used as internal reference (δ, 2.36 ppm).[78]

in agreement with those of Walter and Hayes for the same system. However, Boicelli and co-workers[72] proposed ∼11 water molecules for the first and ∼12 for the second hydration shell. According to their data, free (bulk) water corresponding to the third hydration shell appears inside the micelle when more than 23 molecules of water per phospholipid are added.

The authors of this chapter have studied the NMR chemical shift of water protons in the soybean phospholipid/Triton X-100/toluene system.[78] It was found that at a low water concentration (0.33 to 1%), most of the water is bound. As the water concentration was increased, free water appears and above 4.5% v/v of water, the proton NMR properties of the system actually resemble those of free water (Figure 6).

Recently, a system of reverse micelles formed of soybean phospholipids and Triton X-100 in toluene has been described.[79] The conductivity and optical density of this system were determined to characterize it.[80] The formation of reverse micelles was found to depend on the water concentration. At low water concentrations (0.03 to 3% v/v) and at concentrations above 4.5%, the system is turbid, and between 3 to 4.5% — probably the region of reverse micelles — the system is transparent and stable. If phospholipids were omitted, Triton X-100 in toluene was turbid starting from 1% of water, and increasing amounts of water increased the turbidity, suggesting that Triton X-100 could not aggregate in toluene solution to form reverse micelles without the help of phospholipids. This finding is not surprising in light of the work of Kumar and Balasubramanian[81] with Triton X-100 reverse micelles in cyclohexane. They show that the presence of a long-chain alcohol was necessary to form reverse micelles. It is noteworthy that while these authors required 20% w/w of alcohol in the Triton X-100 mixture in order to solubilize water in stable reverse micelles, only 1% w/w of phospholipids in the Triton X-100 mixture is required for solubilizing the same amounts of water.

In summary, like in reverse micelles of synthetic surfactants,[82,83] water is distributed in the micellar cavity as bound water (in close contact with the polar heads of the amphiphilic molecules) and as bulk water, above a certain water concentration, which is specific for the type of phospholipid and solvent used. While the *size* of the reverse micelles of synthetic detergents is changed with added water,[84] in the case of reverse micelles containing phospholipids, the relation between Wo and the micellar size is not well established.

C. PHOSPHATIDYLCHOLINE GELS

An interesting and novel organization of lecithin in organic solvents has been reported by Scartazzini and Luisi[85] who describe the formation of organogels of lecithin in about 40 different solvents. The gels are formed in a lecithin concentration range between 50 and 200 mM. A Wo for the gel is defined as the critical [water]/[lecithin] ratio at which the gel is initially formed. They have shown that the organogels were formed at a very small and critical Wo (gel) which differed from one solvent to another. Luisi et al.[86] characterized the physicochemical properties of these organogels and proposed a model for these systems. They suggest that lecithin molecules aggregate into long cylindrical reverse micelles as water is added and at a certain concentration the cylindrical micelles start to entangle and form a dynamic network. They point out that organogels are formed only with pure lecithin; any mixture of phospholipids or other impurities prevent their formation.

D. CONCLUSIONS

Phospholipids tend to aggregate in organic media to form reverse micelles with the polar heads localized in the interior of the micelle and the hydrocarbon chains forming the outer shell of the micelle (Figure 7). The structure (spheroid, ellipsoid, laminar, or gel-like), the critical micellar concentration, the aggregation number, and the molecular weight of reverse micelles containing phospholipids depend on the following parameters:

1.   The polarity of the organic solvent
2.   The amount of water solubilized
3.   The existence of guest molecules such as peptides, amino acids, or enzymes
4.   The types of phospholipid present

V. ENZYME ACTIVITY IN REVERSE MICELLES CONTAINING PHOSPHOLIPIDS

A. SOLUBLE ENZYMES
1. General Characteristics of Catalysis in Reverse Micelles Containing Phospholipids

Most of the soluble enzymes that have been studied in reverse micelles containing phospholipids present very similar characteristics as those in synthetic reverse micelles. In

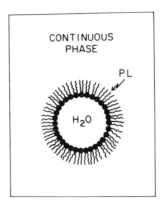

FIGURE 7.   A schematic representation of a phospholipid-containing
reverse micelle.

both systems, enzyme activity depends on the amount of water present in the apolar media.
Table 1 summarizes the studies carried out with soluble enzymes in systems of reversed
micelles containing phospholipids, mainly with lecithin. Only a few of them will be discussed.

Alkaline phosphatase has been studied in reverse micelles of different amphiphile molecules (AOT, PC, PEA, PA) in *n*-heptane.[87] Interestingly, it was found that in all of these
systems, the enzyme was active only when the substrate was trapped in reverse micelles of
AOT. The optimal pH of the reaction in reverse micelles of PC was shifted from pH 10 in
bulk water to pH 10 to 11 in the reverse micelles. The phosphatase activity at the optimal
pH (11) and water content (Wo = 14.8) approached the maximal activity in bulk water.
Varying the water content did not alter substrate specificity or K_m, while V_{max} did depend
on this parameter. The authors discuss the dependence of V_{max} on Wo and suggest that
increasing the free water content in the micellar cavity could increase the water concentration
used for hydrolysis. Nevertheless, the authors fail to discuss why no enzyme activity could
be measured unless the substrate was in AOT reverse micelles. Possibly, the limited water-
solubilizing capacity of phospholipid reverse micelles does not allow enough free water for
the protein mobility required for catalysis.

Superactivity is another interesting phenomenon reported for acid phosphatase[88] and
horseradish peroxidase.[1,89] Both enzymes were incorporated into reverse micelles of egg
lecithin in methanol/pentanol/octane (1:2:37, v/v/v). For the former enzyme, the catalytic
activity at Wo = 11.1 and pH 3 was higher than in bulk water. When the lecithin concentration was below 10 to 30 mM, V_{max} increased about 60-fold, and at higher concentrations
it decreased. Horseradish peroxidase, at Wo = 13 and pH 7, behaved in a similar way. At
low lecithin concentrations, the activity was higher than in the aqueous medium. Increasing
the lecithin concentration decreased the activity and at about 30 mM it was lower than in
an aqueous medium. It is noteworthy that this system contained 7.5% (v/v) of alcohol as
cosurfactant which could explain the superactivity found.

Lactate dehydrogenase has been studied in a system of reverse micelles containing
soybean phospholipids (8.5 mg/ml) and Triton X-100 (15% v/v) in toluene.[90] The activity
of this enzyme in 3.8% (v/v) of water is about 30 times lower than in a totally aqueous
medium. It was found that in the presence of elevated concentrations of the denaturant Gdn-
HCl (1.5 and 2.0 M), the activity increased approximately 20 times, reaching about 60%
of the maximal activity in an aqueous medium. Although Triton X-100 increases the total
amount of water solubilized in the phospholipid-containing micelles,[80] the water content
(Wo ~ 10) is probably still limited in order to express maximal activity. It was suggested

TABLE 1
Soluble Enzymes in Reverse Micelles Containing Phospholipids

Enzyme	Experimental system	Reaction catalyzed	Wo
Acid phosphatase from wheat germ[88]	Egg PC (1 to 50 mM) in methanol/pentanol/octane (1:2:37, v/v)	Hydrolysis of *p*-nitrophenyl phosphate	11.1

At low PC the catalytic activity is higher than in aqueous solution; 60-fold increase in V_{max}; activity decrease with increasing PC.

Alkaline phosphatase from calf intestine[87]	Soybean PC (50 mM) alone or in a mixture with PE and PA or AOT in heptane	Hydrolysis of *p*-nitrophenyl phosphate	10

Optimal pH shift to 11; enzyme is active in all reverse micelles systems just when the substrate is in AOT reverse micelles.

α-Chymotrypsin from bovine pancreas[91]	Dihexanoyl PC or diheptanoyl PC (30 mM in isooctane/hexanol (9:1, v/v)	Hydrolysis of *N*-glutaryl-Phe-NH-NP	3—15

Maximal activity around Wo = 8.

Mitochondrial F_1-ATPase[122]	Asolectin (8.5 mg/ml) in toluene	Hydrolysis of ATP	10

Enzyme becomes resistant to cold denaturation and acquires remarkable thermostability. Catalytic activity is very low in comparison with aqueous media.

Lactate dehydrogenase from bovine heart[90]	Asolectin (8.5 mg/ml) in toluene/Triton X-100 (85:15 % v/v)	Pyruvate + NADH → lactate + NAD$^+$	10

Enzyme activity at Wo = 10 is about 30 times lower than in standard water mixture; 1.5 to 2.0 M guanidine chloride increased the activity by approximately 20 times.

Lipase from *Rhizopus delemar*[92]	Soybean PC (25 mM) in *n*-hexane	Synthesis of triacylglycerol	2.5—15

Maximal activity at Wo = 10. Enzyme is active at pH from 5 to 9 (optimal pH in bulk water is about 5.6).

Lipase from *R. delemar*[94]	Soybean PC (50 mM) + palmitic acid (25 mM) in isooctane	Hydrolysis of *p*-nitrophenyl palmitate	2—5.5

Maximal activity at Wo = 3.3 and pH 5.5.

Lysozyme from egg white[143]	Dihexanoyl PC in cyclohexane/hexanol (9:1, v/v)	Hydrolysis of 4-methylumbelliferyl-*N,N',N''*-tri-acetyl-β-chititrioside	5—17

Enzyme retains activity. The enzyme is in a functional state at pH 5 according to spectroscopic studies.

Peroxidase from horseradish[1,89]	Egg PC (5 to 30 mM) in methanol/pentanol/octane (1:2:37, v/v/v)	Peroxidation of 2-pyrogallol	13

Catalytic activity depends on PC. At low PC the activity is higher than in aqueous solution. With increasing PC the activity decreases.

Trypsin from bovine pancreas[91]	Dihexanoyl PC or dioctanoyl PC (20 mM) in isooctane/hexanol (9:1, v/v)	Hydrolysis of BzPheValArg-NH-NP	3—15
	Diheptanoyl PC (20 mM) in isooctane/hexanol (9:1, v/v)	Hydrolysis of BzArg-OEt	3—15

Maximal activity for both reactions about Wo = 8.

that Gdn-HCl inside the water core of the reverse micelles facilitates the protein-solvent interactions and the protein mobility required for the catalytic cycle (see Chapter 5 for further discussion).

Recently, a study of α-chymotrypsin from bovine pancreas has been reported in a system of dihexanoyl PC in isooctane/hexanol (9:1, v/v).[91] The dependence of the enzymatic activity in the Wo has been studied in the range of 4 to 15 and an optimal value of Wo = 8 for this enzyme has been reported.

As illustrated in Table 1, a wide range of enzymatic activities can take place in reverse micelles containing phospholipids. Generally, it appears that enzyme activity in reverse micelles containing PC depends on the lipid and water concentration present in the system. The reader should notice that all the micellar systems specified in Table 1 contain an alcohol (methanol, pentanol, and hexanol) or a synthetic cosurfactant (AOT and Triton X-100), in addition to the phospholipid, except for the study of the lipase from *Rhizopus delemar*.[92] In this latter case, phosphatidylcholine was solubilized in pure *n*-hexane, but the reaction mixture contained 1,2-diacylglycerol (or an aliphatic alcohol) and fatty acids as the reactants for triacylglycerol synthesis. These compounds could participate in the formation of the reverse micelles. It is important to emphasize once again the fact that reverse micelles of phospholipid in organic solvent can be hydrated to a very limited extent, and this is probably the reason enzymes (alkaline phosphatase, F_1ATPase, and lactate dehydrogenase) display a low activity in such systems which require the presence of a cosurfactant, or alternatively a denaturant, to improve activity.

2. Effect of Wo on Catalytic Activity in Reverse Micelles Containing Phospholipids

The dependence of catalytic activity on the molar ratio of water to surfactant (Wo) for synthetic surfactants is usually represented by a bell-shaped curve. There is an optimum value of Wo for which the activity of the solubilized enzyme is maximum. In many instances, this is also the case for reverse micelles containing phospholipids. Nevertheless, the curves are not always bell-shaped. Bru et al.[93] discuss, in a theoretical study, the localization and catalysis of enzymes in reverse micelles of synthetic detergents. They propose a model which attempts to explain the different shapes of the activity vs. Wo curves depending on the specific location of the enzyme in the interior of the reverse micelle. They consider three basic locations — free water, bound water, and surfactant apolar tails — and their considerations which allow a reasonable fit of experimental results.

In the systems of synthetic surfactants (AOT, CTAB, Brij-96), this relationship has been explored extensively. It was found that at optimal values of Wo with respect to activity, the inner dimension of the micelle corresponds to the size of the enzyme molecule entrapped.[1] Unfortunately, this relationship has not been studied systematically for reverse micelles containing phospholipids. However, there is evidence suggesting that the dependence of enzyme activity on water content is not exclusively a matter of micellar dimension or total content of water.

Morita and co-workers have studied the activity of a lipase from *R. delemar* in phosphatidylcholine reverse micelles in *n*-hexane.[92] They determined synthesis of triacylglycerols in a range of Wo between 2.5 and 15, and they obtained a very shallow bell-shaped curve of enzymatic activity with an optimal value at Wo = 10. In a recent work of Schmidli and Luisi[94] with the same enzyme, the hydrolytic activity was studied in a system of PC and caprylic acid in isooctane. In a range of Wo between 2 and 5.5, an optimal value at Wo = 2 was reported. Hence, two different Wo values were determined for the optimal activity of the same enzyme entrapped in two distinct reverse micellar systems. This could be explained by differences in the amount of water molecules bound to the surfactants in the two systems which would result in different amounts of free water available for catalysis. In addition, differences in the dimensions of the reverse micelles and the nature of the water-surfactant interphase could have pronounced effects on the optimum Wo required for maximum activity.

B. MEMBRANE PROTEINS
1. Historical Perspectives

It is somehow remarkable that although integral membrane proteins would seem very attractive candidates to be studied in reverse micelles, they are seldom studied in these models.

This family of proteins strongly interacts with the hydrophobic region of the bilayer; thus, its incorporation into reverse micelles could yield information regarding protein function and arrangement in biological membranes. From this perspective, the surfactants of choice would be the phospholipids. It has already been discussed that phospholipid reverse micelles may play an important role in some membrane functions such as fusion and flip-flop. Since biological membranes contain integral membrane proteins, it is also relevant to explore how these proteins behave in such systems and alter their properties.

Since the early 1950s, organic solvents have been used to extract very hydrophobic small polypeptides, often called proteolipids, from biological membranes.[95-98] Also during the 1950s, it was found that soluble protein-lipid complexes that formed in water could be extracted into apolar solvents and display catalytic activity.[99,100]

The initial motivations to transfer functional membrane proteins into phospholipid reverse micellar systems arose from the need to have them in a volatile organic solvent for their reassembly into planar bilayers.[101] Most of the work carried out with membrane proteins has involved their transfer from an aqueous media into alkanes containing phospholipids. It is interesting that in the initial studies there was not a clear view of the structure of the lipid-protein complexes, and reverse micelles were thought of as possible structures which could allow protein solubilization in the organic phase.[102,103] This occurred in spite of the fact that studies about reverse micelles,[104] even with soluble proteins,[99,100] had already started to appear.

Early studies had shown that cytochrome *c* could be transferred into isooctane using a mixture of neutral and acid phospholipids.[105] It was thought that the acidic phospholipids neutralized the positive charges of cytochrome *c*, favoring its partition into the solvent. Later, it was found that a low pH or cations, Ca^{2+} being the most efficient, enhanced the partition of the cytochrome *c*-phospholipid complexes.[103] The cytochrome *c* extracts were used to form black lipid membranes, uncovering the potential application of this approach to the reconstitution of membrane functions.[103]

Thus, the charge neutralization principle was applied to detergent-solubilized and partially purified membrane proteins. It was thought that since membrane proteins usually have hydrophylic and hydrophobic regions, their partition into apolar organic solvents would be enhanced by neutralizing the charges of the protein.[101,106] Cytochrome oxidase, the last component of the mitochondrial respiratory chain, was transferred efficiently (90%) into hexane using soybean phospholipids (a mixture containing mainly PC, PE and PS). The organic extract was transparent and displayed similar spectroscopic characteristics as functional detergent-solubilized cytochrome oxidase. The partition of this enzyme depended on the presence of Ca^{2+}. Liposomes formed from the organic extract by evaporating the solvent and rehydrating the protein lipid residue retained enzymatic activity, and so did monolayers at an air-water interphase.[101]

Thereafter, rhodopsin, the visual pigment from bovine retinal rods, was chosen as a model membrane protein to further develop a general strategy to transfer functional protein-lipid complexes into apolar organic solvents for their reconstitution in planar bilayers.[9,10] Two complementary procedures were developed, one to transfer detergent-solubilized and purified membrane proteins and the other to directly transfer proteins from biological membranes into apolar solvents using soybean phospholipids as surfactants. The first procedure allowed the purified membrane protein to be delipidated and then reassembled into vesicles made of a single synthetic lipid in the water phase. Thereafter, these vesicles were sonicated in the presence of organic solvents, a cation was added and the suspension was mixed mechanically, and the two phases were separated by centrifugation. The second procedure is very similar, with the exception that a concentrated pellet of membranes is used instead of the aqueous suspension of lipid vesicles with the reassembled purified protein. Later it was shown that the second procedure can be applied to membrane proteins at any stage of purification from the biological membrane to the purified protein.[107]

The choice of solvent, and the amount of protein and phospholipids, is determined by the preservation of biological activity and the efficiency of extraction. Hexane and other alkanes have turned out to be suitable for the membrane proteins tested to date. Pure ethyl ether could be used to perform a second extraction on the emulsified aqueous phase left after the first extraction. The advantage of this second ether extract is that it has a lower lipid-to-protein ratio ($\sim$200) than the first extraction ($\sim$1000); this allows the experimental control of this ratio from 200, which is close to what is found in the photoreceptor disc membranes, to thousands. Within a certain range, it is possible to control the concentrations of almost all the components present in these extracts[9,10,107] including water (Wo from $\sim$3 to $\sim$10 in the first extract, with even higher values in ether[108,109]).

Rhodopsin can be transferred with good yields ($>$70%) into hexane and ether in the presence of phospholipids using the procedures mentioned. This visual pigment retains its photochemical activity in hexane containing phospholipids, as indicated by its dark and bleached spectra and the kinetics of the metarhodopsin I-II transition. Since it is known that this transition requires the presence of water,[110,111] the fact that it occurs in the apolar media indicated that water was present and accessible to the protein.[112]

In addition, the bleached protein, opsin, can spontaneously recombine in hexane containing phospholipids with either 11-*cis* or 9-*cis* retinal to generate rhodopsin or isorhodopsin, respectively, with a yield of $\sim$90%.[112] This process, which occurs under physiological conditions, is very sensitive to the native state of the protein.[113] Its occurrence in the apolar media indicated that a physiologically relevant biochemical reaction was achieved.

Other membrane proteins have been transferred in their functional state into phospholipid-containing apolar solvents using the procedures described above (Table 2). From them, two examples will be further discussed: the purified reaction centers from photosynthetic bacteria, and some of the proteins from the internal mitochondrial membrane which are directly transferred from the biological membrane into the phospholipid-containing apolar media.

The photosynthetic reaction center protein (RC) purified from photosynthetic bacteria was transferred into hexane and other apolar solvents using soybean phospholipids as surfactants.[114-117] Mg^{2+} was used in the extraction procedure, although Ca^{2+} could replace it. The absorbance spectra in the infrared (IR) and visible regions of the transferred RCs was similar to that of the aqueous detergent-solubilized protein. The RC photoactivity in the apolar media was almost fully restored after adding exogenous ubiquinone (Q_{10}) to replace the one removed from the protein by the solvent.[116] EPR spectroscopy indicated that the light-induced EPR absorbance changes and the low-temperature kinetics of the back reactions in the dark were similar in the organic solvent RC extracts and in reaction centers from chromatophore membranes or when purified. The composition of the RC-lipid complexes in hexane, expressed in molar ratios, were determined to be phospholipid/RC = 4000 to 6000, detergent/RC = 400, H_2O/phospholipid = 40, and Mn^{2+}/RC = 2.

The reaction center-lipid complexes in hexane were able to react with exogenous electron carriers dissolved in the bulk solvent; thus, they were capable of achieving biochemical reactions in the apolar media. As in the case of rhodopsin, their photochemical activity was preserved after solvent removal and rehydration, and they were functionally reconstituted in planar bilayers[114] and in large vesicles.[118]

Since membrane proteins can be transferred as lipid-protein complexes into apolar media directly from biological membranes, it seemed interesting to test if there was any specificity

TABLE 2
Membrane-Bound Enzymes in Reverse Micelles Containing Phospholipids

Enzyme	Experimental system	Reaction catalyzed	Wo
Mitochondrial ATPase[79,109,119]	Soybean phospholipids (8.5 mg/ml) alone or with Triton X-100 (15%, v/v) in hexane or toluene	Hydrolysis of ATP	2—10

Activity could be measured in the organic phase. The enzyme was transferred from SMPs.

SR Ca-ATPase[147]			
Bacteriorhodopsin[123]	Soybean phospholipids (10 mg/ml) in hexane	Light induced isomerization of 11-*cis* retinal to *trans*-retinal	ND
Cytochrome *c* oxidase[124,127]	Soybean phospholipids (10 mg/ml) in hexane or toluene	Oxidation of ascorbate and appearance of reduced cytochrome *c* oxidase; reduction of oxygen in water	2—10

The enzyme was transferred from SMP or purified. Partial activity dependent on water content in organic solvents.

Rhodopsin (bovine)[10,112]	Soybean phospholipids or diphytanoyl PC (10 mg/ml) hexane or ether	Dark bleached spectra; meta I to meta II kinetics and photointermediates	2—30

Spectral properties in the solvent similar as in the native membrane protein. Protein transferred from rod outer segments or purified.

Rhodopsin (squid)[118]	Soybean phospholipids (10 mg/ml) in hexane	Dark bleached spectra	~15

Similar spectral properties as the native protein and meta$_a$-meta$_b$ transition.

Reaction centers[114,115]	Soybean phospholipids (10 mg/ml) in hexane and other alcanes	Dark bleached spectra	~30

In the solvent supplemented with ubiquinone, the protein displayed light-induced optical and EPR absorbance changes as RCs in detergent solutions in water.

Succinate dehydrogenase[109]	Soybean phospholipids (10 mg/ml)	Reduction of dichlorophenol-indophenol	~10

in this process. If so, it would be possible to devise alternative methods of membrane protein purification that would not involve detergents. With this idea in mind, whole mitochondria, or submitochondrial particles (SMPs), were extracted with hexane and ether in the presence of soybean phospholipids. Initially, it was necessary to determine which proteins were extracted and if they were able to withstand the procedure. To do this, the activities of the extracted enzymes were measured after removing the solvent and rehydrating the protein-lipid residue. The rehydration and mechanical dispersion of this residue produce multilamellar liposomes containing the extracted mitochondrial proteins. Gel electrophoresis of these liposomes revealed that most of the mitochondrial proteins were transferred into the apolar solvent in the presence of phospholipids.[119,120,109]

The first enzyme assayed was the F_1F_0 ATPase. Around 20% of the mitochondrial ATPase activity was recovered in the liposomes formed from the organic extracts. The ATPase activity in these liposomes was controlled by the natural ATPase inhibitor. Actually, it was possible to use the liposomes formed from the mitochondrial extracts to estimate the number of ATPases that contained the ATPase inhibitor in its inhibitory site in intact mitochondria under various metabolic states.[119,120]

Thereafter, the activity of three different mitochondrial proteins — the ATPase, cytochrome *c* oxidase, and succinate dehydrogenase — were assayed in the liposomes formed from the organic extracts from SMPs. It was found that the three enzymes were transferred into the apolar media with yields varying between 10 and 30% and retaining their specific activity. Judging from the activity, and the heme and cytochrome content of the liposomes, cytochrome *c* oxidase was transferred with the highest efficiency (30%). In fact, the specific activity of this enzyme in the liposomes was enriched 1.6-fold with respect to SMPs sonicated in the presence of lipids. These results indicated that indeed, in principle, conditions could be designed to preferentially transfer a particular membrane protein, as a protein-lipid complex, into apolar solvents.[109] The enrichment of cytochrome *c* oxidase will be further discussed in the section dealing with the structure of these protein lipid complexes in the apolar media.

An interesting feature about the mitochondrial protein-lipid complexes in apolar solvents was their stability. After 10 d in hexane at 4°C, the three enzymes lost less than 50% of their activity measured in liposomes after removing the solvent and rehydrating. Even after 30 d the activities of succinate dehydrogenase and cytochrome *c* oxidase were higher than 50%.[109] On the other hand, the stability at 70, 80, and 90°C of these two enzymes in the protein-lipid complexes extracted with toluene from SMPs increased orders of magnitude in comparison to their stability in SMPs suspended in water. Their thermal stability was shown to be very dependent on the amount of water present in the apolar media.[122] These results and the thermal stability studies on the ATPase activity measured in the apolar media will be discussed in a separate section on protein stability.[79,122]

Up to this stage, with the exception of rhodopsin,[112] bacteriorhodopsin,[123] and RCs,[115] the activity of membrane enzymes in the protein-lipid complexes had hardly been studied in the apolar media. Taking this into account, the catalytic properties of protein-lipid complexes of cytochrome *c* oxidase and the mitochondrial ATPase were explored in toluene.[124,79]

Cytochrome *c* oxidase from bovine heart SMPs or purified was transferred into toluene containing phospholipids (10 mg/ml toluene) and water at 14 µl (high) and 3 µl/ml toluene (low). In the low water containing system, the spectral characteristics of cytochrome *c* oxidase are similar to those in an aqueous media. The extracted enzyme was reduced by ascorbate and cytochrome *c*; however, its oxidation was highly impaired. Both processes occur at a much slower rate in the organic solvent than in water. To achieve cytochrome *c* oxidase-cytochrome *c* interactions in the organic solvent, it was necessary to extract the two proteins together. A mixture of separately made cytochrome *c* and cytochrome *c* oxidase complexes in the solvent was not functional. The enzyme reduced in the apolar solvent by ascorbate was incapable of forming a complex with CO, but formed a complex with cyanide introduced via inverted micelles. Considering the behavior of cytochrome *c* oxidase in water, the results indicated that in the low-water toluene extracts, electron flow through the redox centers of cytochrome *c* oxidase is arrested before cytochrome a_3. Apparently, a certain amount of water is required for proper enzyme turnover and, when water is scarce, cytochrome *a* can be reduced, but not cytochrome a_3. Indeed, in the cytochrome *c* oxidase extracts containing a high water content, the reduction of ascorbate was higher and it was not possible to detect enzyme reduction.[124]

It is worthwhile mentioning that it was recently suggested that cytochrome *a* is accessible to water molecules,[125] and that electron transfer from cytochrome *a* and Cu_A to the a_3-Cu_B site is associated with water movement.[126] In addition, Bona et al.,[127] using purified cytochrome *c* oxidase extracted into hexane containing phospholipids, confirmed the work of Escamilla et al.[124] They found that the protein-lipid complexes in hexane had a hydrodynamic radius of 420 Å and that, at a high water-to-phospholipid ratio (Wo = 8), both cytochrome *a* and a_3 could be reduced and oxygen uptake observed. In contrast, at a low Wo of 1.8, the rate of cytochrome *a* reduction was reduced and that of cytochrome a_3 was inhibited.

More recently, Escobar and Escamilla[128] have studied the respiratory chain complexes from bovine heart SMPs transferred into isooctane containing soybean phospholipids. They observed a slow electron flux from NADH dehydrogenase to cytochrome *c* oxidase. In the presence of cyanide in the apolar media, which inhibits the activity of cytochrome *c* oxidase, a 34% reduction of this enzyme was attained. Their results indicate that electron flow can occur between all the complexes of the mitochondrial respiratory chain. Although they believe that the complexes are transferred individually into the organic solvent in reverse micelles, these results can be interpreted in a different way which will be discussed in the section on structure.

The ATPase from SMPs was studied in a system composed of toluene, soybean phospholipids (8.5 mg/ml), Triton X-100 (15% v/v), and water (0.5 to 7.4%). This system, called TPT, was developed considering previous results on the hydrolytic activity of F_1-ATPase in toluene-soybean phospholipid extracts[122] which are discussed in the section of soluble enzymes in phospholipid-containing reverse micelles. Triton X-100 was introduced to increase the amount of water solubilized in the apolar media since phospholipid reverse micelles, in general, have a very limited capacity to trap water.[75,122] It has been shown that detergents[1,38] and alcohols[130,131] can be used as cosurfactants to increase the water trapping capacity of reverse micelles. Kumar and Balasubramanian[81,131] had used Triton X-100 in combination with long-chain alcohols to form reverse micelles which could solubilize significant amounts of water. Since this detergent only inhibited the ATPase around 30% when present in water at a concentration of 15%, it was chosen for these studies. When the TPT system contained 3.6 to 5% water, it was completely transparent indicating the presence of reverse micelles.[79]

The ATPase hydrolytic activity in the TPT system started at 0.5% water and increased with increasing amounts of water. At 3.8% water, the ATPase had a similar K_m to that observed in water; however, V_{max} was about 100 times lower. The enzyme remained active in the TPT system up to 91°C when the water content was between 0.5 and 2%. The enzyme withstands much higher temperatures in the TPT system than in water. It is interesting that even though ATPase activity was observed in the TPT system in the absence of added phospholipids,[79] the presence of phospholipids seems to be crucial for thermal resistance.[80]

Centrifugation experiments indicated that the ATPase in the TPT system is in a different compartment than the substrate, and that apparently its catalytic capacity is limited by the reduced amount of water present which restricts its mobility.[122]

2. Structure

It is interesting that all the size determinations of the membrane protein-phospholipid complexes in organic solvents carried out with different techniques are in the same range (150 to 850 Å) (Table 3). In contrast, the values reported for the size of phospholipid reverse micelles without protein vary between 20 and 600 Å. Until now, the structural model that has been invoked for membrane proteins in apolar media containing surfactants is that of inverted micelles, where the hydrophylic regions of the protein are inside the micellar water core and the hydrophobic regions are mainly in contact with the organic solvent[21,23,33] (Figure 8).

The values of the radius of phospholipid reverse micelles under 100 Å are incompatible with the dimensions of the protein lipid complexes. On the other hand, membrane proteins usually have hydrophylic regions that project, more or less, to the aqueous phase on both sides of the membrane. This particular feature could be a disadvantage for their comfortable placement in a single reverse micelle. The bulkier hydrophylic region could be localized in the water core of the inverted micelle, but the second hydrophylic region would have to reach for the water core of another micelle, or the protein would have to bend so that this region could also interact with the water core, which is highly unlikely. Clearly, more work

TABLE 3
Reverse Micelles Containing Phospholipids; A Summary of Dimension Studies

Reverse Micelles Free of Guest Molecules

System	Wo	Characterization	Method	Ref.
EPC (13.3 mM) in benzene	13.8	Calculated water pool diameter: ~40 Å	Light scattering	69
EPC (67 mM) in benzene	<6	Spherical, diameter 600 Å	Light-scattering microscopy and ^{32}P-NMR	145, 146
	>6	Spherical, diameter 600 Å		
EPC (67 mM) in CCl$_4$		Spherical	Light-scattering microscopy and ^{32}P-NMR	145, 146
	<6	Diameter 400 Å		
	>6	Diameter 500 Å		
EPC (67 mM) in cyclohexane	<6	Spherical, diameter 300 Å	Light-scattering microscopy and ^{32}P-NMR	145, 146
	>6	Nonspherical		
EPC (3 to 13 mM) in hexanol	0	No micelle exceeded 270 Å	Light scattering	70
Soybean PL (11 mM) in hexane	15.6	Micellar radius of about 40 Å	Small angle X-ray scattering	21

Proteo-Lipid Extracts of Membrane-Bound Proteins

System	Wo	Protein	Dimension	Method
Soybean PL (12.5 mM) in hexane	40	Reaction center (RC) from *Rhodopseudomonas sphaeroides*, 2.6 μM	Particles containing RC, diameter from 150 to 850 Å	Ultrafiltration[115]
Soybean PL (11 mM) in hexane	15.6	Rhodopsin, 1.95 μM	R_G = 160 Å	Small angle X-ray scattering[21]
Soybean PL (200 mM in hexane	8—10	Cytochrome oxidase, 40 to 43 μM	R_G = 540 Å, R_H = 420 Å	Elastic and quasielastic light scattering[127]

Note: EPC, egg phosphatidylcholine; PL, phospholipids; R_G, radius of gyration; R_H, hydrodynamic radius.

is necessary to establish the structural characteristics of phospholipid reverse micelles, especially when membrane proteins are present.

In the study of Ramakrishnan and co-workers[21] using small-angle X-ray scattering, the radius of soybean phospholipid reverse micelles in hexane was found to be around 22 Å, corresponding to about 100 phospholipid molecules per micelle and a molecular weight of around 75,000 Da. Concomitantly, the size of the rhodopsin-phospholipid complexes in hexane was determined to have an average diameter of 160 Å. Since the size of the protein-lipid complex was incompatible with the size of the reverse micelles, the existence of large aggregates of reverse micelles was suggested. In these aggregates, rhodopsin molecules would form intermicellar bridges (Figure 8). Although this model fits the determined size, other possibilities should be considered, particularly when the capacity of reverse micellar systems to lodge whole cells is considered.[132,133] In addition, it is worth noting that all the membrane proteins have been transferred into apolar solvents containing surfactants from bilayer vesicles or organelles.[132,133] Thus, it is not impossible to think of a different arrangement which would preserve the bilayer and allow the separate hydrophilic domains of the membrane protein to interact with water. This model, presented in Figure 8b, would be consistent with the sizes of the protein-lipid complexes thus far reported. Basically, it

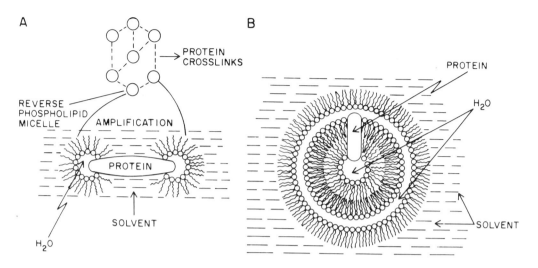

FIGURE 8.   Models of the organization of membrane proteins in apolar solvents containing phospholipids. (A) A model proposed for rhodopsin[21] and used for other membrane proteins. (B) Alternative model which considers the preservation of the bilayer structure and the presence of two aqueous compartments. Only one protein is shown, but many could be present in one vesicle, as is the case for organelles like the mitochondria, or whole cells.

considers that bilayer vesicles, whole organelles, or vesiculated fragments of the organelle may be transferred into the solvent and that the surfactant forms a monolayer that surrounds the vesicle allowing for an internal and an external aqueous compartment. Although the arrangement of cells in apolar media containing surfactants is not known, the system has the capability of housing structures as large as cells, probably keeping the integrity of the membrane, since under certain conditions viability has been demonstrated.[132-133]

This model would explain recent results indicating that it is possible to observe electron transfer through all the complexes of the respiratory chain transferred from SMPs into isooctane containing soybean phospholipids.[128] The authors of this paper[128] believe that the individual complexes are lodged in reverse micelles and collide in the apolar media. We find this explanation unlikely since previously it was found that isolated cytochrome c and cytochrome c oxidase-phospholipid complexes prepared individually were unable to interact in the apolar media. In addition, centrifugation experiments of SMP-phospholipid complexes indicate that the weight of these complexes is not consistent with that of reverse micelles containing one membrane protein.[79]

C. THERMOSTABILITY

An area where enzymes in reverse micelles have found interesting applications is in the study of the relation between enzyme hydration, catalysis, structure, and thermostability. Aside from the interest in basic questions concerning the role of water in enzyme catalysis and structure, this area has become very attractive in the last years due to its biotechnological implications.[5,6,134]

This subject has been investigated using experimental systems containing low amounts of water. One experimental approach used involves resuspending enzymes, at various levels of hydration, directly in organic solvents.[7,135] From these studies it was inferred that one monolayer of water around the enzyme molecule is sufficient to support catalysis. Moreover, enzymes under a limited hydration level placed in organic solvents are strikingly thermostable. It was shown that enzymes like lipases could carry out enzymatic activity at 100°C.[23]

Another approach that is being successfully employed is the incorporation of enzymes within the water cavity of reverse micelles in organic solvents where the water content can be manipulated. Although enzymatic activity has been characterized extensively in reverse micelles containing synthetic detergents,[2,9,138] their thermostability in such systems has barely been explored. However, there are interesting reports about enzyme thermostability in reverse micelles containing phospholipids.

One of the first papers in the literature about enzyme thermal stability in organic solvents was by Austin and co-workers.[137] In this study on the trypsin-membrane interaction in water and organic solvents, they managed to extract an enzyme-phospholipid complex into isooctane. The stability of trypsin in isooctane was investigated at temperatures from −20 to 60°C. They reported a half-life of approximately 28 d at 50°C for the activity of the extracted enzyme in isooctane, while the half-life at 37°C for the enzyme in an aqueous solution was 10 and 60 min in the absence or presence of Ca^+, respectively.

Thereafter, the effect of temperature on catalysis of the soluble enzyme alkaline phosphatase[87] was studied in reverse micelles formed in *n*-heptane from both synthetic and natural detergents (AOT, PE, and PC). This enzyme displays high thermostability in bulk water (stable for 20 min at 60°C); however, in reverse micelles of AOT and PE, the enzyme was inactivated above 45°C, but it was slightly more stable in PC reverse micelles.

More recently, a detailed study of thermostability has been carried out with some proteins from the inner mitochondrial membrane transferred as protein-lipid complexes into organic solvents.[121] The thermostabilities of cytochrome *c* oxidase and of H^+-ATPase transferred from submitochondrial particles (SMPs) into toluene containing soybean phospholipids were studied. In these experiments the enzymes, as protein-lipid complexes, were incubated in toluene at various temperatures for different periods of time. Thereafter, the solvent was removed by evaporation and the activity was measured in an aqueous media. It was found that at 1.3% of water (v/v), the half-life of the ATPase at 70°C was about 11 h, whereas that of cytochrome *c* oxidase was about 100 s. Thermostability of both enzymes increased dramatically by decreasing the water concentration of the apolar phase to 0.03%. At this concentration, the half-life of the ATPase at 90, 80, and 70°C was 5, 48, and 96 h, respectively.

In another paper,[122] the soluble mitochondrial F_1-ATPase from bovine heart was studied in the same system, but its activity was measured in the apolar media. Thermostability and catalytic activity in the organic solvent were correlated with water concentration. At water concentrations between 0.02 and 0.05%, the F_1-ATPase was found to be resistant to cold denaturation and acquired a remarkable thermostability; its half-life at 70°C was more than 24 h. However, at such low concentrations the enzyme was inactive in the solvent. It is unclear if the enzyme was in reverse micelles at the water concentrations (4 to 6%) where a significant activity was observed in the solvent since the solution was very turbid.

In a paper that followed, SMPs were transferred into toluene (85%) that contained phospholipids (8.5 mg/ml) and Triton X-100 (15% v/v), and the ATPase catalytic activity (ATP hydrolysis) in the solvent was studied at high temperatures.[79] At concentrations of water between 0.5 and 2%, the enzyme catalyzed ATP hydrolysis at temperatures up to 91°C. The optimal temperature for hydrolysis was found to be 58°C, independently of the water concentration.

In a study in which Triton X-100 was used as surfactant in different organic solvents (toluene, xylene, and propylbenzene), the thermostability of the mitochondrial ATPase was determined in the presence and absence of phospholipids. The results clearly indicated that, independently of the solvent used, the presence of phospholipids increased enzyme thermostability.[80] Moreover, in the system of Triton X-100 reverse micelles in cyclohexane with 20% of alcohol (octanol or hexanol), attempts to measure lactate dehydrogenase activity totally failed, suggesting again enzyme instability,[138] while when a similar system was used with phospholipids as cosurfactants at a low concentration (1% w/w), the enzyme was active and very stable.[90] Something similar occurred when the activity of lactate dehydrogenase could only be measured for a short time in a system of Brij 96 in cyclohexane. The addition of soybean phospholipids to the system (8.5 mg/ml) markedly improved enzyme activity and stability.[138]

On the other hand, reports about the effect of temperature on enzymatic activity in reverse micelles of synthetic detergents have shown that enzymes are inactivated at 30°C (alcohol dehydrogenase[139,140] and cholesterol oxidase[141]) or at 45°C (polyphenol oxidase[142] and alkaline phosphatase[87]).

From aforementioned data it appears that, while most of the enzymes incorporated into reverse micelles containing phospholipids acquire a remarkable thermostability at elevated temperatures, in reverse micelles of synthetic surfactants they become even more sensitive to temperature than in an aqueous medium. So far, the most important factor that has been shown to be related to enzyme thermostability in organic solvent systems is their water content. However, two features should be pointed out that could contribute to determine thermostability: the nature of the amphiphile molecules that form the reverse micelles (phospholipids vs. synthetic cosurfactants) and the nature of the enzymes that have been studied (soluble or membrane-bound).

It is reasonable to think that, in the synthetic reverse micelles systems, the amount of water necessary to obtain maximal activity (Wo ~ 10) is too high to observe thermostability. Thus, the water content determines enzyme activity as well as its thermostability. In reverse micelles of phospholipids in which water content is very limited, entrapped enzymes acquire remarkable thermostability, although under these conditions usually the catalytic activity is far below that in bulk water. On the other hand, the nature of the surfactant will determine the fraction of free water to which the enzyme is exposed. At the same Wo, reverse micelles formed from mixtures of phospholipids (including charged phospholipids) are likely to have a smaller fraction of free water than those formed from a synthetic surfactant like AOT due to the differences in surfactant hydration.

Finally, it is necessary to make a detailed comparison of the thermostability between a soluble enzyme in phospholipid and in synthetic surfactant reverse micelles. The only data available are with alkaline phosphatase,[87] where the enzyme was slightly more stable in PC than in AOT reverse micelles. Also, a similar study should be conducted with a membrane protein to dissect the contribution to thermostability of the surfactant and the enzyme.

VI. PERSPECTIVES

Reverse micellar systems serve a great number of investigators as an experimental approach in the area of enzymology whether their interest is in structure or catalysis. Most groups have chosen to work with synthetic detergents and soluble proteins; therefore, the knowledge about micellar enzymology in organic solvents has been mainly derived from these systems.[1,2] However, groups interested in reverse micelles as membrane mimetic systems are shifting their attention to natural surfactants such as the phospholipids.[40,124,132,133] Thus, much remains to be done to fully characterize the thermodynamic characteristics of reverse micelles containing phospholipids.

Since many proteins interact with membrane interphases, reverse micelles represent an alternative to characterize these interactions. Actually, a recent model for enzymes in reverse micelles is able to explain how catalysis is differentially affected as the water content in the system is modified depending on the type of protein-interphase interaction.[93] On the other hand, the emerging concepts about the functional relevance of soluble multienzymatic complexes in cell function and the high protein concentration found inside cells raise fundamental quesions as to the state of intracellular water and the influence of its yet mysterious properties on protein stability and catalysis. Phospholipid-containing reverse micelles represent an interesting model where these fascinating questions can be experimentally addressed.

It appears that an important advantage of using phospholipids in the formation of reverse micelles is the increased thermostability they confer to enzymes under conditions of low water content. Enzymes seem to prolong their half-life at elevated temperatures by orders of magnitude in comparison with bulk water. As mentioned earlier, it is important to determine if this is true for both soluble and membrane proteins. Also, it will be important to determine if the presence of a membrane protein affects the organization of surfactants in the hydrophobic environment and to determine their structure. These questions could bear physiological relevance as reverse micelles have been implicated in fusion and flip-flop events in biological membranes.

One of the main disadvantages of reverse micelles containing phospholipids is their limited capacity to solubilize water. While synthetic reverse micelles can solubilize water up to Wo = 60 and still retain stability and transparency, phospholipid-containing reverse micelles are stable only within a narrow Wo range ($\sim$10 without cosurfactant to 30 with a cosurfactant). However, this limitation has been partially overcome by systems that utilize a cosurfactant, which allow up to 60 water molecules per phospholipid.[91]

In the work with cytochrome c oxidase, it was found that, when there was a low water content in the organic solvent containing phospholipids (1 to 3 μl/ml solvent), a step in the catalytic cycle occurs at rates much slower than in aqueous media. Under these conditions it was possible to practically stabilize an intermediary (the half-reduced cytochrome c oxidase, $a^{+2}a_3^{+3}$) and to suspect that water is necessary for enzyme turnover.[124] Therefore, it could be that the transfer to membrane proteins into phospholipid-containing apolar solvents with a low water content could facilitate the study of certain intermediates of catalysis difficult to detect in an aqueous media and to study the participation of water in the catalytic cycle.

REFERENCES

1. **Martinek, K., Levashov, A. V., Klyachko, N., Khmelnitski, Y. L., and Berezin, I. V.,** Micellar enzymology, *Eur. J. Biochem.,* 155, 453, 1986.
2. **Luisi, P. L., Giomini, M., Pileni, M. P., and Robinson, B. H.,** Reverse micelles as hosts for proteins and small molecules, *Biochim. Biophys. Acta,* 947, 209, 1988.
3. **Klibanov, A. M.,** Enzymatic catalysis in anhydrous organic solvents, *Trends Biochem. Sci.,* 14, 141, 1989.
4. **Merkler, D. J., Farrington, G. K., and Wedler, F. C.,** Protein thermostability, *Int. J. Pep. Protein Res.,* 18, 430, 1981.
5. **Zale, S. E. and Klibanov, A. M.,** On the role of reversible denaturation (unfolding) in the irreversible thermal inactivation of enzymes, *Biotechnol. Bioeng.,* 25, 2221, 1983.
6. **Ahern, T. J. and Klibanov, A. M.,** The mechanism of irreversible enzyme inactivation at 100°C, *Science,* 228, 1280, 1985.
7. **Zaks, A. and Klibanov, A. M.,** Enzymatic catalysis in organic media at 100°C, *Science,* 224, 1249, 1984.
8. **Wheeler, C. J. and Croteau, R.,** Terpene cyclase catalysis in organic solvents/minimal water media: demonstration and optimization of (+)-α-pinene cyclase activity, *Arch. Biochem. Biophys.,* 248, 429, 1986.
9. **Darszon, A., Montal, M., and Philipp, M.,** Formation of detergent-free proteolipids from biological membranes: application to rhodopsin, *FEBS Lett.,* 74, 135, 1977.
10. **Darszon, A., Philipp, M., Zarco, L., and Montal, M.,** Rhodopsin-phospholipid complexes in apolar solvents: formation and properties, *J. Membr. Biol.,* 43, 71, 1978.
11. **Martinek, K., Levashov, A. V., Klyachko, N. L., and Berezin, I. V.,** *Dokl. Akad. Nauk SSSR,* 236, 920, 1977.
12. **Wolf, R. and Luisi, P. L.,** Micellar solubilization of enzymes in hydrocarbon solvents, enzymatic activity and spectroscopic properties of ribonuclease in *n*-octane, *Biochem. Biophys. Res. Commun.,* 89, 209, 1979.
13. **Douzou, P., Keh, E., and Balny, C.,** Cryoenzymology in aqueous media: micellar solubilized water clusters, *Proc. Natl. Acad. Sci. U.S.A.,* 76, 681, 1979.
14. **Menger, F. M. and Yamada, K.,** Enzyme catalysis in water pools, *J. Am. Chem. Soc.,* 101, 6731, 1979.

15. **Martinek, K., Levashov, A. V., Khmelnitski, Y. L., Klyachko, N. L., Chernyak, V. Y., and Berezin, I. V.,** Molecular mechanisms of the solubilization of protein (enzymes) in an organic solvent with the aid of a surface-active substance. Colloidal solution of water in an organic solvent as a new microheterogeneous medium for enzymatic reactions, *Dokl. Akad. Nauk. SSSR,* 258, 1488, 1981.

16. **Klyachko, N. L., Levashov, A. V., and Martinek, K.,** Catalysis of enzymes incorporated in reverse micelles of surface-active substances in organic solvents. Peroxidase in the Aerosol OT-water-octane system, *Mol. Biol.,* 18, 1019, 1984.

17. **Barberic, S. and Luisi, P. L.,** Micellar solubilization of biopolymers in organic solvents. V. Activity and conformation of α-chymotrypsin in isooctane-AOT reverse micelles, *J. Am. Chem. Soc.,* 103, 4239, 1981.

18. **Fendler, J. H.,** *Membrane Mimetic Chemistry,* John Wiley & Sons, New York, 1982.

19. **Luisi, O. L. and Magid, L. J.,** Solubilization of enzymes and nucleic acids in hydrocarbon micellar solutions, *CRC Crit. Rev. Biochem.,* 20, 409, 1986.

20. **Grandi, C., Smith, R. E., and Luisi, P. L.,** Micellar solubilization of biopolymers in organic solvents. Activity and conformation of lysozyme in isooctane reverse micelles, *J. Biol. Chem.,* 256, 837, 1981.

21. **Ramakrishnan, V. R., Darszon, A., and Montal, A.,** A small angle X-ray scattering study of a rhodopsin-lipid complex in hexane, *J. Biol. Chem.,* 258, 4857, 1983.

22. **Thompson, K. F. and Gierasch, L. M.,** Conformation of a peptide solubilizate in a reversed micelles water pool, *J. Am. Chem. Soc.,* 106, 3648, 1984.

23. **Chatney, D., Urbach, W., Nicot, C., Vacher, M., and Waks, M.,** Hydrodynamic radii of protein-free and protein-containing reverse micelles as studied by fluorescence recovery after fringe photobleaching. Perturbations introduced by myelin basic protein uptake, *J. Phys. Chem.,* 91, 2198, 1987.

24. **Storch, J. and Kleinfeld, A. M.,** The lipid structure of biological membranes, *Trends Biochem. Sci.,* 10, 418, 1985.

25. **Boicelli, C. A., Conti, F., Giomini, M., and Giuliani, A. M.,** Interactions of small molecules with phospholipids in inverted micelles, *Chem. Phys. Lett.,* 89, 490, 1982.

26. **Chen, S.-T. and Springer, C. S., Jr.,** Hyperfins shift NMR studies of hydrated phospholipid inverted micelles, *Chem. Phys. Lipids,* 23, 23, 1979.

27. **Srivastava, D. K. and Bernhard, S. A.,** Enzyme-enzyme interaction and the regulation of metabolic reaction pathways, *Curr. Top. Cell. Regul.,* 28, 1, 1986.

28. **Srivastava, D. K. and Bernhard, S. A.,** Biophysical chemistry of metabolic reaction sequences in concentrated enzyme solution and in cell, *Annu. Rev. Biophys. Biophys. Chem.,* 16, 175, 1987.

29. **Srivastava, D. K., Smolen, P., Betts, G. F., Fukushima, T., Spivey, H. O., and Bernhard, S. A.,** Direct transfer of NADH between α-glycerol phosphate dehydrogenase and lactate dehydrogenase: fact or misinterpretation?, *Proc. Natl. Acad. Sci. U.S.A.,* 86, 6464, 1989.

30. **Singer, S. J. and Nicholson, G. L.,** The fluid mosaic model of the structure of cell membranes, *Science,* 175, 720, 1972.

31. **Geiger, G. and Singer, S. J.,** The participation of α-actinin in the capping of cell membrane components, *Cell,* 16, 213, 1979.

32. **Ferguson, M. A. J. and Williams, A. F.,** Cell-surface anchoring of proteins via glycosyl-phosphatidyl-inositol structures, *Annu. Rev. Biochem.,* 57, 285, 1988.

33. **Rouser, G. and Fleischer, S.,** Isolation, characterization and determination of polar lipids of mitochondria, *Methods Enzymol.,* 10, 385, 1967.

34. **Seelig, J. and Seelig, A.,** Lipid conformation in model membranes and biological membranes, *Q. Rev. Biophys.,* 13, 19, 1980.

35. **Tanford, C.,** *The Hydrophobic Effect: Formation of Micelles and Biological Membranes,* John Wiley & Sons, New York, 1980.

36. **Israelachvili, J. N., Marcelja, S., and Horn, R. G.,** Physical principles of membrane organization, *Q. Rev. Biophys.,* 13, 121, 1980.

37. **Salem, L.,** The role of long-range forces in the cohesion of lipoproteins, *Can. J. Biochem. Physiol.,* 40, 1287, 1962.

38. **Jacobs, R. E., Hudson, B., and Anderson, H. L.,** A theory of the chain melting phase transition of aqueous phospholipid dispersions, *Proc. Natl. Acad. Sci. U.S.A.,* 72, 3993, 1975.

39. **Israelachvili, J. N., Mitchell, D. J., and Minham, B.-W.,** Theory of self-assembly of hydrocarbon amphiphiles into micelles and bilayers, *J. Chem. Soc. Faraday Trans. 2,* 72, 1525, 1976.

40. **Walde, P., Giuliani, A. M., Boicelli, C. A., and Luisi, P. L.,** Phospholipid-based reverse micelles, *Chem. Phys. Lipids,* 53, 265, 1990.

41. **Luzzati, V. and Tardiev, A.,** Lipid phases: structure and structural transition, *Annu. Rev. Phys. Chem.,* 25, 79, 1974.

42. **Franks, N. P. and Levine, Y. K.,** Low-angle X-ray diffraction, in *Membrane Spectroscopy,* Grell, E., Ed., Springer-Verlag, Berlin, 1981, chap. 7.

43. **Verkleij, A. J. and de Gier, J.,** Freeze fracture studies on aqueous dispersions of membrane lipids, in *Liposomes: From Physical Structure to Therapeutic Applications,* Knight, C. G., Ed., Elsevier/North-Holland, Amsterdam, 1981, chap. 4.

44. **Hauser, H., Pascher, I., Pearson, R. H., and Sundell, S.,** Preferred conformation and molecular packing of phosphatidylethanolamine and phosphatidylcholine, *Biochim. Biophys. Acta,* 650, 21, 1981.

45. **Gruner, S. M., Cullis, P. R., Hope, M. J., and Tilcock, C. S.,** Lipid polymorphism: the molecular basis of nonbilayer phases, *Annu. Rev. Biophys. Biophys. Chem.,* 14, 211, 1985.

46. **Phillips, M. C., Williams, R. M.. and Chapman, D.,** On the nature of hydrocarbon chain motions in lipid liquid crystals, *Chem. Phys. Lipids,* 3, 234, 1969.

47. **Benga, G. and Holmes, R. P.,** Interactions between components in biological membranes and their implications for membrane function, *Prog. Biophys. Mol. Biol.,* 43, 195, 1984.

48. **Nagle, J. F.,** Theory of the main lipid bilayer phase transition, *Annu. Rev. Phys. Chem.,* 31, 157, 1980.

49. **Melchior, D.,** Lipid phase transitions and regulation of membrane fluidity in prokaryotes, *Curr. Top. Membr. Transp.,* 17, 263, 1982.

50. **Cornell, B. A. and Separovic, F.,** Membrane thickness and acyl chain length, *Biochim. Biophys. Acta,* 733, 189, 1983.

51. **Hui, S.-W., Mason, J. T., and Huang, C.,** Acyl chain interdigitation in saturated mixed-chain phosphatidylcholine bilayer dispersions, *Biochemistry,* 23, 5570, 1984.

52. **Klausner, R. D., Kleinfeld, A. M., Hoover, R. L., and Karnousky, M. J.,** Lipid domains in membranes. Evidence derived from structural perturbations induced by free acids and lifetime heterogeneity analysis, *J. Biol. Chem.,* 255, 1286, 1980.

53. **Thomson, T. E. and Tillack, T. W.,** Organization of glycosphingolipids in bilayers and plasma membranes of mammalian cells, *Annu. Rev. Biophys. Biophys. Chem.,* 14, 361, 1985.

54. **Wolf, D. E. and Voglmayer, J. K.,** Diffusion and regionalization in membranes of maturing ram spermatozoa, *J. Cell Biol.,* 98, 1678, 1984.

55. **Karnovsky, M. J., Kleinfeld, A. M., Hoover, R. L., and Klausner, R. D.,** Concept of lipid domains in membranes, *J. Cell Biol.,* 94, 1, 1982.

56. **Chapman, D., Gomez-Fernandez, J. C., and Goni, F. M.,** The interaction of intrinsic proteins and lipids in biomembranes, *Trends Biochem. Sci.,* 7, 67, 1982.

57. **De Kroijff, B., Cullis, P. R., and Verkleij, A. J.,** Non-bilayer lipid structures in model and biological membranes, *Trends Biochem. Sci.,* 5, 79, 1980.

58. **Coolbear, K. P., Berde, C. B., and Keough, K. M. W.,** Gel to liquid-crystalline phase transitions of aqueous dispersions of polyunsaturated mixed-acid phosphatidylcholines, *Biochemistry,* 22, 1466, 1983.

59. **Singer, S. J.,** in *Membrane and Intracellular Communication,* Balian, R., Chabre, M., and Devaux, Ph. F., Eds., North-Holland, Amsterdam, 1981, p. 5.

60. **Cullis, P. R. and De Kruijff, B.,** Lipid polymorphism and functional roles of lipids in biological membranes, *Biochim. Biophys. Acta,* 559, 339, 1979.

61. **Cullis, P. R. and De Kruijff, B.,** The polymorphic phase behavior of phosphatidylethanolamines of natural and synthetic origin. A ^{31}P NMR study, *Biochim. Biophys. Acta,* 513, 31, 1978.

62. **Stier, A., Finch, S. A. E., and Bosterling, B.,** Non-lamellar structure in rabbit liver microsomal membranes. A ^{31}P-NMR study, *FEBS Lett.,* 91, 109, 1978.

63. **De Kruijff, B., Verkleij, A. J., Van Echteld, C. J. A., Gerritsen, W. J., Mombers, C., Noordam, P. C., and de Gier, J.,** The occurrence of lipidic particles in lipid bilayers as seen by ^{31}P NMR and freeze-fracture electron microscopy, *Biochim. Biophys. Acta,* 555, 200, 1979.

64. **Van Venetio, R. and Verkleij, A. J.,** Analysis of the hexagonal II phase and its relations to lipidic particles and the lamellar phase. A freeze-fracture study, *Biochim. Biophys. Acta,* 645, 262, 1981.

65. **Verkleij, A. J.,** *Phospholipid Research and the Nervous System. Biochemical and Molecular Pharmacology,* Horrocks, L. A., Freysz, L., and Toffano, G., Eds., Liviana Press, Padova, 1986, p. 207.

66. **Verkleij, A. J.,** *Electron Microscopic Analysis of Subcellular Dynamics,* Plattner, H., Ed., CRC Press, Boca Raton, FL, 1989.

67. **Noordam, P. C., Van Echteld, C. J. A., De Kruijff, B., and De Gier, J.,** Rapid transbilayer movement of phosphatidylcholine in unsaturated phosphatidylethanolamine containing model membranes, *Biochim. Biophys. Acta,* 646, 483, 1981.

68. **Elworthy, P. H.,** Micelle formation by lecithin in benzene, *J. Chem. Soc.,* 813, 1959.

69. **Elworthy, P. H.,** The structure of lecithin micelles in benzene solution, *J. Chem. Soc.,* 1951, 1959.

70. **Elworthy, P. H. and McIntosh, D. S.,** Micelle formation by lecithin in some aliphatic alcohols, *J. Pharmacol.,* 13, 663, 1961.

71. **Haque, R., Tinsley, I. J., and Schmedding, D.,** Nuclear magnetic resonance studies of phospholipid micelles, *J. Biol. Chem.,* 217, 157, 1972.

72. **Boicelli, C. A., Conti, F., Giomini, M., and Giuliani, A. M.,** Interactions of small molecules with phospholipids in inverted micelles, *Chem. Phys. Lett.,* 89, 490, 1982.

73. **Fung, B. M. and McAdams, J. L.,** The interaction between water and the polar head in inverted phosphatidylcholine micelles, *Biochim. Biophys. Acta,* 451, 313, 1976.

74. **Davenport, J. B. and Fisher, L. R.,** Interaction of water with egg lecithin in benzene solution, *Chem. Phys. Lipids,* 14, 275, 1975.

75. **Poon, P. H. and Wells, M. A.**, Physical studies of egg phosphatidylcholine in diethyl ether-water solutions, *Biochemistry,* 13, 4928, 1974.
76. **Walter, W. V. and Hayes, R. G.**, Nuclear magnetic resonance studies of the interaction of water with the polar region of phosphatidylcholine micelles in benzene, *Biochim. Biophys. Acta,* 249, 528, 1971.
77. **Gotthard, K. and Stelzner, F.**, NMR investigations of the interaction of water with lecithin in benzene solutions, *Biochim. Biophys. Acta,* 363, 1, 1974.
78. **Shoshani, L., Gomez-Puyou, A., Joseph-Nathan, P., and Darszon, A.**, in preparation.
79. **Garza-Ramos, G., Darszon, A., Tuena de Gómez-Puyou, M., and Gómez-Puyou, A.**, Enzyme catalysis in organic solvents with low water content at high temperatures. The adenosinetriphosphatase of submitochondrial particles, *Biochemistry,* 29, 751, 1990.
80. **Fernandez-Velasco, D. A., Garza-Ramos, G., Ramirez-Silva, L. H., Shoshani, L., Tuena de Gómez-Puyou, M., and Gómez-Puyou, A.**, Enzyme catalysis and thermostability at low amounts of water in systems composed of apolar organic solvents, phospholipids, and Triton X-100. The mitochondrial ATPase, in preparation.
81. **Kumar, C. and Balasubramanian, D.**, Studies on the Triton X-100:alcohol:water reverse micelles in cyclohexane, *J. Colloid Interface Sci.,* 69, 271, 1979.
82. **Wong, M., Thomas, J. K., and Nowak, T.**, Structure and state of H_2O in reverse micelles. 3, *J. Am. Chem. Soc.,* 99, 4730, 1977.
83. **Maitra, A.**. Determination of size parameters of water-Aerosol OT-Oil reverse micelles from their nuclear magnetic resonance data, *J. Phys. Chem.,* 88, 5122, 1984.
84. **Zulauf, M. and Eicke, H. F.**, Inverted micelles and microemulsions in the ternary system H_2O/Aerosol-OT/Isooctane as studied by photon correlation spectroscopy, *J. Phys. Chem.,* 83, 480, 1979.
85. **Scartazzini, R. and Luisi, P. L.**, Organogels from lecithins, *J. Phys. Chem.,* 92, 829, 1988.
86. **Luisi, P. L., Scartazzini, R., Haering, G., and Schurtenberger, P.**, Organogels from water-in-oil microemulsions, *Colloid Polym. Sci.,* 268, 356, 1990.
87. **Oshima, A., Narita, H., and Kito, M.**, Phospholipid reverse micelles as a milieu of an enzyme reaction in an apolar system, *J. Biochem.,* 93, 1421, 1983.
88. **Levashov, A. V., Klyachko, N. L., Pshezhetskii, A. V., Berezin, I. V., Kotrikadze, N. G., and Martinek, K.**, Superactivity of acid phosphatase entrapped into surfactant reverse micelles in organic solvents, *Dokl. Akad. Nauk. SSSR,* 289, 1271, 1986.
89. **Kabanov, A. V., Klyachko, N. L., Pshezhetskii, A. V., Namiotkin, S. N., Martinek, K., and Levashov, A. V.**, Kinetic regularities of enzymatic catalysis in systems of surfactant reversed micelles in organic solvents, *Mol. Biol.,* 21, 275, 1987.
90. **Garza-Ramos, G., Darszon, A., Tuena de Gómez-Puyou, M., and Gómez-Puyou, A.**, High concentrations of guanidine chloride activate lactate dehydrogenase in low water media, *Biochem. Biophys. Res. Commun.,* 172, 830, 1990.
91. **Peng, Q. and Luisi, P. L.**, The behaviour of proteases in lecithin reverse micelles, *Eur. J. Biochem.,* 188, 471, 1990.
92. **Morita, S., Narita, H., Matoba, T., and Kito, M.**, Synthesis of triacylglycerol by lipase in phosphatidylcholine reverse micellar system, *J. Am. Oil Chem. Soc.,* 61, 1571, 1984.
93. **Bru, R., Sanchez-Ferrer, A., and Garcia-Carmona, F.**, A theoretical study on the expression of enzymic activity in reverse micelles, *Biochem. J.,* 259, 355, 1989.
94. **Schmidli, P. K. and Luisi, P. L.**, Lipase-catalyzed reactions reverse micelles formed by soybean lecithin, *Biocatalysis,* 3, 367, 1990.
95. **Folch, J. and Lee, M.**, Proteolipids, a new type of tissue lipoproteins. Their isolation from brain, *J. Biol. Chem.,* 191, 807, 1951.
96. **Zahler, P. and Niggli, V.**, The use of organic solvents in membrane research, in *Methods in Membrane Biology,* Vol. 8, Korn, E. D., Ed., Plenum Press, New York, 1977, p. 1.
97. **Lees, M. B., Sakura, J. D., Sapirstein, V. S., and Curtalo, W.**, Structure and function of proteolipids in myelin and non-myelin membranes, *Biochim. Biophys. Acta,* 559, 209, 1979.
98. **Nelson, N., Eytan, E., Notsani, B. E., Sigrist, H., Sigrist-Nelson, K., and Gitler, C.**, Isolation of a chloroplast N,N'-dicyclohexylcarbodiimide-binding proteolipid active in proton translocation, *Proc. Natl. Acad. Sci. U.S.A.,* 24, 2375, 1977.
99. **Hanahan, D. J.**, The enzymatic degradation of phosphatidylcholine in diethyl ether, *J. Biol. Chem.,* 195, 199, 1952.
100. **Widmer, C. and Crane, F. L.**, A lipid soluble form of cytochrome c from the electron transport particle of beef-heart mitochondria, *Biochim. Biophys. Acta,* 27, 203, 1958.
101. **Montal, M.**, Lipid-protein assembly and the reconstitution of biological membranes, in *Perspectives in Membrane Biology,* Estrada-O, S. and Gitler, C., Eds., Academic Press, New York, 1974, p. 591.
102. **Gitler, C. and Montal, M.**, Thin proteolipid films: a new approach to the reconstitution of biological membranes, *Biochim. Biophys. Acta,* 47, 1486, 1972.
103. **Gitler, C. and Montal, M.**, Formation of decane-soluble proteolipids: influence of monovalent and divalent cations, *FEBS Lett.,* 28, 329, 1972.

104. **Mittal, K. L.**, *Micellization, Solubilization and Microemulsions*, Vols. 1 and 2, Plenum Press, New York, 1977.

105. **Das, M. L. and Crane, F. L.**, Proteolipids. I. Formation of phospholipid cytochrome c complexes, *Biochemistry*, 3, 696, 1964.

106. **Montal, M. and Korenbrot, J. I.**, Incorporation of rhodopsin proteolipid into bilayer membranes, *Nature*, 246, 219, 1973.

107. **Darszon, A., Blair, L., and Montal, M.**, Purified rhodopsin phosphatidylcholine complexes in apolar environments, *FEBS Lett.*, 107, 213, 1979.

108. **Darszon, A.**, Rhodopsin-phospholipid complexes in apolar solvents: characteristics and mechanism of extraction, *Vision Res.*, 32, 1443, 1982.

109. **Ayala, G., Nascimiento, A., Gómez-Puyou, A., and Darszon, A.**, Extraction of mitochondrial membrane proteins into organic solvents in a functional state, *Biochim. Biophys. Acta*, 810, 115, 1985.

110. **Walde, G., Durrell, J., and St. George, R. C. C.**, The light reaction in the bleaching of rhodopsin, *Science*, 3, 179, 1950.

111. **Ostroy, S. E.**, Rhodopsin and the visual process, *Biochim. Biophys. Acta*, 463, 91, 1977.

112. **Darszon, A., Strasser, R. J., and Montal, M.**, Rhodopsin phospholipid complexes in apolar environments: photochemical characterization, *Biochemistry*, 18, 5205, 1979.

113. **Hubbard, R. and Ald, G.**, Cis-trans isomers of vitamin A and retinene in the rhodopsin system, *J. Gen. Physiol.*, 36, 269, 1952.

114. **Schonfeld, M., Montal, M., and Feher, G.**, Functional reconstitution of photosynthetic reaction centres in planar lipid bilayers, *Proc. Natl. Acad. Sci. U.S.A.*, 76, 6351, 1979.

115. **Schonfeld, M., Montal, M., and Feher, G.**, Reaction center-phospholipid complex in organic solvents: formation and properties, *Biochemistry*, 19, 1535, 1980.

116. **Kendall-Tobias, M. and Crofts, A. R.**, Reaction centres in hexane, *Biophys. J.*, 25, 54a, 1979.

117. **Packham, N. K., Packham, C., Mueller, P., Tiede, D. M., and Dutton, P. L.**, Reconstitution of photochemically active reaction centres in planar phospholipid membranes, *FEBS Lett.*, 110, 101, 1980.

118. **Darszon, A., Vandenberg, C. A., Schonfeld, M., Ellisman, M. H., Spitzer, N., and Montal, M.**, Reassembly of protein-lipid complexes into large bilayer vesicles: perspectives for membrane reconstitution, *Proc. Natl. Acad. Sci. U.S.A.*, 77, 239, 1980.

119. **Darszon, A. and Gómez-Puyou, A.**, Extraction of mitochondrial protein-lipid complexes into organic solvents: an approach to study the interaction between the ATPase and the mitochondrial ATPase inhibitor protein, *Eur. J. Biochem.*, 121, 427, 1982.

120. **Sanchez-Bustamante, V. J., Darszon, A., and Gómez-Puyou, A.**, On the function of the natural ATPase inhibitor protein in intact mitochondria, *Eur. J. Biochem.*, 126, 611, 1982.

121. **Ayala, G., Tuena de Gómez-Puyou, M., Gómez-Puyou, A., and Darszon, A.**, Thermostability of membrane enzymes in organic solvents, *FEBS Lett.*, 203, 41, 1986.

122. **Garza-Ramos, G., Darszon, A., Tuena de Gómez-Puyou, M., and Gómez-Puyou, A.**, Catalysis and thermostability of mitochondrial F1-ATPase in toluene-phospholipid-low-water-systems, *Biochemistry*, 28, 3177, 1989.

123. **Hwang, S. G., Korenbrot, J., and Stoeckenius, W.**, Structural and spectroscopic characteristics of bacteriorhodopsin in air-water interface films, *J. Membr. Biol.*, 36, 115, 1977.

124. **Escamilla, E., Ayala, G., Tuena de Gómez-Puyou, M., Gómez-Puyou, A., Millän, L., and Darszon, A.**, Catalytic activity of cytochrome oxidase and cytochrome c in apolar solvents containing phospholipids and low amounts of water, *Arch. Biochem. Biophys.*, 272, 332, 1989.

125. **Sassaroli, M., Ching, Y.-C., Dasgupta, S., and Rousseau, D. L.**, Cytochrome c oxidase: evidence for interaction of water molecules with cytochrome a, *Biochemistry*, 28, 3128, 1989.

126. **Kornblatt, J. A., Hoa, G. H. B., and Heremans, K.**, Pressure-induced effects on cytochrome oxidase: the aerobic steady state, *Biochemistry*, 27, 5122, 1988.

127. **Bona, M., Fabian, M., and Sedlák, M.**, Spectral and catalytic properties of cytochrome oxidase in organic solvents, *Biochim. Biophys. Acta*, 1020, 94, 1990.

128. **Escobar, L. and Escamilla, E.**, Behaviour of the respiratory chain in an apolar organic solvents. Cytochrome c oxidase activity in isooctane, submitted.

129. **Keiser, B. A., Varie, D., Barden, R. E., and Wolts, S. L.**, Detergentless water/oil microemulsions composed of hexane, water, and 2-propanol. II. Nuclear magnetic resonance studies, effect of added NaCl, *J. Phys. Chem.*, 83, 1276, 1979.

130. **Khmelnitsky, Y. L., Zharinova, I. N., Berezin, I. V., Levashov, A. V., and Martinek, K.**, Detergentless microemulsions: a new microheterogeneous medium for enzymatic reaction, *Ann. N.Y. Acad. Sci.*, 501, 161, 1987.

131. **Kumar, C. and Balasubramanian, D.**, Spectroscopic studies on the microemulsions and lamellar phases of the system Triton X-100: hexanol:water in cyclohexane, *J. Colloid Interface Sci.*, 74, 64, 1980.

132. **Darszon, A., Escamilla, E., Gómez-Puyou, A., and Tuena de Gómez-Puyou, M.**, Transfer of spores, bacteria and yeast into toluene containing phospholipids and low amounts of water: preservation of the bacterial respiratory chain, *Biochem. Biophys. Res. Commun.*, 151, 1074, 1988.

133. **Pfammatter, N., Ayala, G., and Luisi, P. L.,** Solubilization and activity of yeast cells in water-in-oil microemulsion, *Biochem. Biophys. Res. Commun.,* 161, 1244, 1989.
134. **Zale, S. E. and Klibanov, A. M.,** Why does ribonuclease irreversibly inactivate at high temperatures?, *Biochemistry,* 25, 5432, 1986.
135. **Klibanov, A. M.,** Enzymes that work in organic solvents, *Chemtech,* 16, 354, 1986.
136. **Luisi, P. L.,** Enzymes hosted in reverse micelles in hydrocarbon solution, *Angew. Chem. Int. Ed. Engl.,* 24, 439, 1985.
137. **Austin, P., Dodd, G., and Davis, M.,** Studies on trypsin-membrane interactions in both aqueous and hydrocarbon solvents, *Biochem. Trans. Soc.,* 2, 963, 1974.
138. **Shoshani, L., Gómez-Puyou, A., and Darszon, A.,** unpublished.
139. **Samama, J. P., Lee, K. M., and Biellmann, J. F.,** Enzymes and microemulsions, activity and kinetic properties of liver alcohol dehydrogenase in ionic water-in-oil microemulsions, *Eur. J. Biochem.,* 163, 609, 1987.
140. **Lee, K. M. and Biellmann, J. F.,** Enzyme and organic solvents: horse liver alcohol dehydrogenase in non ionic microemulsions: stability and activity, *FEBS Lett.,* 223, 33, 1987.
141. **Khmelnitski, Y. L., Hilhorst, R., and Veeger, C.,** Detergentless microemulsions as media for enzymatic reactions, cholesterol oxidation catalyzed by cholesterol oxidase, *Eur. J. Biochem.,* 176, 265, 1988.
142. **Sanchez-Ferrer, A., Bru, R., and Garcia-Carmona, F.,** Kinetic properties of polyphenol oxidase in organic solvents. A study in Brij-96 cyclohexane reverse micelles, *FEBS Lett.,* 233, 363, 1988.
143. **Zampieri, G.,** Dissertation Eidgenossische Technische Hochschule (ETH) Nr. 8444, 1987.
144. **Elworthy, P. H. and McIntosh, D. S.,** The interaction of water with lecithin micelles in benzene, *J. Phys. Chem.,* 68, 3448, 1964.
145. **Kumar, V. V., Kumar, C., and Raghunathan, P.,** Studies on lecithin reverse micelles: optical birefringence, viscosity, light scattering, electrical conductivity and electron microscopy, *J. Colloid Interface Sci.,* 99, 315, 1984.
146. **Kumar, V. V., Manoharan, P. T., and Raghunathan, P.,** Nuclear magnetic resonance spin-lattice relaxation in lecithin reverse micelles, *J. Biosci.,* 4, 449, 1982.
147. **Ferreira, S. T. and Verjovski-Almeida, S.,** Fluorescence decay of sarcoplasmic reticulum ATPase. Ligand binding and hydration effects, *J. Biol. Chem.,* 264, 15392, 1989.

Chapter 3

THE MICROSCOPIC STRUCTURE AND DYNAMICS OF WATER AT A SURFACE

Humberto Saint-Martin and Iván Ortega-Blake

TABLE OF CONTENTS

I. INTRODUCTION

The interest in understanding the behavior of a solvent surrounding a nonpolar solute and, in particular, a surface is now well documented, leading to several books devoted to this topic (see, for example, References 1 and 2). The reasons are manifold: basic research in statistical mechanics of liquids to develop analytical approaches for complex problems; the use of new powerful techniques, Monte Carlo and Molecular Dynamics, for envisaging the microscopic role of the solvent in an interesting system; the biochemical interest because of the relevance to biological membranes and enzymic catalysis; and the industrial interest of micelles, microemulsions, and surfaces.

The very general appearance of the phenomena of water solvation of surfaces in physical, chemical, and biological systems, being metallic arrangements or proteins, has led to an enormous amount of research. Particularly, this has permitted the establishment of phase diagrams of mixtures of water, oil, and surfactant. These mixtures can have different forms of aggregation: lamellar, bicontinuous, or intercontiguous phases, and discrete phases as micelles or reverse micelles. The appearance of a particular phase depends on the ratios of the components of the mixture and also on the salinity, temperature, and other environmental factors that can change from one phase to another without apparent change in the macroscopic properties, but it can produce changes in the electric conductivity and it can certainly conduce to quite different phenomena in solvation, catalysis, etc. It is thus very important to understand the mesoscopic description of the system and much theoretical work has been devoted to this, but once it is clear that a particular phase has been attained, for instance reverse micelles, understanding the phenomena will come from a microscopic description.

Microscopic understanding of the phenomena can come from experiments directed to look into the molecular properties and also, with an increasing frequency, from theoretical methods. This work is devoted to an analysis of the information provided by microscopic theoretical methods and the comparison with available experimental work that will permit the elaboration of a molecular picture to start guessing the behavior of the solvent at the interphase.

II. THEORETICAL MICROSCOPIC STUDIES

Very soon after the establishment of quantum mechanics, the straightforward application of physics, the Schrödinger equation, was pursued for the understanding of the microscopic properties of the matter. More than half a century afterward, and due mainly to the advance in computers, it is now possible to obtain molecular and intermolecular properties of small molecules with reliability and criteria to assess this reliability (see, for example, Reference 3). An extension of this idea of obtaining the microscopic behavior from the first principles is to compute the intermolecular potential of a substance, e.g., water, and then to construct a numerical model with a few hundred of these molecules to simulate the substance, e.g., liquid water. This approach has been championed by Enrico Clementi for many years and is now proving to be quite successful and is becoming popular. On the other hand, coming not from quantum mechanics, but from statistical mechanics, is the approach initiated by Bernal, Stillinger, and others of constructing classical potentials for the intermolecular interaction and adjusting their parameters to reproduce known macroscopic properties. Both manners are now used to obtain potentials to be employed in Monte Carlo and Molecular Dynamics simulations. In the Monte Carlo method (see, for example, Reference 4), the molecular movements are performed randomly, using the computed potential energy to sample the configuration space in an intelligent manner. Molecular Dynamics (see, for example, Reference 5) solves Newton's equations of motion with the intermolecular forces derived from the adjusted potential. In both methods, one looks into the statistical averages to obtain the expected values of the molecular properties. Hence it is very important to have a large enough sample, either of the configurational space or of the phase space, to have confidence in the obtained values which are also critically affected by the number of elements considered to simulate the ensemble. Of course, the molecular potential used is crucial since the obtained properties will stem from it.

From our point of view, it is more convenient to develop the intermolecular potential from first principles, mainly because it allows a stepwise refinement of the model that should always conduce to a better representation of the system. For instance, there has been a permanent interest in reproducing liquid water by numerical simulations and many potentials have been used. However, it has been recently shown that the proper reproduction of structural properties requires a very accurate description of the intermolecular interaction;[6-8] that is, the physical model for the water molecule has to include effects such as polarizability and many body corrections. Since reproduction of bulk water is very important for the study of its behavior at the interphase, we will now present a Monte Carlo simulation of liquid water with a large number of waters and made with a very refined potential.

III. BULK WATER

A Monte Carlo simulation was performed on a box of 1668 water molecules and dimensions $\Delta x = 32$ Å, $\Delta y = 32$ Å, and $\Delta z = 48$ Å having periodic boundary conditions. The initial configuration was generated from the combination of several boxes of 343 waters that had attained equilibrium. The large box was allowed to equilibrate for 5×10^5 Monte Carlo steps until the variation in the energy of the ensemble was ± 0.05 kcal/mol. One can obtain from the simulation thermodynamic and structural parameters that can be compared with their experimental counterparts. The potential energy per water molecule obtained in this simulation is 7.96 ± 0.05 which compares rather well with 8.1 ± 0.4, the experimental value.[9] It is interesting to note that a value of 8.2 ± 0.05 was obtained from a simulation with the same potential and 343 waters, showing that there is still a dependence of the molecular properties for this number of waters.

There is one structural parameter which is particularly important: the radial distribution function (rdf). This function is obtained by computing the average number of particular atoms, or molecules, that exist in a shell of radius r centered on the atom we are looking to correlate with, of course normalized to an average isotropic density. For instance, for a crystal, the rdf will have sharp peaks and regions of zero value since there are some distances at which there is a number of atoms, while for most of the distance values there is none. On the other hand, for a homogeneous gas, the distribution of atoms is isotropic and the rdf will have a unit value for all distances except the very close excluded volume. A liquid will have an rdf which is a case intermediate between the above two. The more structure that is in the solvent, the more structure that will appear in the function. A strong advantage of the rdf is that it can be obtained experimentally through X-ray or neutron diffraction, and therefore provides a test for the simulation. As a matter of fact, for a long time there has been a strong desire in reproducing the experimental rdf of liquid water as close as possible. In Figures 1 to 3, we present the rdf for oxygen-oxygen, oxygen-hydrogen, and hydrogen-hydrogen obtained from the simulation as well as their experimental counterparts. We can see that there is quite a good agreement, except for the first peak of the rdf for oxygen-oxygen, possibly due to the lack of vibrations in the physical water model. This is the first time that an *ab initio*-based potential correctly reproduces the second peak which is related to the tetrahedral structure and thus to the water network. The same potential used with 343 waters[7] does not present such an agreement, showing again that for 343 waters there is still an impact on the statistical properties. There are several comparative studies on the qualities of different water potentials (see, for example, References 12 and 13). It is clear now that

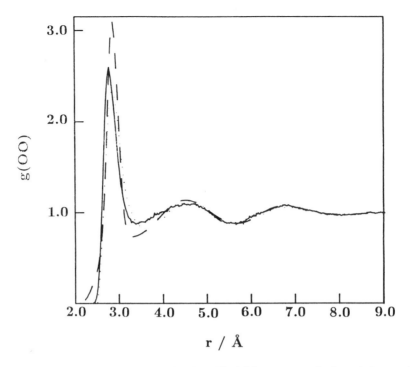

FIGURE 1.   Oxygen-oxygen radial distribution functions. The full line represents the theoretical curve, the dotted line represents the experimental curve of Narten and Levy,[10] and the broken line represents the experimental curve of Soper and Phillips.[11]

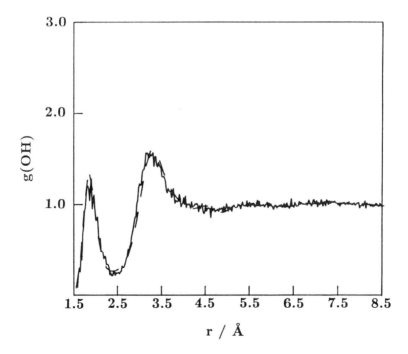

FIGURE 2.   Oxygen-hydrogen radial distribution functions. The full line represents the theoretical curve, and the broken line represents the experimental curve of Soper and Phillips.[11]

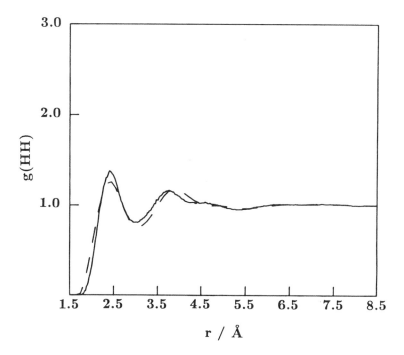

FIGURE 3.   Hydrogen-hydrogen radial distribution functions. The full line represents the theoretical curve, and the broken line represents the experimental curve of Soper and Phillips.[11]

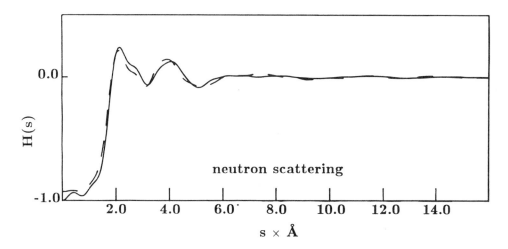

FIGURE 4.   Structure function H for X-ray scattering. The full line represents the theoretical prediction and the broken line represents the experimental curve of Narten.[14]

the only manner to reproduce accurately this structural function is through the refinement of the physical model, and in this respect the first principle's potential is advantageous because of the knowledge of the approximations done and the way to improve them. In addition to this advantage, one can be confident that if the structure of the bulk is well predicted, not adjusted, then similarly good predictions of the water structure in other conditions will be obtained. A more stringent test for the predictions of water structure is in the actual experimental observation, the structure function H. In Figures 4 and 5, we present the theoretical and experimental curves for neutron and X-ray scattering. Again there is very good agreement and an improvement over the 343 waters simulation.[7]

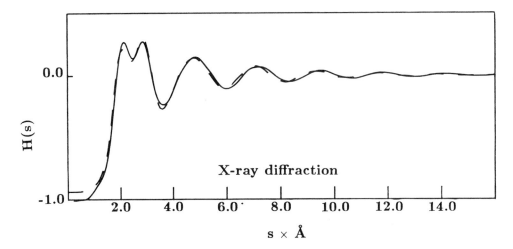

FIGURE 5. Structure function H for neutron scattering. The full line represents the theoretical prediction and the broken line represents the experimental curve of Narten.[14]

From the above results, it is quite clear that this water potential is able to reproduce the energetic and structural properties with a small loss of accuracy if the simulation is done with a reduced number of waters (~300). There is, however, an important deficiency in the potentials fitted to the *ab initio* values: the pressure is very far from the experimental value, and it is clear that this is a value difficult to reproduce in general (see, for example, Reference 15). Hence, with this last concern in mind, we can expect that the structural properties of water are well investigated with highly refined *ab initio*-based potentials. The reason for the improved refinement of the structural properties is the inclusion of physical phenomena in the water model, and since these phenomena are relevant for the discussion at the end of this work, we would like to extend a bit over them here.

Water polarization is obviously an important phenomenon in the water interaction, but there are many water models that are able to reproduce the dimer interaction with a rigid water in both the positive (nuclei) and the negative (electrons) charge posititions. However, when water has to respond collectively, the rigid molecule is forced to do it as a whole molecule, and no intramolecular relaxation of dipoles is allowed. Thus, in spite of a good reproduction of the interaction energy, other properties, such as the dielectric constant, will be impaired. Another important phenomenon is the many-body nonadditivity, i.e., the effect that a third or fourth molecule will have in the interaction of the two or three other molecules. A decade ago it was assumed that for water this correction was small and that its inclusion would be just a fine point. Now it is quite clear that collective properties depend in a relevant manner on these corrections; furthermore, for the liquid state, this correction is much larger than expected.[6] Another feature that has been included in the water potential is the vibrational freedom (see, for example, Reference 16). The current idea is that this will have only a small consequence on the properties of the liquid, but this still has to be clearly established. Charge transfer has been circumvented by flexible analytical forms that can account for their obvious action, even at the pairwise interaction, but it could be that some particular phenomena will require its explicit inclusion. The potential that was used in the bulk simulation just reported, and the lamellar simulations that will be presented afterward, includes polarization and three and four body corrections, not only the induction component, but adjusted to the *ab initio* nonadditive values.

Before entering into the simulation of an interphase, we will present a summary of previous works and an analysis of the general panorama that emerges from them.

IV. THEORETICAL STUDIES OF THE WALL-WATER INTERPHASE

In Table 1, we present a summary of the theoretical works made in the last decade for the wall-water interphase, mainly numerical simulations. Without being exhaustive, this table gives a good glance on the efforts made and the results that emerged from the simulations. First of all, it is clear that the theoretical simulations have become more frequent in recent years and have become more able to yield extensive information about the system. Early work, up to 1985, was performed with models of wall that had some simple forms of repulsion and with poorly refined water potentials. These latter were rigid and nonpolarizable and had proven to yield reasonable predictions of the bulk water structure, but still far from accurate descriptions. Nonetheless, the results are quite clear and consistent. The water density in the neighborhood of the wall has an oscillatory damped behavior, i.e., it shows an increase in density at the wall because of close packing, but this structural phenomenon is buffered pretty fast by the liquid. The density of the bulk is regained in less than 10 Å. Of course, one has to bear in mind the size of the computational cell which is of similar dimensions. There is also a general agreement on the structural effects near the wall, enhanced structure, similar to the ice Ih (where h refers to the normal hexagonal lattice as opposed to ice Ic, with a cubic structure), and an important role of the hydrogen bond. For the dynamical properties, there is no such agreement. Water self-diffusion is proposed to be both faster and slower at the interphase, and the only agreement is that variations are of less than an order of magnitude. A similar pattern appears for the dipole-dipole correlation. In a subsequent phase of the theoretical work, the last 5 years, a search for complexity in the model is characteristic, i.e., simulating the interphase with much more atomic detail and introducing different physical environments, such as charged surfaces or fields. Intermolecular potentials for water, as well as those for the other species now included, were kept simple by necessity. Of course, with much more variability, the modeling of the interphase is now quite complex and the predicted properties show also a certain degree of variability. However, a general result established in a very consistent manner is the absence of structural effects produced by the interphase at distances larger than a few Å (~8). This finding also appears in more recent works where the intermolecular potentials are more refined and will be discussed afterwards. In the vicinity of the interphase, we can see the following effects:

1. The water density presents an oscillatory damped behavior similar to that observed for primitive walls, except for two cases.[27,30] In these cases, the lack of oscillations is attributed to the presence of ions in the simulations, but there are other simulations with ions that show a strong oscillatory behavior.[31,34]
2. All the models with atomic detail of the interphase show a broad boundary, i.e., not a well-defined border, but a rough mixture of the two phases. A very good example can be seen in Figure 6.
3. A general and striking result is observed in the models that include electric fields either through charged surfaces or in a continuous manner. The electric field is almost immediately canceled by water, contrary to the textbook idea that the Gouy-Chapman phase extends over tens of Angstroms.
4. There seems to be a more-packed structure near the wall, similar to ice Ih.
5. The results on the dynamic properties are similar to those reported by the simulations with primitive walls; i.e., there is a discrepancy about the values being increasing or decreasing, but they retain the same order of magnitude as those in the bulk.
6. The inclusion of ions in the model allows us to look into their distribution. However, the results are contradictory, predicting condensation of ions near the wall[34] or repulsion by the surface.[31]

TABLE 1
Theoretical Studies of the Wall Interphase

Year	Model	Potentials	Results	Ref.
1981	Box with periodic conditions in x and y Two hard walls of infinite repulsion on z Δz between walls: 20 Å 150 water molecules Monte Carlo, 3.5×10^8 conf. T = 300 K	Water-water MCY[34] Rigid, nonpolarizable *ab initio* Cubic cutoff	$\rho(z)$, oscillatory damped behavior with 3 peaks; max amplitude = 1.25, max range = 7.5 Å No wetting present Enhanced hydrogen bonding in the surface Orientation of dipole near the wall	17
1983	Box with periodic conditions in x and y Hard repulsive walls, with an "elastic" potential, $\phi(z) = \alpha \exp(\beta z)$ Δz between walls: 20 Å 150 water molecules Molecular dynamics, 25 ps T = 302 K	Water-water ST2[35] Rigid, nonpolarizable Empirical Spherical cutoff, r = 7.4 Å	$\rho(z)$ has an oscillatory, damped behavior, with 2 peaks; max amplitude = 0.2, max range = 5 Å No wetting present Water dipole orientation parallel to the wall Dipole-dipole correlation smaller than bulk rdf shows higher peaks (more structure) close to the wall $D\perp$ in the wall is half of that in bulk $D\parallel$ is not affected D near the wall is 20% less than in bulk Dielectric relaxation in the wall 8.5 GHz and 17 GHz in the bulk	18
1983	Box with periodic conditions in x and y Lennard-Jones walls, with a potential similar to that of water Walls separated until zero pressure is obtained Δz between walls: 25 Å 216 water molecules Molecular dynamics	Water-water ST2[35]	$\rho(z)$ behavior is very dependent on the level of convergence to equilibrium and the number of configurations considered (Δt); is it possible that there will be no oscillation in a large equilibrium sample?; in this one, the behavior is oscillatory, damped with 3 peaks; max amplitude = 0.1, max range = 10 Å No wetting present Water dipole orients parallel to wall Orientational correlation extends to 10 Å $D\perp$ is 30% larger than $D\parallel$ away from the wall, but they are equal at the interphase Hydrogen bonding is only 75% at the interphase	19
1984	Box with periodic conditions in x and y Lennard-Jones wall, simulating paraffin Δz between walls: 24 Å 216 water molecules Molecular dynamics, 72 ps T = 290 K	Water-water ST2[35] Water-water ST2[35] without charges Spherical cutoff, r = 8.2 Å	$\rho(z)$ for a simple liquid (ST2 without charges) shows strong oscillations which extend for a long range, and there is wetting; for water (full ST2) the behavior is oscillatory and damped, max amplitude = 0.2, max range = 7 Å, and there is no wetting Ice Ih-like structure at the wall Orientational structure buffers at 6 Å The molecular orientation is governed by HB interactions Water molecules close to the wall preserve 75% of their HBs	20

TABLE 1 (continued)
Theoretical Studies of the Wall Interphase

Year	Model	Potentials	Results	Ref.
1985	Sphere with restricting forces, r = 9 Å 101 water molecules Molecular dynamics, 30 ps T = 301 K	Water-water TIPS2[36] Rigid, nonpolarizable Empirical Spherical cutoff, r = 8.5 Å	The second peak of the rdf is washed out compared to bulk, which implies loosening of the tetrahedral structure Explicit orientation in the wall Structural preference is not propagated deep into the liquid, and this is assumed to be due to the spherical topology Water molecules close to the interphase sacrifice one HB in order to maximize the total interaction	21
1986	A mixture of 1094 water molecules with 15 sodium octanoate in a 34 Å cubic box Rescaling to maintain constant energy Molecular dynamics, 30 ps (FC) and 72 ps (RC) T = 294 K	Water-water SPC[37] Rigid, nonpolarizable empirical (FC) Lennard-Jones type interaction for other atoms Dihedral potential for rotating heads Truncation of long-range coulombic potential Spherical cutoff, r = 10 Å Charges reduced to 50% (RC) to account for dielectric constant	$\rho(r)$ extends away from micelle with only one peak (RC) and no wetting present Broad boundary in the micelle, very dependent on the potential Same dynamics for water in the interphase and in the bulk, so there is no support for the concept of bound water	22
1986	Box with periodic conditions in x and y Walls with different dielectric constant and charges, 77.3 and 5.0, respectively Δz between walls: 21 Å Monte Carlo, 2×10^5 conf. T = 301 K	Water continuum model Electrostatic ion-ion, ion-wall, and ion-water interactions	The ions are repelled because of the low dielectric constant of the wall	23
1986	Box with periodic conditions in x and y Lennard-Jones type walls Δz between walls: 24 Å 200 water molecules 8 lithium iodine molecules Molecular dynamics, 10 ps T = 283 K	Water-water ST2[35] Lennard-Jones type for Li-I, empirical Spherical cutoff, r = 9.3 Å	$\rho(z)$ has no oscillations near the wall, assumed to be produced by the presence of ions The Lennard-Jones wall is very short-ranged in structure making Ice Ih-like structure near the wall Faster motion near the wall, due to the absence of shear force An influence on the whole lamina increases self-diffusion The water molecules preserve 75% of their HB close to the wall	24
1987	Box with periodic conditions in x and y Hard walls Δz between walls: 20 Å 125 water molecules Monte Carlo, 2.02×10^6 configurations T = 399 K	Water-water TIPS2[36]	$\rho(z)$ oscillatory damped behavior but nonsymmetric; max amplitude = 0.2, max range = 8 Å No wetting present No electric polarization near the surface Ice Ih structure, modest	25

TABLE 1 (continued)
Theoretical Studies of the Wall Interphase

Year	Model	Potentials	Results	Ref.
1987	Box with periodic conditions in x and y Metallic wall with image charge-interactions Gouy-Chapman double layer on one wall Δz between walls: 32 Å 200 water molecules Monte Carlo, 2.0×10^6 conf. T = 399 K	Water-water TIPS2[36]	The field has a minor effect on the structure described in the previous entry The effect of the field is negligible at the neutral wall Hydrogen bonding between waters controls the structure A metallic surface freezes a thin layer of water in a rigid ice Ih structure There is wetting for the metallic surface There are no effects on polarization and angular distribution	26
1988	Mixture of water and sodium octanoate in a 3-D box to form micelles Length of box = 34 Å 1068 water molecules 15 sodium octanoate Molecular dynamics, 50 ps T = 300 K	Water-water SPC[37] Lennard-Jones type for other atoms, with a flexible surfactant Ewald sums for long-range coulombic interactions	The micelles formed are nonspherical but stable $\rho(r)$ outside the micelle does not show any structure, but just an increase up to a plateau The micelle remains partially charged because it is not neutralized completely by the counterion SPC water has a dielectric constant of 65, and the scaling of Reference 22 is not justified The cutoff used for long-range interactions in Reference 22 could yield a difference with this work	27
1988	Periodic boundary conditions in 3-D mixture of: 526 water molecules 52 Na-decanoate 76 decanol Variable size of the unit cell Molecular dynamics, 30 ps T = 300 K	Water-water SPC[37] Lennard-Jones type for other atoms; generic potentials from combination rules Approximation of forces at 3 different ranges: short, with direct cutoff at 7.5 Å medium, updated every 10 steps long, approximated with a charge density	$\rho(r)$, a peak in density is suggested, but there is not enough water for a second one Water penetrates into the bilayers Na^+ also penetrates, but less markedly The interphase is very broad and diffuse There is not a diffuse electrical double layer D inside the membrane is 10 times lower than in bulk D in the interphase is 3 times lower than in bulk Reorientation and rotation of water is slower than in bulk Na^+ concentration is maximum at bulk	28
1989	A mixture of water, monovalent ions, and a double layer through macro ions Statistical mechanics with the RHNC closure	Water: sphere with dipole and quadrupole moments Ions with charge and different sizes Electrostatic interactions	$\rho(z)$ has a maximum close to the wall and depending on the surface charge, it can show oscillations which extend deep into the bulk; these oscillations are also dependent on the size of the macro ions The results are very sensitive to the interaction potential Ice Ih-like structure at the surface The ice structure is not broken by the presence of ions The surface charge is screened extremely fast	29

TABLE 1 (continued)
Theoretical Studies of the Wall Interphase

Year	Model	Potentials	Results	Ref.
1989	Box with periodic conditions in x and y Wall with a repulsive Lennard-Jones type potential Wall with atomic Lennard-Jones in 3 layers Δz between walls: 35 to 50 Å Molecular dynamics	Water-water TIP4P (updated version of TIP2)	Ice Ih-like structure near the surface D‖ is greater in the surface than in bulk The residence time in the surface is somewhat greater than in bulk, but still small For the hydrophobic surface there is no wetting and the deviations from the bulk density do not extend more than 3 molecular layers While in the hydrophobic surface the water molecules preserve 75% of their hydrogen bonds, close to the silica wall they keep only 25%, and the density profile has only 2 peaks	30
1989	Sphere with a radius of 19 Å, having a repulsive wall at 19.4 Å Mixture of: 1000 water molecules 50 Na 50 carboxylic groups Low density to obtain zero pressure Molecular dynamics, 50 ps	Water-water MCY[34] For ions empirical Lennard-Jones Coulombic interactions are scaled (1/$\sqrt{2}$) to account for the dielectric constant	$\rho(r)$ has an oscillatory damped behavior with 3 peaks, max amplitude = 0.5, max range = 8 Å Wetting possible Enhanced hydrogen bonding at the wall surface when ions were considered without charge When ions are charged, the hydrogen bonding at the wall surface is similar to that in bulk, but water molecules in contact with ions have a reduced HB Water structure more densely packed up to 9 Å Zero Na$^+$ concentration at bulk D is reduced around the ions and the interphase Ions and surface compensate effects Strong orientation of water dipoles in the first hydration shells of ions	31
1990	3-D periodic conditions Mixture of "water" and "oil" particles for a total of 39,304 particles Molecular dynamics, 10^5 steps T = E/K	Lennard-Jones truncated at 2.5	$\rho(r)$ oscillations in the interphases of surfactant water Formation of micelles and lamellae Depletion layer after interphase	32
1990	Box with periodic conditions in x and y Repulsive walls Δz between walls: 24 Å 216 water molecules Molecular dynamics, 96 ps T = 302 K	Water-water NEMO[39] *ab initio* polarizable sperical cut-off, r = 8.5 Å	$\rho(z)$ has an oscillatory damped behavior with 3 peaks, max amplitude = 0.75 max range = 10 Å Wetting can be seen from (z) Ice Ih-like structure in the first two layers More structure than with a pair-wise additive potential Induced dipole aligned to the surface Permanent dipole not aligned Molecules at the surface preserve 75% of their hydrogen bonds Polarization of water molecules does not play an important role in the structure	33

TABLE 1 (continued)
Theoretical Studies of the Wall Interphase

Year	Model	Potentials	Results	Ref.
1991	Box with periodic conditions in x and y Wall with gas-crystal potential, neutral Wall with a DC field = 0.1 V/Å Δz between walls: 20 Å 256 water molecules Molecular dynamics, 30 ps, 70 ps T = 298 K	Water-water SPC-FP, SPC plus polarization and vibrations, adjusted to thermodynamic properties Long-range electrostatic interactions are treated with the reaction field geometry	$\rho(z)$ has an oscillatory damped behavior with 2 peaks, max amplitude = 0.5, max range = Å Wetting is seen in both cases, with and without the DC field, but is more marked in the latter The HB structure is broken by the reduction of the O–H distance at the surface, but the HB is enhanced in the bulk; hence there is low entropy at the wall, but high at bulk $D\perp$ is enhanced with respect to bulk water	34
1991	Box with periodic conditions in x and y At one z surface COOH groups and tails The other z surface free Surfactant in hexagonal lattice 200 water molecules 64 surfactant molecules Monte Carlo, 1.8×10^6 conf. T = 300 K	Water-water TIP4[38] Surfactant with generic Lennard-Jones Spherical cutoff with different radii	$\rho(z)$ has oscillatory damped behavior with 2 peaks, and not enough water for a third peak, max amplitude = 0.5, max range = 6 Å Thermal motion of the heads allows water penetration Water dipoles are oriented from the wall to the bulk The first hydration monolayer screens almost completely the membrane force field of the head groups	35

Note: $\rho(z)$ or $\rho(r)$ corresponds to the water density; $D\perp$ represents the diffusion of water perpendicular to the surface; $D\parallel$ represents the diffusion of water parallel to the surface; and D represents the total diffusion.

The results which show agreement in a wide range of surface models and intermolecular potential attain a degree of confiability, but the appearance of discrepancies stresses the fact that many properties are critically dependent on the intermolecular potential.

The most recent works revisit the problem using more refined potentials for water. The results of those simulations are similar to those obtained previously: an oscillatory damped behavior of the density and the ice-like structure at the wall, but, a new physical phenomenon not pointed out by the authors seems to appear: the presupposed bulk density, i.e., that at the middle of the box, where no oscillations are present, shows a decrement suggesting wetting of the surface. A similar result had been observed,[20] but for a simple liquid. This is an important finding since it is a long-range effect and could affect the chemical behavior of water in systems as reverse micelles. In order to shed some light on this point, here we present a simulation of the interphase with a very refined water potential. We want also to clarify the possibility that some of the structure is just a consequence of the lack of equilibration in the simulation, as questioned by Sonnenschein and Heinzinger.[19]

V. WATER STRUCTURE NEAR A FLAT DIELECTRIC WALL

The model considered here is a water box with 343 waters and periodic conditions in x and y. In the z direction, there are two walls with a distance of 22 Å between them. The wall simulates a medium of dielectric constant $\epsilon = 3$ that should resemble the membrane. The water molecules interact with the images of all the molecules which appear in the dielectric slab. The cutoff radius for water-water is 8 Å and for the water-water images is

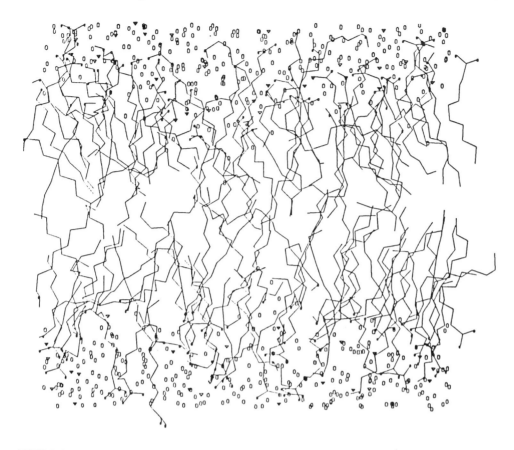

FIGURE 6.   A molecular dynamic simulation result for the interphase water, Na-decanoate, decanol. Projection on the xz plane of a snapshot of the system after 30 ps simulation: decanoate ions and decanol molecules (—), sodium ions (∇), water oxygens ($\bigcirc$), and oxygens in the lipids ($\bullet$). (From Egberts, E. and Berendsen, H. J. C., *J. Chem. Phys.*, 89, 1718, 1988. With permission.)

also 8 Å. There is a zone in the immediate vicinity of the wall (0.5 Å) where the image interaction remains constant, i.e., with the value attained at z = 0.5 Å. This zone has the purpose of considering the atomic range where the image model will fail. In any case, this zone only modulates the water-wall interaction; the value of 0.5 Å yields a strong enough interaction in the closest vicinity, and we will be looking to see if such a large perturbation propagates into the bulk.

A. RESULTS

First of all, we want to mention that there is indeed a very strong dependence of the observed structure on the number of configurations used to attain equilibrium. In our simulation, it was possible to obtain what looked like equilibrium in the potential energy for as many as a million configurations, but if further configurations were allowed to go, one could observe that a very slow convergence rather than equilibrium had been obtained. Unfortunately, in these not quite equilibrated runs, one could see a behavior of the water density and other structural phenomena similar to those reported previously in the literature. In the results that we are reporting now, we threw away 8 millions of configurations and then made statistics on 3.2 millions of configurations; this length of runs had not been considered before.

1. Water Density Profile Along the z Axis

In Figure 7, we can see the water density profile of the simulated lamella; in Figure 8,

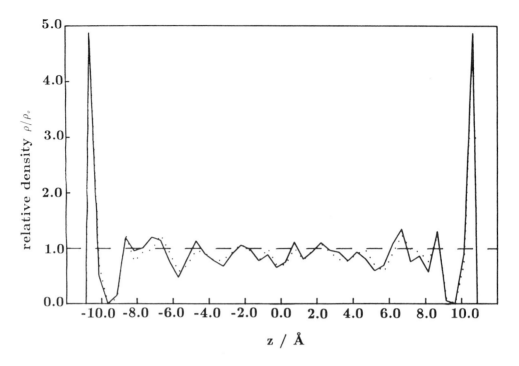

FIGURE 7.   The density of water along the z-axis for the simulation of the lamella (full line). Also shown is the symmetrized function constructed from $[\rho(z) + \rho(-z)]/2$, represented by the dotted line.

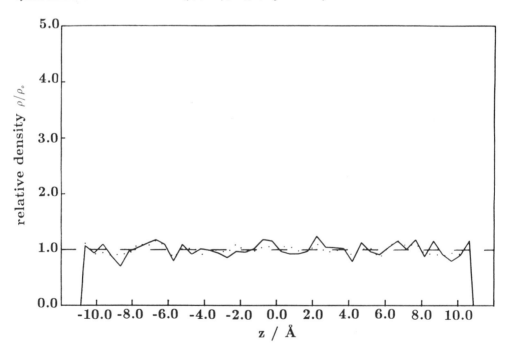

FIGURE 8.   The density of water along the z-axis for the simulation of the bulk (full line). Also shown is the symmetrized function constructed from $[\rho(z) + \rho(-z)]/2$, represented by the dotted line.

the corresponding bulk density is presented; and, in both figures, the symmetrized density obtained from the average value between $\rho(z)$ and $\rho(-z)$ is shown. We can see in Figure 7 that there is a huge peak of almost five times the ordinary density showing a strong close packing produced by the charges' images. However, in spite of this strong surface effect, the bulk density is attained extremely fast (~ 6 Å). It was quite clear during the long equilibration runs that to destroy any local structure, evidenced by the persistence of peaks in $\rho(z)$, required a large number of configurations, indicating an energetic hindrance; this could also be reflected in the need for large relaxation times in molecular dynamics. Apart from being an interesting observation on the capability of water to retain local structure, these findings highlight the need for quite long runs. The symmetrized function does a form of common mode rejection of noise and it is thus more reliable to look into. In this function, we can see that, apart from the fast buffering by the liquid, wetting does indeed appear, in agreement with the other simulations of polarizable models.

2. Hydrogen Bond Pattern at the Surface

A good indication of the energetic properties of water at the surface would be the ability to form hydrogen bonds and the distribution of strengths compared to the bulk. In order to estimate this, we considered the interaction energy of close neighboring waters (less than 3.5 Å); if the energy exceeded -4.0 kcal/mol, we counted them as hydrogen bonded. An analysis of the results as a function of z is presented in Table 2 along with the distribution of hydrogen bond energy values. From this table, we can see that in the bulk there is a rather homogeneous distribution of hydrogen bonds and a constant average energy. In the lamella, there is a strong modification near the wall, as expected, but no increase in the average strength. We can also see a depletion layer of hydrogen bonds, corresponding to the depletion in $\rho(z)$. A more important result is that, on recovering to the bulk, the number of hydrogen bonds is slightly increased while the energy remains almost the same. This result is in agreement with the observation that wetting could be occurring, and the increased ability to form hydrogen bonds is a reflection of the reduced density.

3. Dipole Orientation

The dipole orientation in lamella has been one of the most common phenomena looked into in the theoretical simulation. A strong orientation of the dipole will suggest quite a different electric environment than that at the bulk. Here we want to remark that, in a model with a rigid molecule, the permanent dipole should orient by the orientation of the whole molecule, hence being subject to structural hindrance, but in the polarizable models, there is an induced dipole that can orient without the concomitant movement of the molecule. In Figures 9 and 10, we present the dipole orientation of the lamella and the bulk, respectively, that is counting the projection of the dipole along the z-axis. An isotropic distribution will give a zero value which is observed in the bulk's simulation. In the simulation of the lamella, we can see that, at a few angstroms from the wall, the dipole orientation is the same as that in the bulk.

4. Dipole Value and Dipole Correlation

A much-desired parameter of the simulation is the dielectric constant as a function of the z coordinate since this will let us know to some extent what sort of chemical environment will be occurring. One can obtain the dielectric constant as the response to some field and some model that involves several approximations. The value will be obtained from the expectation values from the dipole and the dipole-dipole correlation. Here we will be looking to these expectation values along the z axis and compare them to those in the bulk. This will permit us to see any difference that could appear without introducing the approximations of the reaction field model, but, of course, this prevents us from obtaining numerical values for the dielectric constant.

TABLE 2
Hydrogen Bonding in Bulk Water and in Water
Between Dielectric Plates

	Bulk		Lamella	
z (Å)	N_{hb}	E_{hb} (kcal/mol)	N_{hb}	E_{hb} (kcal/mol)
−10.63	1.82	−4.56	0.80	−4.43
−10.13	1.78	−4.54	0.00[a]	−3.92
−9.64	1.80	−4.55	0.00	0.00
−9.15	1.81	−4.54	1.74	−4.54
−8.65	1.83	−4.55	1.64	−4.49
−7.66	1.87	−4.53	1.96	−4.73
−7.17	1.71	−4.52	1.78	−4.69
−6.67	1.85	−4.57	2.00	−4.57
−6.18	1.80	−4.57	2.08	−4.58
−5.68	1.67	−4.53	1.93	−4.65
−5.19	1.78	−4.56	2.04	−4.63
−4.70	1.77	−4.58	2.02	−4.57
−4.20	1.83	−4.57	2.00	−4.53
−3.71	1.73	−4.53	2.07	−4.57
−3.21	1.82	−4.54	2.10	−4.60
−2.72	1.88	−4.55	1.98	−4.60
−2.22	1.82	−4.55	1.86	−4.56
−1.73	1.74	−4.53	2.00	−4.60
−1.24	1.61	−4.53	1.93	−4.59
−0.74	1.83	−4.54	1.99	−4.59
−0.25	1.84	−4.54	2.15	−4.60
0.25	1.92	−4.56	2.07	−4.60
0.74	1.88	−4.55	2.11	−4.60
1.24	1.71	−4.58	2.10	−4.61
1.73	1.71	−4.55	1.96	−4.53
2.22	1.88	−4.57	2.09	−4.54
2.72	1.82	−4.66	2.18	−4.56
3.21	1.72	−4.55	2.10	−4.55
3.71	1.73	−4.61	2.14	−4.59
4.20	1.65	−4.53	2.04	−4.57
4.70	1.71	−4.60	2.02	−4.59
5.19	1.64	−4.55	1.89	−4.56
5.68	1.70	−4.55	1.96	−4.55
6.18	1.67	−4.55	2.13	−4.60
6.67	1.72	−4.59	1.89	−4.57
7.17	1.74	−4.56	1.80	−4.58
7.66	1.85	−4.55	2.01	−4.62
8.16	1.75	−4.57	1.58	−4.58
8.65	1.70	−4.57	1.73	−4.67
9.15	1.55	−4.53	1.03	−4.41
9.64	1.84	−4.53	0.00	0.00
10.13	1.83	−4.57	0.47	−4.18
10.63	1.77	−4.55	0.86	−4.45

Note: N_{hb} corresponds to the number of hydrogen bonds, and E_{hb} corresponds to the average strength.

[a] These values are averages over 3.2×10^6 configurations; there were very few which had a hb in this position, so that the average N_{hb} is 2×10^{-4}; nevertheless, the average strength of those few is −3.92 kcal/mol.

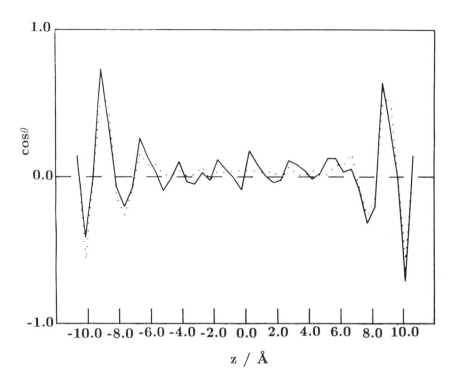

FIGURE 9. The dipole orientation of water along the z-axis for the simulation of the lamella (full line). Also shown is the symmetrized function constructed from $[f(z) + f(-z)]/2$, represented by the dotted line.

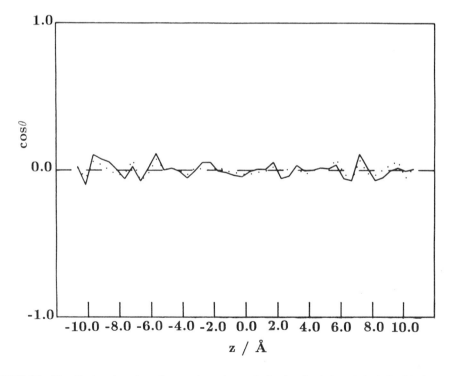

FIGURE 10. The dipole orientation of water along the z-axis for the simulation of the bulk (full line). Also shown is the symmetrized function constructed from $[f(z) + f(-z)]/2$, represented by the dotted line.

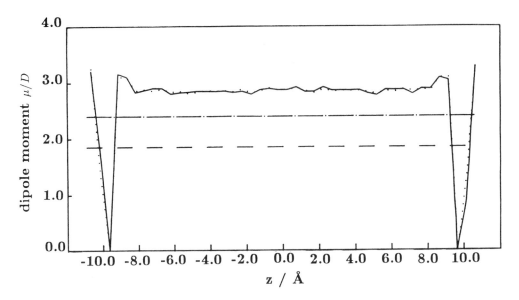

FIGURE 11.   The average dipole value of water along the z-axis for the simulation of the lamella (full line). The broken line shows the dipole of the water molecule in the gas phase, and the dash-dot line shows the dipole of the MCHO model for a single water molecule. Also shown is the symmetrized function constructed from $[f(z) + f(-z)]/2$, represented by the dotted line.

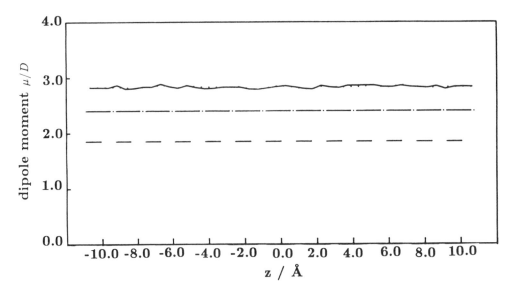

FIGURE 12.   The average dipole value of water along the z-axis for the simulation of the bulk (full line). The broken line shows the dipole of the water molecule in the gas phase, and the dash-dot line shows the dipole of the MCHO model for a single water molecule. Also shown is the symmetrized function constructed from $[f(z) + f(-z)]/2$, represented by the dotted line.

The result for the dipole value of the lamella and the bulk are presented in Figures 11 and 12, and the dipole correlations are presented in Figures 13 and 14. We can see again that there are no long-range modifications — just at a few angstroms from the wall — and there are again bulk properties.

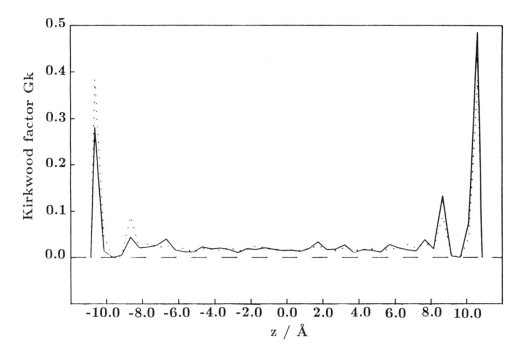

FIGURE 13.   The dipole-dipole correlation of water along the z-axis for the simulation of the lamella (full line). Also shown is the symmetrized function constructed from $[f(z) + f(-z)]/2$, represented by the dotted line.

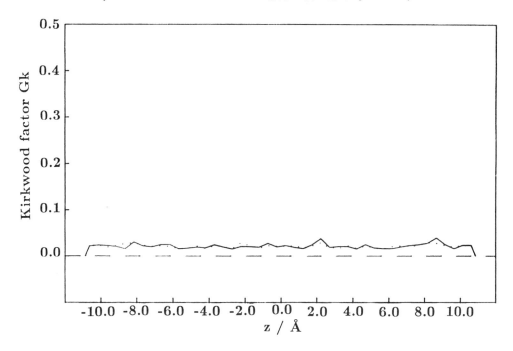

FIGURE 14.   The dipole-dipole correlation of water along the z-axis for the simulation of the bulk (full line). Also shown is the symmetrized function constructed from $[f(z) + f(-z)]/2$, represented by the dotted line.

VI. DISCUSSION

Here we have presented a detailed view of the efforts made by theoretical molecular methods to study the interphase between water solvent and a wall. We have seen the inherent problems of the methodology: increasing complexity of the system vs. refinement of the intermolecular potential. We have also seen how the efforts over a decade have converged to a structural picture, but not yet to the dynamical one. The most striking and consistent result is the very short range of the structural changes produced by the wall, always less than 10 Å. Sometimes this observation goes as far as suggesting that electric perturbances as Gouy-Chapman double layer are also short ranging, a contradiction to the textbook ideas on these phenomena. The shortness of the perturbation is indeed an important matter. A decade ago, Clegg[42] proposed that water inside a cell was so structured, because of the long-range effect of protein surfaces and membranes, that a completely different biochemistry than that observed *in vitro* was working. Even without taking this radical position, the idea of considerable water structure and relative long-range effect of surfaces has been a prevailing one. The theoretical results suggest quite a different picture. How do they relate to experimental evidence?

In Table 3, we present a summary of some recent experimental results that can be compared to the theoretical findings. The first thing that we observe from the experimental panorama is a strong discrepancy on the water properties assumed to exist near the surface. Steytler et al.[43] point out that a general problem of the experimental information is that to separate the effects due to molecules close to the surface from those more distant, this has to weigh in the conclusions drawn from the experiments. One should also remember that the wall-water conditions can change, e.g., from reverse micelle to an intercontiguous phase, by environmental changes that could be produced by experimental manipulation.

From an analysis of Table 3, one can see that there is indeed experimental evidence supporting the general idea behind this review, i.e., that structural effects of the solvent are constrained to very short distances from the wall. Steytler et al.,[43] from the analysis of neutron diffraction studies, stress the idea that structural changes have to be restricted to few angstroms. Similar support can be obtained from Raman spectroscopy,[52] where it is also pointed out that structural changes could be coming from interstitial water rather than from direct surface effects. This last observation is particularly interesting, taking into account that theoretical simulations show a broad undefined boundary, i.e., a roughness of the membrane that will produce such interstices. Rheology and proton NMR[53] also support the idea with the finding of similar mobility of water in bulk and surface. Finally, there are several observations that in an indirect manner support the idea. There seems to be a normal pH value in reverse micelles,[55] i.e., one can expect a bulk dielectric constant for most of the water in the micelle. There seems to be no marked modification of enzyme structure when observed by circular dichroism.[55]

However, there is also a substantial interpretation of experiments along the idea of a strongly modified structure of the solvent at the interphase. Piculell et al.[47-49] report, after NMR studies, a strongly modified water mobility near the surface and the existence of a bound water which exchanges very slowly. None of the theoretical simulations predict anything similar since variations in water diffusion are always small and water exchange is rapid. Low mobility of water was also reported by quasielastic neutron scattering,[46] but it was associated with assumed huge ionic concentrations, a condition that has not been explored by simulations. Dielectric response studies[51] have also been used to advance a tremendous modification of the water structure inside a reverse micelle, leading to the proposal of a general ice-like phase. However, one should note that the reduction in dielectric relaxation observed in this work was also obtained in a molecular dynamics simulation[18] where only a small water layer suffered from structural modifications.

Experimental observations which are ascribed to the interfacial region are rather scarce, and again contradictory. Neutron diffraction suggests enhanced hydrogen bonding[43] whereas fluorescence picosecond studies suggest a more aprotic environment.[44] Unfortunately, more recent reports on these lines, to the best of our knowledge, have not appeared.

TABLE 3
Some Experimental Work on Physical Properties of the Water-Wall Interphase

Year	System	Method	Results	Ref.
1983	Silica samples Spherisorb and Gasil	Neutron diffraction	There is an implicit problem in separating the effect due to the molecules close to surface and those more distant It is suggested, cautiously, that the interfacial region has enhanced hydrogen bonding There appears to be some local ordering of the water molecules in the interfacial region The interfacial region which is strongly hydrogen bonded and ice-like is restricted to few angstroms, <10 Å	43
1984	Reverse micelles	Fluorescence picosecond studies	More aprotic environment near the surface	44
1985	Micelles and microemulsions	Neutron small-angle scattering; differences between protonated and deutrated molecules	Intercontiguous structures of water and oil could be produced by salt	45
1985	Micelles and microemulsions	Quasielastic neutron scattering	In small water droplets, the local concentration of ion may be extremely high (12 M) This high concentration would be responsible for the decrease in the diffusion constant of water: four times less than bulk	46
1986	Protein solutions	Spin relaxation ^{17}O and ^{2}H	There is a class of water that exchanges very slowly	47
1986	Colloidal silica	Same	A marked reduction in water diffusion	48
1986	Same	Low-frequency dispersion	Reduction of the water radial diffusivity by two orders of magnitude and of the lateral by one order	49
1986	Micelles and microemulsions	Dielectric response	Water behaves quasi-frozen in the microemulsion phase A dielectric relaxation lower than that observed in free water; 40 GHz vs. 99 GHz	50
1986	Micelles and microemulsions	Dielectric response	The temperature dependence of the dielectric response suggests an ice-like structure of water in a micelle Due to higher connectivity of the hydrogen bond, dipole relaxation decreases by six orders of magnitude, comparable to liquid crystal	51
1989	Lamellar and inverted micelles	Raman spectrum O–H bond stretching	Near multilamellar phospholipids there is a small fraction of the water molecules in crystalline structure The shape of the O–H bonds depends on supramolecular structure adopted by phospholipids rather than on the surface interactions In colloidal particles, the Raman spectrum of water is almost identical to bulk The presence of small interstices also prevents water from crystallizing, and the Raman spectra depend on the size of the water domains rather than the nature of surface It is believed that reduced long-range intermolecular vibrational coupling is the origin of the Raman change	52

TABLE 3 (continued)
Some Experimental Work on Physical Properties of the Water-Wall Interphase

Year	System	Method	Results	Ref.
1989	Bilayer aqueous colloids	Rheology H-NMR	The mobility of water seems to be the same as that in the bulk The mobility seems to decrease with micelle formation Thus, there are two types of water, but they are not sharply defined	53
1990	AOT micelles	Thermochemistry	Water in a micelle has a freezing temperature well below zero H of freezing decreases with decreasing amount of water The unfreezable water is strongly bound to AOT and prevents ordinary ice Water in the micelle is present mostly in a bound form; hydration water Solubility drastically decreases with increasing amount of water Enzyme activity increases with decreasing amount of water Thermal stability of the protein increases with decreasing amount of water	54
1991	Reversed micelles with enzyme	^{31}P-NMR circular dichroism enzyme activity	The ionization behavior of the phosphate groups is similar to that in bulk; no anomaly of the micellar pH No marked perturbation on the enzyme structure by micellar water The physical properties of water in micelles of AOT or lecithin are very similar, suggesting that special properties are more determined by physical constraint that by chemical composition	55

VII. CONCLUSIONS

A combined theoretical and experimental effort during the last decade in determining the properties of water at an interphase has advanced our understanding of this crucial problem. Even if there is no complete agreement, we feel that the fact that water attains the bulk properties, or very similar ones, after a few angstroms is emerging. The only exception observed so far, and with the most refined potentials, is wetting, which has shown a longer range of action than the dimensions of the boxes used in the simulations — just above 10 Å. However, this leaves some critical observations as open questions. Why does water in a reversed micelle have a low freezing point?[54] Why are solubility and enzyme thermal stability affected?[54] And mainly, why are catalytic reactions so much affected?

One could think that water crystallization is hindered by a local surface structure that does not allow for the ice topography; this is quite possible, mainly through the idea of interstitial water. One could think that a physical phenomenon such as wetting could conduce to local ionic distributions that would change the microscopic picture so far obtained. Further theoretical and experimental work is needed and will certainly help us to understand an important and intriguing problem.

REFERENCES

1. **Luisi, P. L. and Straub, B. E., Eds.,** *Reverse Micelles, Biological and Technological Relevance of Amphiphilic Structures in Apolar Media,* Plenum Press, New York, 1984.

2. **Degrogio, V. and Corti, M., Eds.,** *Physics of Amphiphiles — Micelles, Vesicles and Microemulsions,* North-Holland, Amsterdam, 1985.

3. **Sauer, J.,** Molecular models in *ab initio* studies of solids and surfaces: from ionic crystals and semiconductors to catalysts, *Chem. Rev.,* 89, 199, 1989.

4. **Binder, K., Ed.,** *Topics in Current Physics: Application of the Monte Carlo Method in Statistical Physics,* Springer-Verlag, Heidelberg, 1984.

5. **Ciccoti, G., Frenkel, D., and McDonald, I. R.,** *Simulation of Liquids and Solids, Molecular Dynamics and Monte Carlo Methods in Statistical Mechanics,* North-Holland, Amsterdam, 1987.

6. **Gil-Adalid, L. and Ortega-Blake, I.,** Four body nonadditivity in liquid water, *J. Chem. Phys.,* 94, 3748, 1991.

7. **Saint-Martin, H., Medina-Llanos, C., and Ortega-Blake, I.,** Nonadditivity in an analytical intermolecular potential: the water-water interaction, *J. Chem. Phys.,* 93, 6448, 1990.

8. **Niesar, U., Corongiu, G., Clementi, E., Kneller, G. R., and Battachayra, D. K.,** Molecular dynamics simulation of liquid water using the NCC ab-initio potential, *J. Phys. Chem.,* 94, 7949, 1990.

9. **Weast, R. C. and Astle, M. J., Eds.,** *CRC Handbook of Chemistry and Physics,* 63rd ed., CRC Press, Boca Raton, FL, 1983.

10. **Narten, A. H. and Levy, H. A.,** Liquid water: molecular correlation functions from X-ray diffraction, *J. Chem. Phys.,* 55, 2263, 1971.

11. **Soper, A. K. and Phillips, M. G.,** A new determination of the structure of water at 25°C, *Chem. Phys. Letts.,* 107, 47, 1986.

12. **Watanabe, K. and Klein, M.,** Effective pair potentials and the properties of water, *Chem. Phys.,* 131, 157, 1989.

13. **Cieplak, P. and Kollman, P.,** A new water potential including polarization: application to gas-phase, liquid and crystal properties of water, *J. Chem. Phys.,* 92, 6755, 1990.

14. **Narten, A. H.,** Liquid water: atom pair correlation function from neutron and X-ray diffraction, *J. Chem. Phys.,* 56, 5081, 1972.

15. **Levitt, M.,** Molecular Dynamics of macromolecules in water, *Chemica Scr.,* 29A, 197, 1989.

16. **Lie, G. C. and Clementi, E.,** Molecular Dynamics simulation of liquid water with an ab-initio flexible water-water interaction potential, *Phys. Rev. A,* 33, 2679, 1986.

17. **Jönsson, B.,** Monte Carlo simulations of liquid water between two rigid walls, *Chem. Phys. Letts.,* 82, 520, 1981.

18. **Marchesi, M.,** Molecular Dynamics simulations of liquid water between two walls, *Chem. Phys. Letts.,* 97, 224, 1983.

19. **Sonnenschein, R. and Heinzinger, K.,** A molecular dynamics study of water between Lennard-Jones walls, *Chem. Phys. Letts.,* 102, 550, 1983.

20. **Lee, C. Y., McCammon, J. A., and Rossky, P. J.,** The structure of liquid water at an extended hydrophobic surface, *J. Chem. Phys.,* 80, 4448, 1984.

21. **Belch, A. and Berkowitz, M.,** Molecular Dynamics simulations of TIPS2 water restricted by a spherical hydrophobic boundary, *Chem. Phys. Letts.,* 113, 278, 1985.

22. **Jönsson, B., Edholm, O., and Teleman, O.,** Molecular Dynamics simulation of a sodium octanoate micelle in aqueous solution, *J. Chem. Phys.,* 85, 2259, 1986.

23. **Bratko, D., Jönsson, B., and Wenneström, H.,** Electrical double layer interactions with image charges, *Chem. Phys. Letts.,* 128, 449, 1986.

24. **Spohr, E. and Heinzinger, K.,** A molecular dynamics study of an aqueous LiI solution between Lennard-Jones walls, *J. Chem. Phys.,* 84, 2304, 1986.

25. **Valleau, J. P. and Gardner, A. A.,** Water-like particles at surfaces. I. The uncharged unpolarizable surface, *J. Chem. Phys.,* 86, 4162, 1987.

26. **Gardner, A. A. and Valleau, J. P.,** Water-like particles at surfaces. II. In a double layer and at a metallic surface, *J. Chem. Phys.,* 86, 4171, 1987.

27. **Watanabe, K., Ferrario, M., and Klein, M.,** Molecular Dynamics study of sodium octanoate micelles in aqueous solution, *J. Phys. Chem.,* 92, 819, 1988.

28. **Egberts, E. and Berendsen, H. J. C.,** Molecular Dynamics simulation of a smectic liquid crystal with atomic detail, *J. Chem. Phys.,* 89, 1718, 1988.

29. **Patey, G. N. and Torrie, G. M.,** Water and salt water near a charged surface: a discussion of some recent theoretical studies, *Chemica Scr.,* 29A, 39, 1989.

30. **Rossky, P. J. and Lee, S. H.,** Structure and dynamics of water at interfaces, *Chemica Scr.,* 29A, 93, 1989.

31. **Linse, P.,** Molecular Dynamics study of the aqueous core of a reversed micelle, *J. Chem. Phys.,* 96, 4992, 1989.

32. **Smit, B., Hilbers, P. A. J., Esselenk, K., Rupert, L. A. M., Vanos, N. M., and Schlijper, A.,** Computer simulations of a water interface in the presence of micelles, *Nature,* 348, 624, 1990.

33. **Wallqvist, A.,** Polarizable water at a hydrophobic wall, *Chem. Phys. Letts.,* 165, 437, 1990.

34. **Zhu, S. B. and Robinson, G. W.,** Structure and dynamics of liquid water between plates, *J. Chem. Phys.,* 94, 1403, 1991.

35. **Makovsky, N. N.,** Structure of the surfactant water interface, *Mol. Phys.,* 72, 235, 1991.

36. **Matsuoka, O., Clementi, E., and Yoshimine, M.,** Water dimer potential surface, *J. Chem. Phys.,* 64, 1351, 1976.

37. **Stillinger, F. H. and Raman, A.,** Improved simulation of liquid water by molecular dynamics, *J. Chem. Phys.,* 60, 1545, 1974.

38. **Jönsson, B., Wenneström, H., and Halle, B.,** Ion distribution in lamellar liquid crystals. A comparison between results from Monte Carlo simulations and solutions of the Poisson-Boltzmann equation, *J. Phys. Chem.,* 84, 2179, 1980.

39. **Berendsen, H. J. C., Postma, J. P. M., van Gunsterson, W. F., and Hermans, J.,** Interaction models for water in relation to protein hydration, in *Intermolecular Forces,* Pullman, B., Ed., Reidel Dordrecht, 1981, 331.

40. **Jörgensson, W. L., Chandrasekar, J., Madura, J. D., Impey, R. W., and Klein, M. L.,** Comparison of simple potential functions for simulating liquid water, *J. Chem. Phys.,* 79, 926, 1983.

41. **Wallqvist, A. and Karlström, G.,** A new non empircal force field for computer simulations, *Chemica Scr.,* 29A, 131, 1989.

42. **Clegg, J. S.,** Alternative views on the role of water in cell function, in *Biophysics of Water,* Franks, F. and Mathias, S., Eds., John Wiley & Sons, Chichester, 1982.

43. **Steytler, D. C., Dore, J. C., and Wright, C. J.,** Neutron diffraction studies of the water in meso and micropores, *Mol. Phys.,* 48, 1031, 1983.

44. **Rodgers, M. A. J.,** Picosecond studies of rose bengal fluorescence in reverse micellar systems, in *Reverse Micelles, Biological and Technological Relevance of Amphiphilic Structures in Apolar Media,* Luisi, P. L. and Straub, B. E., Eds., Plenum Press, New York, 1984, 165.

45. **De Geyer, A. and Tabony, J.,** Evidence for intercontiguous structures in concentrated microemulsions, neutron small angle scattering results, *Chem. Phys. Lett.,* 113, 83, 1985.

46. **Tabony, J.,** Quasi elastic scattering measurements of molecular motions in micelles and microemulsions, *Chem. Phys. Lett.,* 113, 75, 1985.

47. **Piculell, L.,** Water spin relaxation in colloidal systems. I. ^{17}O and ^{2}H relaxation in dispersions of colloidal silica, *J. Chem. Soc. Faraday Trans.,* 82, 387, 1986.

48. **Piculell, L. and Halle, B.,** Water spin relaxation in colloidal systems. II. ^{17}O and ^{2}H relaxation in protein solutions, *J. Chem. Soc. Faraday Trans.,* 82, 401, 1986.

49. **Halle, B. and Piculell, L.,** Water spin relaxation in colloidal systems. III. Interpretation of the low frequency dispersion, *J. Chem. Soc. Faraday Trans.,* 82, 415, 1986.

50. **Nimtz, G., Binggeli, B., Börngen, L., Marquardt, P., Schäffer, U., and Zorn, R.,** Dielectric and structural properties of a water-oil emulsion at the gel microemulsion transition, *Europhys. Lett.,* 2, 103, 1986.

51. **Marquardt, P. and Nimtz, G.,** Critical dielectric behaviour of micellar water below the cloud point of a water-oil microemulsion, *Phys. Rev. Letts.,* 8, 1036, 1986.

52. **Lafleur, M., Pigeon, M., Pézolet, M., and Caillé, J. P.,** Raman spectrum of intersticial water in biological systems, *J. Phys. Chem.,* 93, 1522, 1989.

53. **Matsumoto, T., Ito, D., and Kohno, H.,** Interaction between amphiphiles and water molecules in concentrated bilayers aqueous colloids, *Colloid Polym. Sci.,* 267, 946, 1989.

54. **Luisi, P. L., Häring, G., Maestro, M., and Rialdi, G.,** Proteins solubilized in organic solvents via reverse micelles: thermodynamic studies, *Thermochim. Acta,* 162, 1, 1990.

55. **Peng, Q. and Luisi, P. L.,** The behaviour of proteases in lecithin reverse micelles, *Eur. J. Biochem.,* 188, 471, 1990.

Chapter 4

MECHANISTIC ENZYMOLOGY IN ANHYDROUS ORGANIC SOLVENTS

Alan J. Russell, Sudipta Chatterjee, Igor Rapanovich, and James G. Goodwin, Jr.

TABLE OF CONTENTS

I. INTRODUCTION

In recent years the use of enzymes in anhydrous organic solvents has dramatically increased.[1-3] When placed in organic media, enzymes exhibit exciting and useful novel properties, such as greatly enhanced thermostability,[4] profoundly altered specificity,[5] and the ability to catalyze new reactions and improve others.[6] In addition, organic solvents have many advantages over water as reaction media for industrial processes.[2] The advantages of using enzymes in organic solvents include, but are not limited to, increased ease of product recovery, shifting of thermodynamic equilibria, enhanced solubility of hydrophobic substrate molecules, and reduced microbial contamination of biocatalytic reactors. Consequently, enzymatic catalysis in organic solvents holds the promise of making significant contributions to modern industry. The increased use of enzymes in anhydrous organic media has also been matched by an increase of publication in this area. A computer search of the literature from 1967 to 1980 using the keywords ''enzymes'' and ''organic solvent'' scored 122 hits, whereas the same search just 10 years later recorded 422 hits. The publication bonanza has of course inspired many review articles which discuss the properties and applications of enzymes in anhydrous systems.[1-3]

Nonaqueous enzymology, however, is not without its own problems. Perhaps the most severe problem is that at present there is relatively little information available regarding how solvents interact with the protein and how these interactions are propagated into effects on enzyme activity, substrate specificity, and structural stability. Much of the published work to date has attempted to demonstrate the structural integrity of enzymes in the presence of anhydrous organic solvents.[7-9] This pioneering research has not, however, explained the drastically reduced activities of some enzymes in nonaqueous media[10] and as such the information required for optimization of activity must still be generated. This paper will describe what is currently known about the mechanism of enzymes in organic solvents, how the solvent can be used to regulate enzyme activity[11] and specificity,[12] and how we have chosen to investigate the dependence of enzyme mechanism on solvent.

It has been argued that when enzymes are placed in nonaqueous solvents, they are suspended in a shell of water and thus are not truly functioning in the organic media. Although the enzyme molecules are undoubtedly surrounded by about a monolayer of water (5%, $w_{water}/w_{protein}$), the total concentration of water in the system is usually less than 0.01% ($v_{water}/v_{solvent}$). Given these values, it is not unreasonable to refer to the enzyme as functioning in an essentially anhydrous system. It is, however, worthwhile to note that the location and activity of the water molecules present are of profound importance,[13] and much research is currently in progress to elucidate the role that water plays in controlling the activity, specificity, and stability of enzymes suspended in anhydrous solvents. We also wish to stress that in this paper we shall not consider the use of enzymes in aqueous/nonaqueous mixtures. The effect of cosolvents on enzyme activity and specificity in predominantly aqueous solutions has been studied intensely for over 20 years[14] and is not directly pertinent to the current discussion.

II. STRUCTURAL INTEGRITY OF ENZYMES IN ORGANIC SOLVENTS

The most obvious question regarding the use of an enzyme in an organic solvent is whether the protein exists in its native conformation (the conformation which results in activity in aqueous media). If the structure of the enzyme is equivalent in a variety of solvents, then information pertaining to kinetics and mechanism which is generated in organic solvents would be of considerable use. Any structural changes caused by suspension in anhydrous solvents are of particular interest for two reasons. First, when comparing the activity and specificity of an enzyme in different solvents, we must ensure that we compare "apples with apples": conformational changes must be understood before relationships between enzyme function and solvent properties can be elucidated. Second, if an enzyme retains its catalytic power in an organic solvent, while undergoing a significant structural alteration, then the new enzyme is of interest in its own right.

Direct physical studies of enzymes suspended in organic solvents have shown conclusively that the overall structure of the enzyme is the same in water and in anhydrous solvents.[7-9] This finding should be of no surprise based on our knowledge of protein structure and basic physical chemistry.

When a solid particle, consisting of enzyme, salt, and water molecules, is placed in anhydrous organic solvents, the activity of the enzyme (and by implication the structure of the enzyme) will be dependent on the level of interaction between the solvent and the protein molecules. Specifically, if water and salt molecules surrounding the enzyme prevent any interaction of protein and solvent (by virtue of the high dielectric of water, only one or two molecules of water are sufficient to effectively screen electrostatic interactions between bulk solvent and a point on the protein surface[15]) then the protein will have a conformation equivalent to that in a crystalline environment. This will be similar to the structure in aqueous solution. Protein crystals are solid particles which are water-saturated and rigid (if they were not rigid, crystallography would be impossible), and it is generally accepted that the crystal structure resembles closely the solution structure (as demonstrated by numerous nuclear magnetic resonance experiments). If a particle of enzyme in an anhydrous organic solvent does not interact with the solvent (a situation which can only exist if the protein is somehow screened from the solvent), then a single protein molecule will be in a similar situation to a molecule within a crystal.

Only when the solvent has the ability to interact with the protein will the protein structure be affected. This can occur in two ways. First, the dielectric of water may not be sufficient to screen electrostatic interactions between solvent and protein. This will only be the case for very polar nonaqueous solvents, and even then the effect of the solvent will be minimal so long as the water molecules surrounding the protein remain in place. Second, the most dramatic effect on activity will occur when the water and salt molecules become detached from the protein and enter the bulk solvent. The ability of a solvent to strip water molecules in this way will be related to the hydrophilicity of the solvent.[10] When the solvent is more hydrophilic than the surface of the protein interacting with water, then, by definition, it will be energetically favorable for the water to enter bulk solvent. Another driving force for the removal of water from the surface of the enzyme will be the increase in entropy which results from the release of bound water. The free energy of these processes may itself be sufficient to overcome the activation energy of unfolding, and the protein may take up a new conformation in which hydrophilic side chains interact with each other rather than with the solvent. More seriously, the protein will no longer be protected from the bulk solvent and the relatively low dielectric of organic solvents will result in dramatic changes in the delicate balance of electrostatic interactions within the protein. This will undoubtedly inactivate the enzyme. Indeed, attenuation of intramolecular electrostatic interactions by more polar solvents will depopulate hydrogen-bonded conformers and permit others which are favored sterically but disfavored electrostatically.

If one accepts this two-step hypothesis — where, first, water is stripped from the enzyme (a function of the hydrophilicity of the solvent) and then the solvent can penetrate and interact directly with the protein (the interaction being most serious, in terms of protein conformation, when the solvent has a low dielectric) — then the worst solvents will be those that are hydrophilic in nature. Polar solvents will have the ability to disrupt hydrogen bonds and ionic interactions and will favor conformations with large dipole moments. Thus, our current knowledge predicts that enzyme structure in organic solvents should be dependent on the nature of the solvent. Indeed, intuitively one would predict that hydrophobic organic solvents, which are more hydrophobic than the surface of the enzyme, will enable the enzyme to retain its native conformation. All the kinetic and structural studies performed to date appear to support this hypothesis fully.

The first direct demonstration of structurally intact enzymes in organic solvents was performed using electron paramagnetic resonance (EPR).[7,9] It should be noted that these experiments were performed on immobilized enzyme. The relatively low mobility of a spin label on alcohol dehydrogenase indicated that the enzyme had not unfolded. Although EPR can only investigate the environment of the spin label, the data suggest that immobilized alcohol dehydrogenase can maintain its native conformation in organic solvents. An enzyme does not have to be immobilized in order to retain its native conformation in a solvent. In an examination of the structure of α-lytic protease directly suspended in anhydrous acetone, Klibanov and colleagues have shown that the micro-environment of the active-site histidine is identical in water and in organic solvents.[8] Solid-state ^{15}N-nuclear magnetic resonance (NMR) was used to probe the presence of a hydrogen-bonded network at the active site of the enzyme. The chemical shift of the ^{15}N-labeled histidine is extremely sensitive to its environment and was shown to be equivalent to enzyme dissolved in buffer, suspended in acetone, and crystallized from aqueous solution.[8]

In conclusion, there is no doubt that the structure of an enzyme in anhydrous and relatively hydrophobic organic solvents is equivalent to the structure of an enzyme in water. The notion that the enzyme may exist in a variety of structural forms (only one of which is active) is also not supported by experimental data. The EPR and NMR data show no evidence of alternate structures and they support the view that the structure of each molecule of enzyme in the system is similar.

A. INACTIVATION OF ENZYMES SUSPENDED IN ORGANIC SOLVENTS

Inactivation of enzymes suspended in organic solvents will be of two types: reversible and irreversible. If the solvent causes an irreversible inactivation of the enzyme, then when the protein is resuspended in water, the enzyme will no longer function. Given the fact that a wide variety of enzymes exhibit activity in organic solvents, it is clear that inactivation of either type will be relatively slow. For instance, we have shown that alkaline phosphatase, an enzyme commonly utilized in immunodetection assays, is slowly inactivated by exposure to anhydrous hydrophilic organic solvents. The rate of inactivation is dependent on the nature of the solvent, but the half-life is typically in excess of 20 h.

Alkaline phosphatase was lyophilized from 0.01 M glycine buffer (pH 9.8), suspended in solvent (dimethylformamide, tetrahydrofuran, acetonitrile, or acetone), and shaken for either 5, 20, or 36 h in sealed vials. After incubation the enzyme was separated from solvent by centrifugation, followed by brief lyophilization, and resuspended in buffer. The activity of the enzyme was then assayed using *p*-nitrophenylphosphate (2 mM). The results of this study are shown in Figure 1. Since the enzyme is not completely inactivated by the solvent, there is the potential for the enzyme to exhibit activity in nonaqueous media (assuming that there is no reversible inhibition). We are currently optimizing the activity of alkaline phosphatase with chemiluminescent substrates for use in organic solvents, enabling an increase in sensitivity of detection since water quenches the luminescent signal. It seems clear that solvents will inactivate proteins in a number of different ways, and our main interest should be in predicting how a given solvent will interact with a given protein. We are currently surveying the stabilities of many unrelated proteins in organic solvents in an attempt to correlate structural stability to solvent physical properties. The solvent properties which are of particular relevance are hydrophobicity, dielectric constant, polarity, dipole moment, and viscosity.

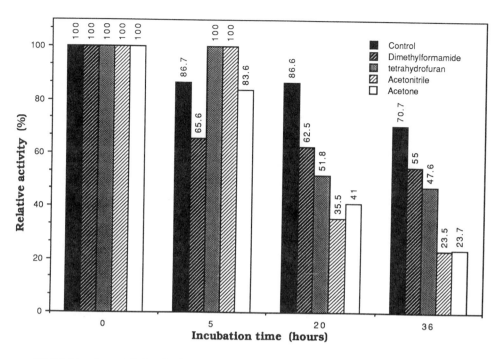

FIGURE 1. Irreversible inactivation of alkaline phosphatase in hydrophilic organic solvents (see text for detailed description).

III. MECHANISTIC INTEGRITY OF ENZYMES IN ORGANIC SOLVENTS

The preceding discussion leaves little doubt that enzymes which are suspended in organic solvents are structurally intact, and thus our focus must now be on how biocatalysts function in anhydrous systems. Indeed, comparisons of the activity of an enzyme in different solvents are of limited use unless the enzyme has the same mechanism in each solvent. The mechanistic integrity of a number of enzymes suspended in organic solvents has been studied using a variety of steady-state kinetic techniques.

In 1966 measurements were made on the activity and behavior of lyophilized enzymes suspended in essentially anhydrous organic solvents.[16] Dastoli and Price showed that chymotrypsin[17] and xanthine oxidase[18] retained their native activity in organic solvents. Further, they showed that not only did the oxidation of crotonaldehyde by xanthine oxidase follow Michaelis-Menten kinetics, but that the enzyme also retained up to 90% of its native activity when lyophilized and suspended in anhydrous organic solvents. Measurements of V_m (the maximal velocity) and K_m (the Michaelis constant) were performed in 14 anhydrous solvents. Interestingly, there was no definite correlation between the catalytic constants for the enzyme in organic solvents and the hydrophobicity of the solvent. In addition, the authors reported that the activity data could not be correlated to the viscosity, dipole moment, or dielectric constant of the solvent. They do, however, report that *"it appears likely that catalytic activity of undissolved enzymes would be of general occurrence rather than restricted to a few enzymes."* Unfortunately, the importance of these initial experiments was not recognized, and the field of nonaqueous enzymology lay essentially dormant for another two decades until Klibanov revolutionized the field with the demonstration that enzymes were actually more thermostable in organic solvents than in water.[19]

The discovery by Klibanov that serine proteases were active in organic solvents and could catalyze transesterification (alcoholysis) reactions has enabled detailed mechanistic studies of enzyme function in these environments. The mechanism of serine proteases is understood at the molecular level, and thus we are able to test whether these enzymes behave similarly when in different solvents. In water, peptide and synthetic ester substrates are hydrolyzed by serine proteases via the acyl-enzyme mechanism (for a detailed description, see Fersht[20]). The enzyme and substrate first associate noncovalently to form the enzyme-substrate complex. This is followed by attack of the hydroxyl group of the active-site serine residue upon the carbonyl of the substrate forming the tetrahedral intermediate in which the oxyanion of the substrate is stabilized by the backbone chain of the protein. The collapse of this intermediate leads to the release of the amine or alcohol and the formation of the covalent acyl-enzyme intermediate. After hydrolysis, resulting from the attack of water, the enzyme product complex is formed which then collapses to release free enzyme and product. In the case of transesterification, water is replaced by an alcohol as the nucleophile in the final step of the reaction.

The initial studies in this area focused on straightforward determinations of the specificity constants for subtilisin, chymotrypsin, and porcine pancreatic lipase in a variety of solvents. It was shown that each enzyme obeyed standard Michaelis-Menten kinetics in the range of substrate concentrations attainable and that the effect of varying ester concentrations at different concentrations of alcohol was the same in water as in organic solvents.[11,21] The latter experiment is, of course, a test for the acyl-enzyme mechanism. One of the most significant problems with all these experiments was that the values for k_{cat} and K_m could not be separated since the K_m was so high that substrate solubility became limiting. Indeed, since the plot of substrate concentration vs. rate was linear, the highest substrate concentrations used were at best only 5% of K_m. Nevertheless, the primary and secondary plots imply that the acyl-enzyme mechanism is followed in organic solvents. Further evidence is already mounting that this is indeed the case.

Linear free-energy relationships (LFER) have long been recognized as a useful tool for investigating enzyme mechanism. Utilizing a series of parasubstituted phenyl acetates in a transesterification reaction with hexanol, Kanerva and Klibanov[22] have determined the dependence of the specificity constant for the serine protease subtilisin on the substituent constant (σ) of the substrate. Linear correlations were obtained in a total of six solvents (including water). Such a Hammett analysis, which quantitates the charge distribution of the transition state, is extremely sensitive to the mechanism of the enzyme. Since the ρ values (which are characteristic of the reaction rather than the substrate) for subtilisin in water, tetrahydrofuran, acetone, butyl ether, acetonitrile, and *tert*-amyl alcohol are almost identical, it can be assumed that the transition-state structure of the enzyme in each of these solvents is the same. This is further evidence that serine proteases utilize the acyl-enzyme mechanism when suspended in anhydrous solvents. An enzyme which is not limited by substrate solubility in organic media is peroxidase. Ryu and Dordick[23] have determined V_m and K_m for a number of substrates in a variety of solvents. These results have shown that LFERs exist not only between the catalytic efficiency and solvent hydrophobicity, but also between activity and substrate hydrophobicity. This important finding reminds us that when comparing activity of enzymes with a variety of substrates in different organic solvents, we must not be tempted to assume that all changes in activity are the result of changes on the enzyme. More recently, Klibanov and colleagues[24] have investigated kinetic isotope effects of labeled substrates on subtilisin. Their findings have demonstrated once more a *"marked independence of the transition state structure of the enzymatic process on the nature of the reaction medium."*

Although this mass of evidence suggests the formation of an acyl-enzyme intermediate in the transesterification reaction, the stable intermediate has never been detected or isolated from an enzyme in an organic solvent. Thus it remains necessary to prove that it is formed and to determine the individual rate constants for the reaction (see below). We are addressing this problem in a straightforward manner. As already described, the acyl-enzyme intermediate is formed via a covalent link between the substrate and the active serine residue, Ser221 for subtilisin, the serine hydroxyl being particularly nucleophilic because of its location in the active site of the protein. Protein engineering has been used to study the importance of this amino acid side chain, replacing it with a thiol side chain, forming thiolsubtilisin from subtilisin. The kinetic[25] and structural[26] properties of thiolsubtilisin in aqueous solution have been studied for over two decades. Thiolsubtilisin is structurally indistinguishable from native subtilisin, and thus comparisons of the enzymes yield much information about the importance and nucleophilicity of Ser221 in the native protein. If an acyl enzyme were formed by subtilisin in organic solvents and the change in solvent does not affect nucleophilicity, then the ratio of $(k_{cat}/K_m)_{thiol}/(k_{cat}/K_m)_{native}$ should be the same in aqueous and nonaqueous media. Preliminary data indicate that this is indeed the case in some hydrophobic solvents.[27] Our approach, which simply compares the activity of thiolsubtilisin and subtilisin in different solvents, will test whether Ser221 plays an important role in biocatalysis in organic solvents. Perhaps even more interesting is that once we have established a database consisting of $(k_{cat}/K_m)_{thiol}/(k_{cat}/K_m)_{native}$ ratios for each solvent, we will be able to dissect which solvent properties affect Ser221 directly. This research is currently in progress in our laboratory.

IV. NEED TO DETERMINE INDIVIDUAL RATE CONSTANTS

As stated in Section I, the potential applications of organic phase enzyme catalysis in industry are significant. Enzymes are insoluble in almost all organic solvents and therefore act as heterogenous catalysts in these systems. Although an advantage in industry, the heterogeneity of the system has made detailed kinetic investigations of enzymes in solvents difficult if not impossible. For instance, the external diffusional limitations for an enzyme particle can be reduced by rapid agitation; however, it is much harder to evaluate the role of internal diffusion in such systems. Of course, using immobilized enzymes and proteins on nonporous supports has alleviated many problems associated with determining the role of internal diffusion in such processes.[3,28]

It has already been shown that measuring the apparent kinetic constants for enzyme activity in organic solvents and correlating these properties with those of the solvent yields important information about enzyme structure and function.[12,23,29] These analyses were performed with experimentally determined values for k_{cat}, K_m, or k_{cat}/K_m. These are all apparent kinetic constants and, thus, are a function not only of the individual rate constants for a particular mechanism, but also any other factors affecting rate (such as diffusion). In order to understand how a solvent interacts with an enzyme, and thus determine how to use solvent to control enzyme activity and specificity, we must first understand how every form of the enzyme is affected by solvent. Such a complete characterization of an enzyme requires that the mechanism of the enzyme is established and that the free energy of each form of the enzyme has been determined. This enables a free-energy profile of the enzyme-catalyzed reaction to be formed. Once such a profile exists for the reaction in a variety of environments, the energetic consequence of placing an enzyme in an organic solvent can be deduced. Using protein engineering, this approach has been used with great success in determining the role played by individual amino acids in biocatalysis.[30] For instance, Fersht and colleagues have dissected the catalytic machinery of tyrosyl t-RNA synthetase and produced an energy profile of the reaction. In a series of classic protein engineering experiments, they have been able to deduce which amino acid side chains are important in each step of the reaction.[31] These investigations of "protein physiology"[32] have laid the ground rules for the research path which we are currently following.

Instead of mutating an enzyme, we propose to alter its environment and determine the effect on the energy profile. When making such comparisons of reaction profiles, one must ensure that not only is the mechanism of the enzyme the same in differing environments, but also that the rate-determining step is equivalent in all cases. These requirements necessitate an initial investigation of the effect of solvent on the mechanism of the enzyme and the determination of all the individual rate constants during the reaction.

V. METHOD FOR DETERMINATION OF RATE CONSTANTS FOR ENZYMES IN ORGANIC SOLVENTS

The transesterification reaction catalyzed by subtilisin (a serine protease) in anhydrous organic solvents has the following proposed reaction scheme:

$$E + S_1 \Leftrightarrow ES \Rightarrow EA + P_1 \underset{+S_2}{\Rightarrow} E + P_2$$

where E, S, and P are enzyme, substrate, and product, respectively. It should be noted that for this mechanism the total transit time is the sum of the individual steps, thus:[21]

$$\frac{[E_0]}{v} = \frac{k_{-1} + k_2}{k_{+1}[S_1]k_2} + \frac{1}{k_2} + \frac{1}{k_3[S_2]} \tag{1}$$

where k_{-1} and k_{+1} are the rate constants for the reverse and forward reactions of the first step, k_2 and k_3 are the rate constants for the second and third steps, respectively. $[E_0]$ is the total enzyme concentration, $[S]$ is the concentration of substrates, and v is the initial rate of the reaction.

A. STEADY-STATE ISOTOPIC TRANSIENT KINETIC ANALYSIS

Many techniques used in the study of heterogenous catalysis have not yet been applied to immobilized enzyme systems. The use of steady-state isotopic transient kinetic analysis (SSITKA) in the determination of residence times is of particular relevance to nonaqueous enzymology. The steady-state isotopic transient technique pioneered by Happel et al.[33-35] and Biloen et al.[36] has permitted the monitoring of important kinetic parameters on solid heterogeneous catalysts under steady-state reaction conditions. Experimentally the decay or development of isotopic species is monitored (typically by mass spectrometry) at steady state after switching between reactant isotopes in the feed stream without perturbing the reactor conditions; a typical normalized isotopic transient is displayed in Figure 2. For a homogeneous surface reaction, the rate can be written as:

$$\text{Turnover frequency (TOF)} = \theta/\tau_{site}$$

with θ and τ_{site} being the coverage of the active catalytic sites and average surface lifetime of reaction intermediates, respectively. Unlike conventional steady-state methods, SSITKA is able to deconvolute the reaction rate into contributions due to coverage of intermediates vs. those due to the reactivity of the reaction intermediates. This technique is very powerful since it permits one to address the nature of groups of reaction sites. For conventional transient techniques[37] which attempt a similar decoupling of the reaction rate, the analysis is complicated by the depletion of reaction intermediates during the transient. Thus, for SSITKA, k, the rate constant, corresponds inversely to the area under the normalized transient curve in Figure 2 and is closely related to $1/\tau_{site}$ if readsorption can be neglected or accounted for. The steady-state surface concentration of reaction products and intermediates ($N_{i(s)}$) can

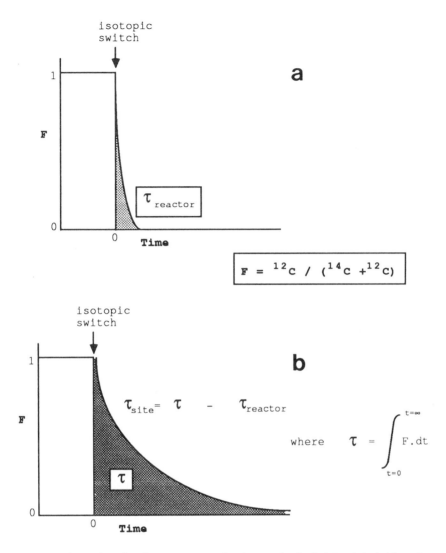

FIGURE 2.   Isotopic transients for a heterogeneous catalyst (see text for detailed description): (a) non-interacting tracer and (b) interacting species in a heterogenous catalytic system. The decay curves shown assume that the reactor is of a plugged-flow type.

be calculated by integrating the transient curves (after correction for reactor holdup) and applying the formula:

$$N_{i(s)} = (A)^* \text{(rate of } i \text{ formation)} \qquad (2)$$

where A is the area under the normalized transient response curve.

Proper analysis of data acquired by SSITKA also permits one to quantify the heterogeneity of the reaction sites. Currently, the most powerful method is that developed by de Pontes et al.[38] which performs an inverse Laplace transformation of transient data to obtain an *a priori* distribution function for k and consequently for the strength of the sites. The relationship has the form:

$$\text{TOF} = \theta_0 \int_0^\infty [ke^{-kt}f(k)]dk \qquad (3)$$

the activity distribution function $f(k)dk$ being the probability that the intrinsic activity lies between k and $k + dk$. Recently, work from the laboratories of Goodwin at the University of Pittsburgh has been presented for ammonia synthesis over a commercial iron catalyst[39] showing a comparison of reactivity distribution functions based on the Temkin theoretical model and on the transformation method developed by de Pontes et al. It could be inferred that the apparent nonuniformity of the catalyst was derived from the basic structure of the surface rather than from adsorbate-induced surface segregation or aggregation. A fit of the Temkin model to the SSITKA results yielded reasonable values for the Bronsted transfer coefficient, the active site density of the working catalysts surface, and a nitrogen adsorption affinity window that is commensurate with that implied by the *a priori* method of de Pontes et al.

Further work has also demonstrated the usefulness of SSITKA for a wide variety of investigations of heterogeneously catalyzed reactions, including CO hydrogenation,[40] CO oxidation,[41] the oxidative coupling of methane,[42,43] CO + NO reaction,[41] ammonia synthesis,[41,44] and the partial oxidation of propylene to acrolein.[46]

To date, all applications of SSITKA have been to gas-phase reactions on solid metal and metal oxide catalysts. However, it would appear, based on calculations, to be applicable to liquid-phase and enzyme-catalyzed reactions.

B. APPLICATION OF SSITKA TO ENZYME CATALYSIS

SSITKA analysis of enzymes in organic solvents will be possible if the choice of enzyme, substrate, solvent, and reactor enables discrimination between the residence time without interaction in the reactor and residence time on the active sites. In order to minimize the distribution of residence times in the reactor, a plug-flow reactor is preferred and, in this case, is acceptable due to the low incidence of secondary reaction of the primary product. Such a reactor will also make it easier to retain the immobilized enzyme in the reactor without having to resort to complicated filtering and sampling systems. Such systems would be difficult to employ in this instance due to the short residence times and the small scale of the process dictated by the method.

In order to determine easily the residence time on the enzyme site, it should be at least 5% of the non-interactive residence time in the reactor. For 50 mg of catalyst in a 5-ml bioreactor, there would be approximately 1×10^{17} sites ($3 \, \mu M$). Temperature can be altered to control residence time (residence time increasing with decreasing temperature). Using subtilisin and N-acetyl-L-tyrosine methyl ester, catalyst residence times between 1 and 60 s (these values are from current estimates of k_{cat} in organic solvents) for the entire reaction can be achieved. If the residence time in the reactor (independent of whether enzyme is present or not) is below 1 min, then the determination of catalyst residence time would be relatively straightforward.

If flow through the reactor is F (ml/min), the volume of the reactor is V (ml), and the ester substrate concentration in the feed is $[S]_{in}$, then the dilution rate, D, will be given by $D = F/V$. Assuming that the reactor is converting only 5% of substrate, then, at steady state, the concentration of ester substrate in the reactor $[S]$ will be constant and approximately equal to that of output. Note that $([S]_{in} - [S])$ will determine the limit of sensitivity. Thus:

$$\frac{d[S]}{dt} = D^*[S]_{in} - D^*[S] - \left(\frac{k_{cat}[E]_0[S]}{K_m + [S]} \right) \qquad (4)$$

where k_{cat} is the turnover number for the enzyme and K_m is the Michaelis constant of the ester for the enzyme.

At steady state the rate of change of substrate concentration is zero (d[S]/dt = O), and hence:

$$D^*([S]_{in} - [S]) = \left(\frac{k_{cat}[E]_0[S]}{K_m + [S]}\right) \tag{5}$$

In our system, $K_m >> [S]_{in} > [S]$, since in organic solvents K_m is high and low substrate concentrations can be utilized. Thus:

$$D^*([S]_{in} - [S]) = \left(\frac{k_{cat}[E]_0[S]}{K_m}\right) \tag{6}$$

Since $k_{cat}[E]_0$ is merely the maximal rate at a given enzyme concentration (V_{max}), then:

$$[S] = \frac{D \cdot [S]_{in}}{\left(\dfrac{V_m}{K_m}\right) + D} \tag{7}$$

Obviously, the critical parameters in defining the amount of substrate converted to product in a nonaqueous system are the dilution rate, D, and the specificity factor of the enzyme. (V_m/K_m). If (V_m/K_m) is greater than D, then [S] will be less than $[S]_{in}$ (or if (V_m/K_m) < D, then [S] < $[S]_{in}$) and product is detectable in each case. Of course, if (V_m/K_m) is much greater than D, then [S] tends to zero and there is complete conversion. In addition, if (V_m/K_m) is much less than D, then [S] will tend to $[S]_{in}$ and there is no conversion at all. Operating conditions can be chosen such that, at steady state, approximately 0.1 to 10% of substrate would be converted to product. In a typical experiment using subtilisin in anhydrous octane, a flow rate which can be maintained accurately with a peristaltic pump is about 5 ml min^{-1}. For a volume of the reaction mixture of 5 ml, the dilution rate, D, would be 1 min^{-1} and the reactor residence time would be 1 min. To achieve a 1% conversion to product at this dilution rate, then D = V_m/K_m × 99. The important question is whether V_m/K_m for the enzyme can be approximately 1.7 × 10^{-4} s^{-1}, and whether under these conditions k_{cat} is less than 0.2 s^{-1} (since above this value the residence time on the enzyme will be insignificant compared to the reactor residence time). It is expected that k_{cat} for subtilisin with the chosen substrate will be approximately 0.1 s^{-1},[1] within the above guidelines. Although it is more difficult to estimate K_m, we expect a value of approximately 5 to 10 mM. Thus, with an enzyme concentration of 3 μM, V_m/K_m would be within the above guidelines.

Since radiolabeled substrates can be utilized, if K_m for the substrate is an order of magnitude higher than expected, then the substrate conversion can be reduced to 0.1% and detection would still be feasible. By optimizing flow rate, enzyme concentration, temperature, and percent conversion for the substrate chosen, it should be possible to generate isotopic transient-kinetic data. The residence time of an inert tracer in a perfect reactor would give the result shown in Figure 2a (where F is the fraction of labeled inert over total). Enzymes in organic solvents function like many conventional catalysts in that, depending on reaction conditions, the rate constants for the reaction can be relatively small. Since residence time is merely the reciprocal of a rate constant, this translates into a significant residence time of a labeled substrate on the active site of the enzyme (assuming that the substrate only binds to the active site). The hold-up of substrate on the enzyme in the reactor exhibits itself as a gradual decay in the amount of unlabeled product as shown in Figure 2b. Using standard data analysis techniques, it is possible to extract the residence time by integration. In addition, if the reaction is pseudo-first order, then for a plot of ln F (corrected for reactor hold-up) vs. time, the slope of the line will be the rate constant for the reaction.

TABLE 1
Values of Parameters Used in Feasibility Calculations

Flow (ml min^{-1})	Total area (m^2)	Volume (ml)	Conversion (%)	[S]$_{in}$ (mM)	K$_m$ (mM)
1	1	1	0.1	0.01	1
2	5	2	1	0.1	10
4	10	5	5	1	100
6	25	10	10	10	
8	50				
10	100				

Note: See text for detailed description.

As stated above, a reactor **will** have a residence time of its own, and it is necessary to ensure that the catalyst residence time is measurable with respect to the former. Thus, the τ (residence time) shown in Figure 2b is actually comprised of two residence times:

$$\tau_{total} = \tau_{enzyme} + \tau_{reactor} \qquad (8)$$

where τ_{enzyme} is the residence time on the enzyme active site. However, $\tau_{reactor}$ can be easily accounted for using an inert tracer.

We have investigated the feasibility of biocatalytic SSITKA by taking a range of reasonable values for all significant parameters and calculating which combination of these enables the differentiation between reactor and substrate-enzyme residence times. The volume (V) of the reactor was allowed to vary between 1 and 5 ml, with the flow rate (F) being set between 0.1 and 5 ml min^{-1}. Since the enzyme will be immobilized, the concentration which can be achieved is a function of the surface area of the support. Glass supports (which do not alter in size in the presence of organic solvents) are available with surface areas ranging from 10 to 200 m^2 g^{-1}. In order to calculate the enzyme concentration (E$_0$), the surface area of one enzyme molecule (S$_e$) was assumed to be 10,000 Å^2. Then, assuming a total available surface area of (S$_a$):

$$[E_0] = \frac{(S_a)}{(V \cdot S_e \cdot N)} \qquad (9)$$

The values of the parameters used in calculating the residence time for reactor and enzyme are given in Table 1. All possible combinations of these parameters were used in calculating k$_{cat}$ with the following equation:

$$k_{cat} = \frac{D \cdot ([S]_{in} - [S]) \cdot (K_m + [S])}{([E_0 \cdot [S])} \qquad (10)$$

The only variable which cannot be controlled is the K$_m$. Thus, for the experiments to be successful, for each value of K$_m$ there must be a set of conditions under which the enzyme and reactor residence times are distinguishable. Computer analysis, selected results of which are presented in Table 2, reveals that for each K$_m$ there is a reactor configuration which should enable determination of enzyme residence time. The program was designed to list only reactor configurations which resulted in an enzyme residence time (1/k$_{cat}$) of greater than 30% reactor residence time (1/D), and a turnover number (k$_{cat}$) of less than 20 min^{-1}. These are very conservative estimates since an enzyme site residence time of 1% of reactor residence time can be distinguished. There were more than 600 configurations which were identified as acceptable for the parameter ranges used.

TABLE 2
A Selection of Reactor Conditions Under Which Turnover Number Can Be Distinguished from Dilution Rate

V (ml)	F (ml/min)	S_a (m²)	Conv. (%)	$[S]_{in}$ (mM)	K_m (mM)	D (min⁻¹)	k_{cat} (min⁻¹)
1	1	1	1	10	1	1	0.66
2	1	5	1	1	1	0.5	0.24
1	1	5	10	1	1	1	2.54
1	1	5	5	0.01	1	1	0.64
10	2	1	0.1	1	1	0.2	0.24
2	4	5	0.1	1	10	2	0.53
10	4	5	0.1	1	10	0.4	0.53

Note: The reactor volume (V) and flow (F) are assumed to be between 1 and 10 ml and 1 and 4 ml/min, respectively. The surface area available for the attachment of enzyme is given by S_a. The degree of conversion of substrate to product is given by Conv., the initial substrate concentration being $[S]_{in}$. The Michaelis constant (K_m) has been given reasonable values for the enzymes being utilized. The dilution rate (D) for the reactor and the turnover number (k_{cat}) for the enzyme have been calculated using Equation 10.

The use of SSITKA in the determination of the rate constants for enzyme catalysis in nonaqueous media has not been reported previously. The demonstration that this technique is applicable to such systems will represent a significant advance in the methodologies available for the study of biocatalysts in organic solvents. One of the drawbacks of non-aqueous enzymology is the unpredictability of activity in different solvents. At present, no detailed kinetic investigation (designed to separate macro- and microscopic kinetic constants) has been performed using enzymes in organic solvents. The presence of such data will increase our understanding of nonaqueous enzymology. Indeed, in aqueous systems, an enzyme is not considered fully characterized until the individual kinetic constants have been determined.

A reaction of interest is the transesterification of *N*-acetyl-L-tyrosine methyl ester (ATM), catalyzed by subtilisin in a variety of organic solvents. In order to rapidly switch the isotopic content of the substrate feed, we have designed a small-scale (5-ml total volume) plugged-flow bioreactor. The drawback of a continuous stirred tank reactor (which is often utilized with enzymes in organic solvents) is that the concentrations of enzyme which are attainable are much lower in such a system. The only difference between the biocatalytic reactor described and more conventional SSITKA reactors is that the catalyst will be a complex biological molecule. Importantly, since the enzyme is immobilized, the reactor could be used in either an aqueous or a nonaqueous mode. In the choice of volume and flow rate, it will be necessary to ensure that the residence time of substrate on the enzyme is at least 5% of the residence time in the reactor. To measure the residence time in the reactor, a labeled inert tracer molecule (dodecane) will be added with the substrates. It is, of course, critical that the residence time of the substrate for the enzyme be distinguishable from that of the tracer. For instance, if the tracer residence time is on the order of minutes, then a substrate/enzyme residence time on the order of milliseconds will be undetectable. The time scale of the substrate and tracer residence times must be similar.

We are collecting data on the isotopic composition of reactor contents both before and after switching from unlabeled to labeled substrate. The use of substrates labeled at different positions in transient-kinetic experiments gives residence times (and thus rate constants) for separate segments of the reaction. In Figure 3, the segment of the reaction for which a residence time can be calculated is highlighted for each of the different substrates we are

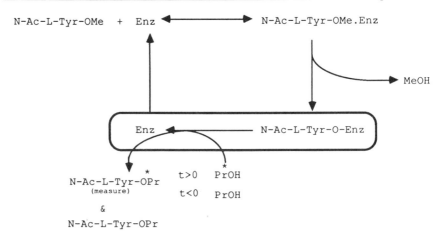

1.

Residence times and site concentration measured when using N-Ac-L-Tyr-OMe*

t>0   N-Ac-L-Tyr-OMe* + Enz ⟷ N-Ac-L-Tyr-OMe*.Enz

t<0   N-Ac-L-Tyr-OMe

MeOH*
(measure)
&
MeOH

Enz ⟵ N-Ac-L-Tyr-O-Enz

PrOH

N-Ac-L-Tyr-OPr

2.

Residence times and site concentration measured when using N-Ac-L-Tyr-OMe*

t>0   N-Ac-L-Tyr-OMe* + Enz ⟷ N-Ac-L-Tyr-OMe*.Enz

t<0   N-Ac-L-Tyr-OMe

MeOH

Enz ⟵ N-Ac-L-Tyr-O-Enz*

N-Ac-L-Tyr-OPr & N-Ac-L-Tyr-OPr*   PrOH
(measure)

3.

Residence times and site concentrations measured when using PrOH*

N-Ac-L-Tyr-OMe + Enz ⟷ N-Ac-L-Tyr-OMe.Enz

MeOH

Enz ⟵ N-Ac-L-Tyr-O-Enz

N-Ac-L-Tyr-OPr*   t>0   PrOH*
(measure)   t<0   PrOH
&
N-Ac-L-Tyr-OPr

FIGURE 3.   Determination of residence times with biocatalytic SSITKA (see text for detailed description). The mechanism shown is for the esterase (Enz)-catalyzed alcoholysis of N-acetyl-L-tyrosine methyl ester (N-Ac-L-Tyr-OMe) with propanol (PrOH) to form N-acetyl-L-tyrosine propyl ester (N-Ac-L-Tyr-OPr). The position of the radiolabel on either of the substrates is shown with an asterisk (*). The section of the reaction for which the residence time will be quantified for each substrate combination has been circled.

using. The individual residence times for the substrates on the enzyme in a number of different solvents are currently being determined in our laboratory. The data we obtain will enable the elucidation of which reaction steps are affected by changing the bulk solvent. By using a series of similar solvents, we are able to delineate the effect of changing, for instance, hydrophobicity without significantly changing, for instance, dielectric constant. For example, we will study the reaction described in dodecane, decane, octane, heptane, and hexane. The individual steps of the reaction can then be analyzed for their "susceptibility" to various bulk solvent properties. **It should also be noted that SSITKA is not limited to the substrates and enzyme that we have chosen. Theoretically, it should be possible to optimize the same type of analysis for any enzyme-substrate pair in organic solvents. Thus, if our approach is successful, it would be possible to use the same analysis to investigate specificity as a function of solvent.**

VI. CONTROL OF ACTIVITY OF ENZYMES BY SOLVENT

Having addressed the issue of the structural and mechanistic integrity of enzymes in the absence of aqueous solvents, it should be clear that a full characterization of an enzyme should include an analysis of how variation of solvent will affect activity and specificity. Although nonaqueous enzymology is still in its infancy, there are a number of hypotheses which have already been proposed which relate solvent properties to the activity of an enzyme. It is generally agreed that enzymes are more active in very hydrophobic solvents; the debate begins when we consider why increasing hydrophilicity appears to drastically reduce the activity of enzymes.

A favored hypothesis has suggested that hydrophilic solvents can strip water from the surface of the enzyme. Studies on the rehydration of lyophilized proteins have indicated that water increases the mobility of proteins, and this increase in mobility can be correlated to an increase in activity. Thus, it has been proposed that the removal of water is equivalent to the removal of the lubricant which enables enzymes to exhibit activity in nonaqueous surroundings. Evidence supporting this view is that the addition of compounds which can mimic the hydrogen-bonding properties of water tend to increase the activity of enzymes in solvents. The only way to study this phenomenon directly is to quantify the mobility of protein and correlate this to activity. Although the studies on the rehydration of proteins are important, they do not address the issue of how mobility affects catalysis in organic solvents. EPR has been used to study the mobility of proteins suspended in anhydrous solvents and the effect of the addition of water to such systems.[7] Interestingly, although the addition of water to lyophilized alcohol dehydrogenase did result in an order of magnitude increase in activity, there was no change in flexibility as measured by EPR. This implies that flexibility is actually not of central importance when considering how a solvent can tailor the activity of an enzyme. This argument is also supported by the fact that some enzymes are actually more active in organic solvents than in water (although they are still more rigid); an example of such an enzyme is dehalogenase which exhibits 120% of its native activity when suspended in an anhydrous solvent.[45] There can be little doubt that for proteins in which flexibility plays a key role in aqueous solution, the effect of a solvent on mobility will be important. What remains unclear is whether the activity of many enzymes in water or in organic solvents is actually controlled in a rate-limiting fashion by mobility.

It has been argued that rather than mobility being of central importance in determining rate, the nature of water which is associated with the protein is crucial.[13] This immediately raises a key question: Is it possible to measure accurately the number of water molecules associated with a protein when suspended in an organic solvent? Most researchers agree that experimental determinations of the number of "bound water molecules" is extremely difficult, and the methods used so far are crude at best. Until more refined techniques, such as solid-state NMR, can be adapted to measure the exact number of water molecules associated with the surface of an enzyme, it will be difficult to correlate how water controls the rate of biocatalytic processes. Consideration of the activity of the water associated with the enzyme has also been proposed as a possible way to predict the response of an enzyme to a given solvent.[46] The idea that comparisons of enzyme activity in different solvents should be made at equivalent activities of water should be accepted, although the question as to why different solvents can change the activity of water associated with an enzyme is also important. In this chapter we have already described a two-step hypothesis which would explain the response of an enzyme to different solvents. However, unless the dependence of each of the rate constants for a given multistep reaction upon solvent properties can be determined, it will remain difficult to understand or predict how a solvent will interact with a protein. It is for this reason that we have proposed the use of SSITKA for the analysis of enzyme function in organic solvents.

VII. CONTROL OF SELECTIVITY OF ENZYMES BY SOLVENT

Selectivity is the ubiquitous hallmark of all biological processes. Indeed, it is the specificity of enzymes which make them such remarkable catalysts and the finely tuned bindings sites of antibodies which allow the body to defend itself against an almost infinite variety of antigens. Over the years there has been much interest in manipulating the selectivity of enzymes so that they can catalyze reactions with "unnatural" substrates.[34] Until recently, the only way to change selectivity has been to alter the protein itself either by chemical modification[47] or site-directed mutagenesis.[48] Although protein engineering is a powerful tool for the protein chemist, it requires that the gene encoding the protein of interest has been cloned and that a detailed understanding of its native structure is available.[49] Thus, in many cases changing the structure of an enzyme is not feasible, and it is necessary to search for alternative approaches.

It has been shown that when proteins are placed in anhydrous media, the substrate specificity of the enzyme may be altered and even reversed.[6] Since proteins function in organic solvents, this enables us to address the question of to what extent biological specificity can be tailored by the reaction medium. We shall now discuss a number of examples demonstrating that selectivity is a strong function of the solvent.

Subtilisin (SN) and chymotrypsin (CHT) are catalytically active in organic solvents. In aqueous solution these enzymes catalyze the hydrolysis of some amino acid esters.[50] In organic solvents one can replace water, which acts as the nucleophile in the hydrolysis reaction, with an alcohol and catalyze an alcoholysis reaction (this has been commonly referred to as a transesterification). In water, CHT is 10^5 times more active in the hydrolysis of N-acetyl-L-phenylalanine ethyl ester than with N-acetyl-L-serine methyl ester. In octane, however, the enzyme shows a fivefold preference for the hydrophilic substrate.[11] Thus, placing CHT in nonaqueous media has reversed its specificity from hydrophobic to hydrophilic substrates. This reversal also holds for SN and porcine pancreatic lipase. Interestingly, the rate of reaction of CHT with hydrophilic amino acid ester substrates is greater in organic solvents than in water.

One may be tempted to think that enzymes in organic solvents have merely lost all semblance of specificity and, for this reason, function with unusual substrates. In fact, nothing could be further from the truth. There have been several demonstrations of the high degree of specificity of enzymes in organic solvents. These include asymmetric reductions catalyzed by alcohol dehydrogenase,[51] regioselective acylation of secondary hydroxyl groups in sugars catalyzed by lipase,[52] enzymochemical regioselective oxidation of steroids with SN,[53] and regioselective oxidation of phenols catalyzed by polyphenol oxidase.[54]

One of the elusive goals of protein chemists has been to alter predictably the enantioselectivity of enzymes. Recently, it has been observed that proteases in organic solvents have relaxed stereospecificity, enabling us to study the role of solvent characteristics on enantioselectivity.[12] We discovered that a simple biophysical model could be used to relate the hydrophobicity of a solvent to the enantioselectivity of subtilisin when placed in that solvent: the more hydrophobic the solvent, the less selective the enzyme and vice versa. The result is exactly what one would expect, since stereoselectivity in SN is determined by binding into a hydrophobic binding pocket. The binding reaction is driven by the release of entropic energy upon the expulsion of ordered water molecules from the pocket. In hydrophilic solvents the release of water from this binding region will be more favored than in hydrophobic solvents, and as a result the enzyme will be more selective. Thus, we have demonstrated the predictable and rational control of enantioselectivity of enzymes by changing the reaction medium rather than the enzyme. Our model also predicts that, for substrates with more bulky side chains, SN will have a greater difference between enantioselectivity in water and organic solvents. This is indeed the case: when alanine is replaced by phenylalanine in butyl ether there is a tenfold increase in enantioselectivity in water.

Further experiments on the enantioselectivity of proteins in various organic solvents have shown how dramatic the solvent effect can be. It has been possible to use SN for the enzymatic resolution of racemic amines on the millimole scale.[2] Studying the acylation of α-methylbenzylamine as a function of the solvent, it was shown that in 3-methyl-3-pentanol the ratio of activity with the S-amine to the R-amine was 7.7, enough to resolve the racemic amine to an enantiomeric excess of 85%. In the same solvent it was possible to enzymatically resolve α-methyltryptamine to >95% enantiomeric excess. The enzyme in toluene (and at least six other solvents) showed no selectivity. This demonstrates that rather than changing the enzyme in order to resolve different racemic mixtures, it is now possible to change the reaction medium and in this way alter the effectiveness of the biocatalyst. Hence, instead of screening many enzymes for the selectivity required, one can now alter the environment of a few such enzymes in order to manipulate their activities.

Klibanov has now proposed that protein flexibility and mobility should be taken into account when attempting to predict enantioselectivity and activity of enzymes in organic solvents.[55] In the reaction of vinyl butyrate and an alcohol, catalyzed by subtilisin, there is a correlation between enantioselectivity and the dipole moment, dielectric constant, and partition coefficient of the solvent. It is argued that as the solvent allows for increased flexibility in the enzyme (which occurs in high dielectric solvents), the specificity will decrease because the enzyme will be more "forgiving". The role of molecular flexibility in proteins suspended in organic solvents has already been discussed in terms of activity. Briefly, no direct correlation has been found between flexibility and activity of subtilisin and alcohol dehydrogenase in organic solvents.[7,56] Undoubtedly, flexibility will play an important role in controlling specificity and activity, although whether flexibility is the most important parameter remains to be seen.

Biological selectivity is, of course, not limited to enzymes. Indeed, one of the most striking examples of specificity is in the interaction between an antibody and antigen. Work at the Massachusetts Institute of Technology (MIT) has investigated whether the binding characteristics of an antibody for its antigen are retained when placed in an organic solvent.[57] We chose to study the binding of a monoclonal antibody, 2E11 (produced in collaboration with Professor S. Tannenbaum and colleagues), to 4-aminobiphenyl, which is an extremely potent carcinogen released in cigarette smoke. We first immobilized 2E11 on porous glass beads and then measured the dissociation constant for binding to the hapten in water and a variety of organic solvents. Porous glass beads were chosen since other supports would swell or shrink in different solvents. We discovered that in water the K_d is 2.3 μM, while in acetonitrile it is 23 μM. Thus, 2E11 does retain its ability to bind hapten in an organic solvent. The strength of binding was related to the nature of the solvent. The hydrophobic antigen undoubtedly binds in a hydrophobic binding pocket on the surface of the antibody. As the hydrophobicity of the solvent increases, the strength of binding decreases; this is of course equivalent to the variation of stereospecificity of subtilisin with solvent described above. In order to demonstrate that the binding of 2E11 to 4-aminobiphenyl was specific and occurring at the same site in water and organic solvents, we performed three further experiments.

First, in the presence of a 300-fold molar excess of 2-aminobiphenyl, there is no change in the binding of 4-aminobiphenyl to the antibody in acetonitrile; in other words, the antibody has absolute specificity for 4-aminobiphenyl in the presence of 2-aminobiphenyl. In water the antibody is not as specific, although the binding is tighter, and under conditions of our experiment there is a 20% reduction in binding of 4-aminobiphenyl. Second, we showed that immobilized bovine serum, glucose isomerase, or boiled and denatured antibody showed no detectable binding for 4-aminobiphenyl in acetonitrile. Finally, we demonstrated that the FAB fragment of 2E11 was sufficient for specific binding of 4-aminobiphenyl, indicating that 4-aminobiphenyl probably binds to the same site in water and organic solvents.

The binding properties of monoclonal antibody 2E11 in organic solvents are no longer the only demonstration of antibody-hapten specificity in anhydrous media. The binding of triazine herbicides to antibodies in organic solvents has also been investigated.[58] A direct correlation between solvent hydrophobicity and strength of binding was apparent once again. A correlation between solubility of hapten in a solvent and strength of binding in that solvent was also observed.

The examples described above indicate that "solvent engineering" is a novel method for manipulating biological specificity. Rather than changing the biological molecule in order to alter its properties, we can now change the reaction medium. The scope and mechanisms of the solvent engineering approach to rational modification of enzyme specificity is currently under investigation in many laboratories.

VIII. CONCLUSIONS

We have described what is currently known about the mechanisms of enzymes in organic solvents and how solvents can be used to tailor the properties of proteins. Obviously there is much work still to be done. We must elucidate how solvents interact with the surface of enzyme particles and why some enzymes exhibit extraordinary activity in organic solvents (dehalogenases, for example) while others operate almost a million times less efficiently in solvents as in water (serine proteases, for example). Mechanistic enzymology requires a detailed understanding of the relationship between the structure of a biocatalyst and its function. Indeed, when a new enzyme activity is discovered, there is a well-defined scientific path to follow in order to elucidate the catalytic chemistry. Initially, and at the most basic level, a method is required which can yield quantitative information about the rate constants for the reaction in question. After further studies on the specificity of activity and the structure of the enzyme, strategies can be designed to optimize activity under conditions where that activity may be most useful. In fact, we think nothing of testing the activity of one enzyme on hundreds of different substrates and at many different temperatures and pHs in order to develop a clear picture of how the enzyme works. Why then are we not equally keen to analyze the behavior of an enzyme in a number of different solvents? One reason is probably that until recently quantitative mechanistic studies on enzymes in anhydrous environments have been limited to correlations between specificity constants (actually V_{max}/K_m's) and solvent properties. However, the wealth of information obtained in these studies, and those described above, has shown that enzymology in organic solvents is already an important subject in its own right. Further, there can be little question that the more we learn about how an enzyme interacts with bulk solvent the more we will known about how biocatalysts function both in and out of water. This information will enable the design of new biocatalytic processes in extreme environments.

ACKNOWLEDGMENTS

The research on alkaline phosphatase, thiolsubtilisin, and biocatalytic SSITKA described above is funded by the American Chemical Society, Washington, DC (PRF 22019-G4), the Army Research Office, Raleigh Durham, NC (DAALO3-90-M-0276), and the National Science Foundation, Washington, DC (BCS-9057312).

REFERENCES

1. **Zaks, A. and Russell, A. J.,** Enzymes in organic solvents; properties and applications, *J. Biotechnol.,* 8, 259, 1988.
2. **Klibanov, A. M.,** Enzymic catalysis in anhydrous organic solvents, *Trends Biochem. Sci.,* 14, 141, 1989.
3. **Dordick, J. S.,** Enzymatic catalysis in monophasic organic solvents, *Enzyme Microb. Technol.,* 11, 194, 1989.
4. **Zaks, A. and Klibanov, A. M.,** Enzymatic catalysis in organic media at 100°C, *Science,* 224, 1249, 1984.
5. **Margolin, A. L., Tai, D.-F., and Klibanov, A. M.,** Incorporation of D-amino acids into peptides via enzymatic condensation in organic solvents, *J. Am. Chem. Soc.,* 109, 7885, 1987.
6. **Zaks, A. and Klibanov, A. M.,** Substrate specificity of enzymes in organic solvents vs. water is reversed, *J. Am. Chem. Soc.,* 108, 2767, 1986.
7. **Guinn, R. M., Skerker, P. S., Kavanaugh, P., and Clark, D. S.,** Activity and flexibility of alcohol dehydrogenase in organic solvents, *Biotech. Bioeng.,* 37, in press.
8. **Burke, P. A., Smith, S. O., Bachovchin, W. W., and Klibanov, A. M.,** Demonstration of structural integrity of an enzyme in organic solvents by solid-state NMR, *J. Am. Chem. Soc.,* 111, 8290, 1989.
9. **Clark, D. S., Creagh, L., Skerker, P., Guinn, R. M., Prausnitz, J., and Blanch, H.,** Enzyme structure and function in water-restricted environments, *ACS Symp. Ser.,* 392, 104, 1989.
10. **Zaks, A. and Klibanov, A. M.,** The effect of water on enzyme action in organic media, *J. Biol. Chem.,* 263, 8017, 1988.
11. **Zaks, A. and Klibanov, A. M.,** Enzymatic catalysis in nonaqueous solvents, *J. Biol. Chem.,* 263, 3194, 1988.
12. **Sakurai, T., Margolin, A., Russell, A. J., and Klibanov, A. M.,** Control of enzyme enantioselectivity by the reaction media, *J. Am. Chem. Soc.,* 110, 7236, 1988.
13. **Halling, P. J.,** High affinity binding of water by proteins is similar in air and organic solvents, *Biochim. Biophys. Acta,* 1040, 225, 1990.
14. **Clement, G. E. and Bender, M. L.,** The effect of aprotic dipolar organic solvents on the kinetics of chymotrypsin catalyzed hydrolyses, *Biochemistry,* 2, 836, 1963.
15. **Russell, A. J.,** Protein engineering of pH dependence of subtilisin BPN', Ph.D. thesis, Imperial College of Science and Technology, University of London, 1987.
16. **Dastoli, F. R., Musto, N. A., and Price, S.,** Reactivity of active sites of chymotrypsin suspended in an organic medium, *Arch. Biochem. Biophys.,* 115, 44, 1966.
17. **Dastoli, F. R. and Price, S.,** Catalysis by xanthine oxidase suspended in organic media, *Arch. Biochem. Biophys.,* 118, 163, 1967.
18. **Dastoli, F. R. and Price, S.,** Further studies on xanthine oxidase in nonpolar media, *Arch. Biochem. Biophys.,* 122, 289, 1967.
19. **Zaks, A. and Klibanov, A. M.,** Enzymatic catalysis in organic media at 100°C, *Science,* 224, 1249, 1984.
20. **Fersht, A. R.,** in *Enzyme Structure and Mechanism,* W. H. Freeman & Co., New York, 1985, 118.
21. **Zaks, A. and Klibanov, A. M.,** Enzyme-catalyzed processes in organic solvents, *Proc. Natl. Acad. Sci.,* 82, 3192, 1985.
22. **Kanerva, L. and Klibanov, A. M.,** Hammett analysis of enzyme action in organic solvents, *J. Am. Chem. Soc.,* 111, 6864, 1989.
23. **Ryu, K. and Dordick, J. S.,** Free energy relationships of substrate and solvent hydrophobicities with enzymatic catalysis in organic media, *J. Am. Chem. Soc.,* 111, 8026, 1989.
24. **Adams, K. A. H., Chung, S.-H., and Klibanov, A. M.,** Kinetic isotope effect investigation of enzyme mechanism in organic solvents, *J. Am. Chem. Soc.,* 112, 25, 1991.

25. **Philipp, M., Tsai, I.-H., and Bender, M. L.,** Comparison of the kinetic specificity of subtilisin and thiolsubtilisin toward *n*-alkyl *p*-nitrophenyl esters, *Biochemistry,* 18, 3769, 1979.
26. **Tsai, I.-H. and Bender, M. L.,** Conformation of the active site of thiolsubtilisin: reaction with specific chloromethyl ketones and arylacryloylimidazoles, *Biochemistry,* 18, 3764, 1979.
27. **Russell, A. J. and Chatterjee, S.,** Manuscript in preparation.
28. **Russell, A. J. and Klibanov, A. M.,** Inhibitor-induced enzyme activation in organic solvents, *J. Biol. Chem.,* 263, 11624, 1988.
29. **Kanerva, L. and Klibanov, A. M.,** Hammett analysis of enzyme action in organic solvents, *J. Am. Chem. Soc.,* 111, 6864, 1989.
30. **Fersht, A. R., Knill-Jones, J. W., Bedouelle, H., and Winter, G. P.,** Reconstruction by site-directed mutagenesis of the transition-state for the activation of tyrosine by the tyrosyl-t-RNA synthetase: a mobile loop envelopes the transition state in an induced fit mechanism, *Biochemistry,* 27, 1581, 1988.
31. **Fersht, A. R., Shi, J. P., Wilkinson, A. J., Blow, D. M., Carter, P., Waye, M. M. Y., and Winter, G. P.,** Analysis of enzyme structure and activity by protein engineering, *Angew. Chem.,* 23, 467, 1985.
32. **Knowles, J. R.,** Tinkering with enzymes: what are we learning?, *Science,* 236, 1252, 1987.
33. **Happel, J., Suzuki, I., Kokayeff, P., and Fthenakis, V.,** Multiple isotope tracing of methanation over nickel catalyst, *J. Catal.,* 65, 59, 1980.
34. **Happel, J., Cheh, H. Y., Otarod, M., Ozawa, S., Severdia, A. J., Yoshida, T., and Fthenakis, V.,** Multiple isotope tracing of methanation over nickel catalyst. II. Deuteromethanes tracing, *J. Catal.,* 75, 314, 1982.
35. **Otarod, M., Ozawa, S., Yin, F., Chew, M., Cheh, H. Y., and Happel, J.,** Multiple isotope tracing of methanation over nickel catalyst. III. Completion of ^{13}C and D tracing, *J. Catal.,* 84, 156, 1983.
36. **Biloen, P., Helle, J. N., van den Berg, F. G. A., and Sachtler, W. H. M.,** On the activity of Fischer-Tropsch and methanation catalysts: a study utilizing isotopic transients, *J. Catal.,* 81, 450, 1983.
37. **Biloen, P. and Sachtler, W. H. M.,** The mechanism of Fischer-Tropsch synthesis, *Adv. Catal.,* 30, 165, 1981.
38. **de Pontes, M., Yokomizo, G. H., and Bell, A. T.,** A novel method for analyzing transient response data obtained in isotopic tracer studies of CO hydrogenation, *J. Catal.,* 104, 147, 1987.
39. **Nwalor, J. U. and Goodwin, J. G., Jr.,** reported at the Pittsburgh-Cleveland Spring Symp., April 6 to 8, 1988.
40. **Iyagba, E. T., Nwalor, J. U., Hoost, T. E., and Goodwin, J. G., Jr.,** The effect of chlorine modification of silica-supported Ru on its CO hydrogenation properties, *J. Catal.,* 123, 1, 1990.
41. **Gallaher, G., Goodwin, J. G., Jr., and Blackmond, D. G.,** personal communication, 1990.
42. **Peil, K., Goodwin, J. G., Jr., and Marcelin, G.,** An examination of the oxygen pathway during methane oxidation over a Li/MgO catalyst, *J. Phys. Chem.,* 93, 5977, 1989.
43. **Peil, K., Goodwin, J. G., Jr., and Marcelin, G.,** Surface concentrations and residence times of intermediates on SM_2O_3 during the oxidative coupling of methane, *J. Am. Chem. Soc.,* 112, 7863, 1990.
44. **Nwalor, J. U., Goodwin, J. G., Jr., and Biloen, P.,** Steady-state isotopic transient kinetic analysis of iron-catalyzed ammonia synthesis, *J. Catal.,* 117, 121, 1989.
45. **Thiele, J., Muller, R., and Lingens, F.,** Enzymatic dehalogenation of 4-chlorobenzoate by 4-chlorobenzoate dehalogenase from *Pseudomonas* sp. CBS3 in organic solvents, *Appl. Microb. Biotech.,* 27, 577, 1988.
46. **Halling, P. J.,** Organic liquids and biocatalysts: theory and practice, *Tibtech.,* 7, 50, 1989.
47. **Ottensen, M. and Ralston, G.,** The ionization behaviour of subtilisin type Novo, *C. R. Trav. Lab. Carlsberg,* 38, 457, 1972.
48. **Wells, J. A. and Estell, D. A.,** Subtilisin — an enzyme designed to be engineered, *Trends Bioch. Sci.,* 13, 291, 1988.
49. **Alvaro, G. and Russell, A. J.,** Rational modification of enzyme catalysis by engineering surface charge, *Methods Enzymol.,* 202, 620, 1991.
50. **Kezdy, F. J. and Bender, M. L.,** The kinetics of the chymotrypsin-catalyzed hydrolysis of *p*-nitrophenyl acetate, *Biochemistry,* 1, 1097, 1962.
51. **Grunwald, J., Wirz, B., Scollar, M. P., and Klibanov, A. M.,** Asymmetric oxidoreductions catalyzed by alcohol dehydrogenase in organic solvents, *J. Am. Chem. Soc.,* 108, 162, 1986.
52. **Therisod, M. and Klibanov, A. M.,** Facile enzymatic preparation of monoacylated sugars in pyridine, *J. Am. Chem. Soc.,* 109, 3977, 1987.
53. **Riva, S., Chopineau, J., and Klibanov, A. M.,** Protease-catalyzed regioselective esterification of sugars and related compounds in anhydrous dimethylformamide, *J. Am. Chem. Soc.,* 110, 584, 1988.
54. **Kazandjian, R. Z. and Klibanov, A. M.,** Regioselective oxidation of phenols catalyzed by polyphenol oxidase in chloroform, *J. Am. Chem. Soc.,* 107, 5448, 1985.
55. **Fitzpatrick, P. A. and Klibanov, A. M.,** How can the solvent affect enzyme enantioselectivity, *J. Am. Chem. Soc.,* in press.

56. **Dordick, J.,** Enzyme design for non-aqueous media, *Proc. AIChE Annual Meeting,* San Francisco, 1990, 108E.
57. **Russell, A. J., Trudel, L. J., Skipper, P. R., Tannenbaum, S. R., Groopman, J. D., and Klibanov, A. M.,** Antibody-antigen binding in organic solvents, *Biochem. Biophys. Res. Commun.,* 158, 80, 1989.
58. **Stocklein, W., Gebbert, A., and Schmid, R. D.,** Binding of triazine herbicides to antibodies in anhydrous organic solvents, *Anal. Lett.,* 23, 1465, 1990.

Chapter 5

WATER AND ENZYMES IN ORGANIC SOLVENTS

Georgina Garza-Ramos, D. Alejandro Fernández-Velasco, M. Tuena de Gómez-Puyou, and Armando Gómez-Puyou

TABLE OF CONTENTS

I. INTRODUCTION

Traditionally, the function of enzymes has been studied in water; this is not surprising, since under physiological conditions enzymes work in water. However, there are numerous reports that indicate that enzyme catalysis may be observed in systems in which the predominant component is an organic solvent of low polarity or a supercritical fluid. Although in these nonconventional reaction mixtures water is not the most abundant component, a certain and, in many cases, a defined amount of water must be introduced in order to achieve enzyme catalysis. Indeed, in no case has catalysis been observed in the complete absence of water, albeit the amount at which catalysis begins to be detected is lower than that required to form a monolayer of water molecules around the protein.

The purpose of this chapter is to describe the behavior of enzymes in systems that are predominantly formed with organic solvents of low polarity, but the main emphasis will be on how, in such systems, the amount of water or the solvent that is in contact with the enzyme affects enzyme behavior. Accordingly, a brief review of the effect of water on some of the properties of proteins is considered to be in line prior to the discussion of the characteristics of enzymes in systems in which the main component is an apolar organic solvent.

II. WATER AND ENZYME CATALYSIS

Enzymes are structures that have been shaped by evolution to carry out catalysis in water. However, catalysis depends on the three-dimensional structure of the protein, and thus it seems more than a coincidence that water is an essential component in the formation of the three-dimensional structure of proteins. In the classical work of Kauzmann,[1] attention was called to the importance of hydrophobic interactions in the protein core of globular proteins, water being the driving force for the formation of these interactions. However, water has other functions: water molecules diminish charge-charge interactions, cover the surface of proteins, and are expelled when monomers come together to form multimeric enzymes. Thus, water molecules are active participants in the processes involved in the folding and assembly of proteins, and they also confer the protein distinct structural features.

The importance of water in protein function and structure is more easily appreciated if it is recalled that proteins are not static entities. Indeed, there is overwhelming evidence indicating that proteins are highly dynamic molecules.[2,3] Numerous crystallographic and spectroscopic studies have shown that protein motions vary from large-amplitude alterations of protein structure to localized vibrations of specific atoms. The various movements that occur in proteins are well-suited for whatever function a protein carries out and they cover a large number of processes. Since proteins exist in water, it is worthwhile to ask to what extent protein movements and function depend on the interaction of the protein with the surrounding media. Large amplitude movements, such as hinge-bending in the arabinose binding protein[4] and loop motions that result in the coverage of the active site of lactate dehydrogenase[5] or triose phosphate isomerase,[6] definitely involve large adjustments between the protein and water. Indeed, from an overall examination of crystallographic data of enzymes, with and without ligands, it is apparent that during catalysis with the majority of the enzymes, there is an important interplay between protein groups and the solvent. In fact, during catalysis, all enzymes undergo conformational changes; hence, in a catalytic cycle, there is movement of protein groups into or away from the surrounding water, and this necessarily implies the formation of new arrangements of water molecules with distinct protein groups.

The time lapses of the various types of protein motions have been calculated. Local vibrations of the main-chain and side-chain atoms are in the picosecond range, whereas the movement of protein domains are in a time range of microseconds or longer.[7] Water relaxation times are in the picosecond range. Thus, water provides an excellent milieu for proteins in the sense that it supports the smooth functional protein mobility by rapidly adjusting to whatever conformational changes the protein must undergo.

Although the dynamic nature of protein structures and the dependence of catalytic activity on protein mobility is well-documented, a substantial amount of work at the molecular level is needed to fully understand the role of water on protein function. A landmark in the understanding of protein-water interactions was made by Careri et al.,[8] Rupley et al.,[9] and Finney and Poole.[10,11] They determined the events that take place upon progressive hydrations of dry lysozyme. In the experimental protocols, progressive hydration of the protein was achieved by placing the dry protein (either in films or powders) in chambers of different controlled humidity. At distinct hydration levels, the authors made infrared (IR), nuclear magnetic resonance (NMR), and electron spin resonance (ESR) spectroscopic analyses, and heat capacity measurements. The point of full hydration was estimated by the change in heat capacity that occurs when all elements in the surface of the protein interact with water; this is at a hydration level of 0.32 g of water per gram of protein (about 300 water molecules per molecule of lysozyme[8]). From their data, as increasing amounts of water molecules contacted the enzyme, a dynamic picture of increasing protein mobility became apparent. At 0.07 g of water per gram of protein, water interacts with charged groups and acid groups ionize. Subsequently, at 0.25 g of water per gram of protein, polar groups are hydrated,[12] and according to Poole and Finney,[11] NH and CO groups hydrate at 0.26 and 0.32 g/g, respectively. Apolar groups are the last to hydrate. Of relevance in the experiments is that enzyme activity starts to appear when the content of water is 0.2 g/g and increases as the lysozyme is further hydrated. The increase in activity runs parallel to the mobility of an included noncovalently bound spin probe.[9] In these works, a central point relevant to enzyme function became apparent: full coverage of the protein by water molecules is not necessary for enzyme activity.

A factor that importantly affects the flexibility of proteins, and hence catalysis, is the shielding effect of protein-bound water molecules of polar and charged protein groups, as this results in a lowering of their interaction energy. According to Finney and Poole,[10] as water is added to lysozyme to a hydration level of 0.07 to 0.12 g/g, the protein acquires a certain flexibility. As water is progressively added, there is a significant increase in protein motions until, at a level of 0.2 g of water per gram of protein, the protein acquires the necessary flexibility to carry out catalysis, albeit at low rates.

Bone and Pethig,[13,14] taking advantage of the characteristic dipole movement of water molecules, investigated their interaction with the protein by dielectric relaxation spectroscopy. Their findings indicate that at low hydration levels, water molecules tightly bind to sorption sites with relaxation times similar to those of the protein. At higher levels of hydration, water molecules become weakly bound and become increasingly rotationally restricted as more water molecules are introduced. At higher hydration levels, bound water molecules exhibit an effective dipole movement comparable to that of bulk water. According to the authors, the protein undergoes increasing motion until at around 0.2 g/g; water acts as a "plasticizer" that effectively supports protein movements linked to the catalytic process.[13]

Recently, the activity of ethanol oxidase spread on DEAE-cellulose or glass beads and exposed to atmospheres of variable humidity was studied. Interestingly, in the work of Barzana et al.,[15] the substrate ethanol was in the gas phase. Enzyme catalysis increased with the activity of water (relative humidity at equilibrium), but again it was found that the enzyme-catalyzed oxidation of ethanol could take place at water contents that were below the monolayer value.

III. ENZYMES IN ORGANIC SOLVENTS

In addition to the aforementioned studies of enzymes at various levels of hydration, there are numerous studies that deal with the properties of enzymes in systems of low water content and in which the predominant component is an apolar organic solvent. There are essentially two variations of this type of systems. In one, enzymes at various levels of hydration are dispersed or suspended in a solvent of low polarity. In a second type of system, the enzymes are trapped in the water pool of reverse micelles. Although in both the main component is an organic solvent, there are striking differences in the disposition of the enzyme and of the physicochemical characteristics of the two systems; hence, they will be discussed separately.

A. ENZYMES DISPERSED IN ORGANIC SOLVENTS

Given that water is the natural habitat of enzymes, it is remarkable that enzymes suspended or dispersed in organic solvents exhibit catalytic activity. Even though the enzymes are in an insoluble state, Klibanov[16] and Dordick[17] have shown that in such conditions several different and unrelated enzymes can carry out catalysis. This suggests that a substantial number of enzymes have the potential to function under these apparently harsh conditions. Although the characterization of the structure and function of enzymes suspended in organic solvents is far from complete, Klibanov[16] indicated that for attaining enzyme catalysis in such conditions, certain requirements have to be fulfilled. An important point relates to the amount of water; Zaks and Klibanov[18] determined the activity rates of several enzymes dispersed in organic solvents at different amounts of water, and found that, for all, the activity increased as the amount of water in the system was raised. However, they also found that the plot of activity vs. the amount of water varied with the enzyme used and the solvent employed.

For the case of the solvent, it was observed that less water was required for enzyme activity as the hydrophobicity of the solvent was increased. The authors explained these observations on the grounds that with the less polar solvents, water partitioned to the enzyme to a larger extent than with the more polar solvents. Thus, an important point emerged: enzyme activity depends on the amount of water in contact with the enzyme and not on the total amount of water in the system. Even though there are striking experimental differences, the results of Zaks and Klibanov[18,19] are in accordance with those of Rupley et al.[8,9] and Finney and Poole[10,11] in powders or films of lysozyme at different levels of hydration. It is important to note that from their respective results, the same conclusion was reached: at the monolayer level of water molecules per enzyme molecule or less, the enzyme carried out catalysis. In other words, bulk water is not necessary for activity.

A point that requires extensive study is that not all enzymes exhibit the same sensitivity to increasing amounts of water. This was observed in the aforementioned studies of Zaks and Klibanov and also in those of Bone.[14] In the former, it could be that proteins are affected differently by the solvent or that different enzymes necessitate distinct hydration levels in order to acquire the necessary flexibility. In any case, it is likely that the different water requirements reflect intrinsic features of the protein.

Another interesting feature of enzymes in organic solvents is that they can catalyze reactions that in conventional water mixtures are not easily detected. It was reported that pancreatic lipase could catalyze esterification, acyl exchange, thiotransesterification, aminolytic, and oximolytic reactions.[20] This evidently shows that enzyme catalysis in nonconventional systems may be of importance in biotechnological processes. However, it also raises basic biochemical problems. A factor that should be considered in these observations is that the equilibrium constant of reactions may be strikingly different in aqueous and nonaqueous media. With molecules such as pyrophosphate and ATP, it has been shown that the equilibrium constant of hydrolysis shifts in several orders of magnitude by modifications in the organization of water by solvents, such as dimethylsulfoxide,[21,22] or in wet chloroform.[23] In fact, spontaneous synthesis of pyrophosphate and ATP by soluble inorganic pyrophosphatase and soluble mitochondrial ATPase, respectively, may be observed.[21,22] In reverse micelles, it was shown that lipase catalyzed the esterification of glycerol[24] and that the end point of the reaction depended on the amount of water. These findings relate to those of Han et al.,[25] who found that the equilibrium constant of lipase-catalyzed hydrolysis of triglycerides increased with increasing concentrations of water.

There is another feature of the system with enzymes dispersed in organic solvents. Some of the enzymes studied normally utilize water soluble substrates; however, due to the nature of the system, once an enzyme is dispersed or suspended in an organic solvent, it can easily utilize substrates that are soluble in the solvent. Thus, with an enzyme dispersed in an apolar organic solvent, conversion of strictly water-soluble substrates may present technical difficulties. On the other hand, it has the advantage that in an organic solvent, enzymes readily utilize substrates that are difficult to handle in water media. This may represent biochemical and technological advantage, particularly if it is considered that as the enzyme is insoluble in apolar organic solvent, it can easily be separated from the substrate or product of the reaction.

B. ENZYMES IN REVERSE MICELLES

Another type of reaction mixture that is predominantly formed with organic solvents and in which the activity of numerous enzymes has been studied is that known as the reverse micelle system or microemulsions.* Surfactants dissolved in organic solvents can form spherical structures with water entrapped in the interior; in this arrangement, the polar heads of the surfactant are oriented towards the water pool, whereas the hydrophobic chains contact the organic solvent. Since enzymes can be entrapped in reverse micelles, many studies have been made to probe their characteristics in this novel environment and to explore their potentiality in biotechnology.[26,27]

With respect to the study of the role of water in the structure and function of enzymes, one of the attractive features of reverse micelles is that the size of the water pool is determined by the ratio of water molecules to surfactant molecules; this is commonly referred to as Wo. Hence, at a constant concentration of surfactant, the size of the micelle depends on the amount of water introduced. For one of the most commonly used systems — water/*bis*(2-ethylhexyl) sodium sulfosuccinate (AOT)/isooctane — at Wo of 5, the diameter of the water pool is around 20 Å and may increase to about 80 Å at Wo of 50. This is of importance, since the amount of water introduced determines the amount and the type of water that an enzyme sees (vide infra).

For measurements of enzyme activity, we generally introduce substrates and enzymes dissolved in water in a two-step procedure to a mixture of solvent and surfactant followed by stirring; i.e., more frequently, enzyme reactions are started by adding the enzyme to the organic solvent that already contains the substrate or vice versa. In these systems, a water-soluble enzyme is presumed to lie in the water phase of the micelle (Figure 1). According to their physicochemical properties, the substrates and the enzymes distribute in the water phase, the interphase of the micelle, or the organic solvent.

1. Arrangement of Enzymes in Reverse Micelles

The majority of the enzymes studied in reverse micelle systems have been water-soluble proteins; these include monomeric and multimeric enzymes.[26,27] However, integral membrane proteins such as cytochrome oxidase,[28] rhodopsin,[29] the Ca-ATPase of sarcoplasmic

---

* Some authors use the term microemulsions, particularly when referring to structures at high Wo. Here the term reverse micelles will be used.

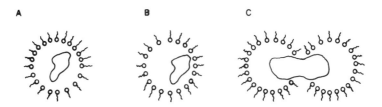

FIGURE 1.   Models for the arrangement of different types of enzymes in reverse
micelles. (A) A soluble enzyme; (B) an enzyme that normally is attached to the surface
of a biological membrane; and (C) an integral membrane protein. For further details,
see text.

reticulum,[30] and the ATPase of submitochondrial particles[31] have also been transferred to
reverse micelle systems. Proteins that localize in the surface of a biological membrane or
between two membrane surfaces, such as cytochrome c[32] and myelin basic protein,[33,34] have
also been studied in reverse micelles. In consequence, it may be asked how these three
different types of enzymes arrange themselves in reverse micelles. Up to date there is no
precise answer to this question, but several models have been proposed. For a water-soluble
enzyme, perhaps the simplest case, it is visualized that the enzyme lies in the water pool
of the micelle surrounded by water molecules. For the myelin basic protein[33,34] and cytochrome c[32] that normally lie between two membranes, there is evidence that indicates that
in reverse micelles these proteins contact the internal surface of the micelle (Figure 1). In
this respect, Chatenay et al.[35] reported that surfactant and myelin basic protein molecule
compete for interfacial water and that the protein modified the properties of lamellar phases.[36]
In view of the charges that line the interior layer of a reverse micelle, it may be considered
that a water-soluble protein may attach to the polar surface of the micelles. Indeed, if the
dimensions of an average protein are taken into account, this would not seem to be an
unlikely possibility, particularly at low Wo. Such an attachment could modify some of the
properties of the enzyme. On the other hand, it has been suggested that with proteins that
normally exist at the surface or between membrane surfaces, i.e., cytochrome c and myelin
basic protein, the reverse micelle system closely mimics their biological location.

   The arrangement of membrane proteins such as rhodopsin, cytochrome oxidase, the
mitochondrial ATPase, and the Ca-ATPase is more difficult to visualize. These proteins
possess a hydrophobic portion that is buried in the apolar milieu of their membranes. It has
been suggested[28,37] that in reverse micelles, the polar moiety(ies) of the protein is in contact
with the water pool of the micelle, whereas the hydrophobic portion contacts the organic
solvent (Figure 1). For the case of a membrane enzyme that has two hydrophilic moieties
on opposite sides of the biological membrane, such as cytochrome oxidase, it had to be
postulated that each of these portions was in the interior of two different micelles. Other
arrangements of this type of protein are suggested (see Chapter 2).

2. Enzyme Activity in Reverse Micelles

   Illustrative of the potential of the reverse micelle system in enzymology is that an
overwhelming number of enzymes have been transferred to reverse micelle systems.[26,27]
Under optimal conditions many of these enzymes exhibit rates of catalysis that in many
instances are similar to those observed in standard aqueous mixtures. Enzyme kinetics in
these systems have been extensively studied. However, some points deserve comment. For
a reaction to take place, substrate-filled micelles must collide and fuse with enzyme-filled
micelles. The rate of successful collisions between micelles have been determined to be in
the microsecond range.[38,39] Thus, at first sight it would appear that collisions between micelles
would not limit enzyme activity. However, the study of enzyme kinetics in reverse micelles
presents a large degree of complexity.[40,41] It has been calculated that the concentration of

micelles in such systems may be in the m*M* range; commonly, enzymes are introduced in n*M* concentration. Thus, when a substrate is introduced in the μ*M* range, there will be filled and empty micelles. Moreover, as the amount of water added distributes in all micelles, the concentration of substrate in the substrate-filled micelle should be corrected by a factor that takes into account the number and the amount of water in the empty micelles. Attempts to design methods that allow the kinetic analysis of enzymes in such systems have been described[41,42] (see Chapters 6, 7, and 8).

Enzymes entrapped in reverse micelles may also convert substrates that are in the interphase of the micelle.[25,42,43] In addition, there are reports that show that in reverse micelles enzymes may convert hydrophobic substrates;[44,45] for this particular type of substrates, their partitioning into the water phase should be considered.

Obviously, the evaluation of kinetic data with hydrophobic substrates or those that lie in the interphase may present difficulties. However, the problems may not be more complex than those faced in conventional water systems with enzymes that transform substrates that arrange in aggregates or liposomes, i.e., phospholipases, or with enzymes that use substrates that have to cross a biological membrane for conversion.

Thus, the overall data indicate that enzymes in reverse micelles are able to transform substrates with widely different solubility properties. This is in contrast to the systems in which enzymes are dispersed in organic solvents (see above) and in which the conversion by enzymes may be limited by the solubility properties of the substrate.

An interesting experiment that illustrates an additional potential of enzymes in reverse micelles is that described by Chopineau et al.[46] This was based on the fact that the existence and the properties of reverse micelles depend principally on the ratio of water molecules to surfactant molecules. The latter authors observed that β-D-glucosidase entrapped in reverse micelles formed with octanol, octyl-β-D-glucoside, and water could catalyze the hydrolysis of the detergent. Thus, as catalysis progressed, the system underwent changes of its physicochemical properties. According to the authors, the enzyme through its catalytic action was able to modify its own microenvironment.

3. Water and Enzyme Catalysis in Reverse Micelles

One of the most relevant characteristics of enzymes in reverse micelles is that their activity is drastically affected by the amount of water in the system. At Wo around 5, the activity of many enzymes is several-fold lower than in all water media, but as the amount of water is increased, activity may reach, and even exceed, that which is observed in all aqueous mixtures.[26,27] In several instances, it has been observed that maximal activity is attained at ratios of water molecules to detergent molecules of about 10. It is of importance to explore the mechanisms involved in this commonly observed phenomenon, as very likely the ascertaining of the mechanisms will provide insight into the relation between water and enzyme activity.

As to how water affects catalysis in reverse micelles, as well as in other different systems, there are two important factors: the amount of water and the "type of water". The characteristics of the water pool of reverse micelles has been explored by several techniques.[47-51] Wong et al.,[48] by ^{1}H and ^{23}Na-NMR spectroscopy of reversed micelles formed with AOT in heptane, found that the spin-lattice relaxation times and spin-spin relaxation rates of ^{1}H of water decreased markedly up to 1% water and became much slower as water was increased. Along the same line, Maitra[50] analyzed by NMR the chemical shifts of water protons at various amounts of water in AOT reverse micelles. Their data were consistent with the idea that at low water concentrations, water exists largely in the bound state. In fact, it was calculated that the first six molecules of water are bound more firmly than the subsequent ones.[50,52] In addition, Tsujii et al.[49] investigated the viscosity of the water pool in reverse micelles formed with dodecylammonium propionate in benzene by ^{13}C-NMR techniques;

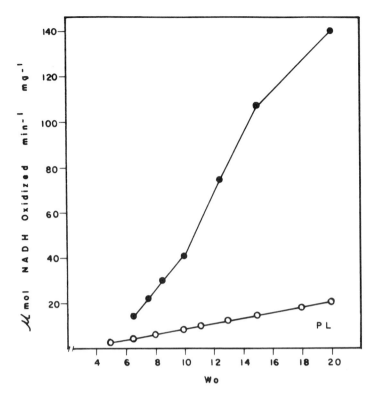

FIGURE 2.   Activity of bovine heart lactate dehydrogenase in systems formed with phospholipids or cetyltrimethyl ammonium bromide (CTAB). At the indicated Wo in a final volume of 1.0 ml, the activity of lactate dehydrogenase was measured in a system formed with toluene, phospholipids, and Triton X-100;[31] the concentration of the reagents in the water phase was 20 mM phosphate buffer (pH 7.4), 30 mM pyruvate, and 7.5 mM NADH. The activity was also measured in a system formed with *n*-octane, CTAB, and hexanol.[42] In this case, the water phase contained 20 mM phosphate buffer (pH 7.4), 2.0 mM pyruvate, and 136 nmol NADH.

they found that at low water/detergent ratios the apparent viscosity of the water pool corresponded to that of a 78% aqueous glycerol solution.

The overall data indicate that properties of water in reverse micelles differ from those of bulk water, but that as content of water of the micelle is increased, the differences become smaller. Thus, when the activity of an enzyme is measured at increasing water concentrations, it is difficult to distinguish to what extent activity is affected by a net increase in the amount of water or to changes in the overall properties of the water pool.

In this respect, a particular class of reverse micelles that merits comment is that formed with phospholipids (for review see Walde et al.[53]). Briefly, studies on the hydration shell of phosphatidylcholine in benzene showed that either 2 to 3 or 9 to 11 molecules of water per polar head form a first hydration shell; a further 9 to 11 water molecules make up a second shell.[54] In addition, NMR data showed that there is one water molecule tightly bound to the polar head of the surfactant in rapid exchange with the rest of the water molecules.[55] In contrast, in reverse micelles formed with synthetic detergents at a Wo of 10, the properties of water approach those of bulk water.

Thus, it was of interest to compare the activity of the same enzyme in systems formed with synthetic detergent and phospholipids. It was found that in a wide range of Wo, the activity of heart lactate dehydrogenase was significantly higher in the former (Figure 2). Most likely, this different activity reflects the effect of the properties of water on the enzyme.

In consequence, it would seem that by varying the components of the micelle, it is possible to affect the activity of the enzyme. Moreover, it was found that in reverse micelles, denaturants may increase the activity of lactate dehydrogenase,[56] suggesting that perturbations of the enzyme environment could lead to a high expression of catalysis. This phenomenon will be discussed below in more detail.

In dealing with enzyme activity in bulk and bound water, it is worth noting that there is a long-standing discussion as to whether enzymes in living cells function in bound or bulk water. In this latter respect, it is interesting to note that Swezey and Epel[57] assayed the activity of five enzymes at saturating substrate concentrations in electrically permeabilized cells. Thus, assays were made in conditions that were equivalent to the intracellular milieu. They observed that the activity of the permeabilized cells was significantly lower than in homogenates. Although several factors may account for their data, it is possible that the state of intracellular water may be related for at least part of the results. If indeed intracellular water is mostly in the ''bound'' state, micellar enzymology will be a valuable model in the study of enzymes under ''physiological'' water conditions.

Not many studies have been carried out on the structure of proteins in reverse micelles, particularly from the point of view of correlating activity with structure. However, the data on myelin basic protein[33] and lysozyme[58] indicated that they are less unfolded in reverse micelles than in all aqueous solutions. This has not been observed for ribonuclease,[59] but for α-chymotrypsin[60] and trypsin,[61] the data suggest that the enzyme is in an environment of relatively low polarity.

IV. ENZYME THERMOSTABILITY

The studies with enzymes placed in contact with low amounts of water show that in these conditions enzyme catalysis is severely restricted. This has been observed with enzymes at various levels of hydration, with enzymes dispersed in organic solvents, and with enzymes placed in reverse micelles. It has been proposed that the cause of the low activity is the low flexibility that the enzyme has in such conditions.[13,19,31] In other words, it is visualized that in a low-water environment, catalysis is diminished due to an impairment in the interplay of protein groups with the surrounding medium. Therefore, several groups and ourselves have addressed the question of whether, with a limiting amount of water, properties of enzymes other than catalysis could be affected; one of these is enzyme thermostability.

As a background, it is worthwhile recalling that at room temperatures thermostable enzymes form thermophiles, in comparison to their mesophilic counterparts, are rather rigid molecules.[62] This is the result of intramolecular interactions and forces that increase the stability of the protein. With respect to the relation between protein rigidity and enzyme catalysis and thermostability, it is illustrative that the specific activity of thermophilic enzymes is lower than that of mesophilic enzymes at temperatures in the 25°C range. Thus, the low intrinsic flexibility of thermophilic enzymes, as compared to their mesophilic counterparts, would account for both high thermostability and low catalytic rates at relatively low temperatures.

The thermostability of lipases,[63] terpene cyclase,[64] α-amylase,[65] and chymotrypsin[66] dispersed in organic solvents has been explored; in addition, the thermostability of cytochrome oxidase[67] and the mitochondrial ATPase[31,68] has been studied in systems formed with phospholipids (with and without Triton X-100) in toluene. For all these markedly different enzymes, the results indicate that at low amounts of water, the thermostability of the enzymes is much higher than in aqueous media. In these experiments, it was also found that as the amount of water in the system is increased, the enzymes become more susceptible to thermal inactivation. In consonance with these findings, differential scanning calorimetry studies of ribonuclease, cytochrome *c*, and lysozyme in reverse micelles (AOT/isooctane)

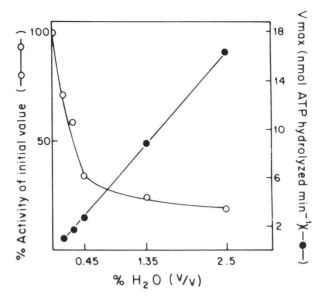

FIGURE 3. Thermostability and activity of soluble mitochondrial ATPase in systems formed with phospholipids in toluene. V_{max} was calculated from Lineweaver-Burk plots with different concentrations of Mg-$(\tau^{32}P)$ATP at each of the water concentrations indicated. Stability was assayed by incubating the enzyme transferred to the toluene/phospholipid system with the indicated concentrations of water for 1 min. At that time aliquots were withdrawn to assay the remaining ATPase activity after transferring the enzyme to all water media. (Reprinted with permission from Garza-Ramos, G., Darszon, A., Tuena de Gómez-Puyou, M., and Gómez-Puyou, A., *Biochemistry,* 28, 3177, 1989. Copyright 1989, American Chemical Society.)

showed that as Wo of the micelles increased, stability decreased.[69] Therefore, it would appear that the amount of water in contact with the enzyme is instrumental in thermal inactivation and denaturation, i.e., the higher the amount of water in contact with the enzyme, the lower the capacity of the enzyme to resist high temperatures.

In systems formed with phospholipids in organic solvents, it was found that enzymes resisted temperatures of 70 to 90°C for significant lengths of time.[31,67,68] In contrast, in reverse micelles formed with synthetic surfactants, enzymes unfold[69] or lose activity in the 40 to 50°C temperature range. Hence, the phospholipid system is of interest in the sense that enzymes exhibit a higher thermostability. Also it is noted that in this system rates of enzyme catalysis are lower than in reverse micelles formed with synthetic detergents (see Figure 2).

In the phospholipid system, an inverse relation between enzyme catalysis and thermostability has been observed with cytochrome oxidase and the mitochondrial ATPase. The half-life of the former at 70°C is about 11 h,[67] but in the same conditions, the enzyme fails to show catalytic activity (electron transport from ascorbate to oxygen) at 25°C; interestingly, electron transport is arrested after reduction of hema a.[28] With the mitochondrial ATPase, it was observed that at water concentrations in which the enzyme is strikingly thermostable, its capacity to hydrolyze ATP at 25°C is barely detectable.[31,68] The relation between rates of catalysis and thermal stability as a function of water concentration is shown in Figure 3 for the mitochondrial ATPase. As the amount of water in contact with the enzyme increases, there is a progressive rise in the ability of the enzyme to hydrolyze ATP but the stability of the enzyme gradually decreases. A similar behavior has been observed for lipase dispersed in organic solvents[63] and to a lower extent with cytochrome oxidase in reversed micelles.[28,67]

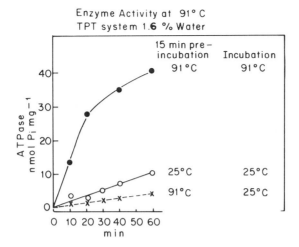

Enzyme Activity at 91°C
TPT system 1.6% Water

FIGURE 4.   ATPase activity of submitochondrial particles transferred to a system of toluene/phospholipids/Triton X-100. After transferring the particles to a system made up with toluene/phospholipids/Triton X-100 and 0.32% water (v/v), the mixture was incubated for 15 min at the temperatures shown; after that time a water solution that contained the substrate was added to start the hydrolytic reaction, and the incubation continued at the indicated temperatures. At different times aliquots were withdrawn to assay the amount of ATP hydrolyzed. The addition of the water solution to start the reaction yielded a final concentration of 1.6% water (v/v), with 10 mM Tris-HCl (pH 7.4), 3 mm Mg_2, and 3 mM ($\tau^{32}P$)ATP (in the water space). (Reprinted with permission from Garza-Ramos, G., Darszon, A., Tuena de Gómez-Puyou, M., and Gómez-Puyou, A., *Biochemistry,* 29, 751, 1990. Copyright 1990, American Chemical Society.)

The effect of water on enzyme stability and catalysis may be explained in terms of protein rigidity or flexibility. As the amount of water in contact with the enzyme is progressively increased, there will be an enhancement of protein flexibility, and, hence, a higher capacity to carry out catalysis. However, the increased flexibility as induced by water* has a cost; i.e., the enzyme becomes more susceptible to thermal inactivation or denaturation.

V. ENZYME CATALYSIS AT HIGH TEMPERATURES

Once it was found that enzymes in organic solvent systems with low amounts of water exhibit a high thermostability, it was natural to ask if in such conditions enzymes could work at temperatures that in all water media are detrimental for enzyme activity. Zaks and Klibanov[63] were the first to observe enzyme catalysis at surprisingly high temperatures; i.e., lipase suspended in organic solvents with 0.015% water catalyzed transesterification of tributyrin and heptanol at a temperature of 100°C. The amount of water was central for enzyme catalysis; if the amount of water in the system was raised, a decrease in the rate of activity took place.

Along the same line, it was explored if at high temperatures the mitochondrial ATPase could catalyze ATP hydrolysis in systems made up of toluene, phospholipids, and Triton X-100.[31] It was found (Figure 4) that at a water concentration of 1.6% (v/v), the enzyme could catalyze ATP hydrolysis at a temperature of 91°C. Catalysis at this temperature was also critically dependent on the amount of water; at higher or lower water concentrations, lower or no hydrolysis at all was detected.

It is worth pointing out that enzyme catalysis at high temperatures was observed in two

---

* Indeed, water has been referred to as a plasticizer[13] or a lubricant.[18]

different experimental systems (enzymes dispersed in organic solvents and in reverse micelles) and with two markedly different enzymes (lipase and the mitochondrial ATPase) that convert different substrates. Thus, it is evident that in low-water systems, the temperature at which many enzymes can carry out catalysis is much higher than in conventional aqueous media. It is also clear that the temperature at which an enzyme can work depends on the amount of water.

The following is a scheme that considers the properties of enzymes in contact with variable amounts of water. It is based on the idea that as the amount of water in contact with the enzyme is increased, it becomes more flexible.

Low water   Excess water

H_2O H_2O

$E_1 \longleftrightarrow E_2 \longleftrightarrow E_3$

Thermostability   High $\longleftrightarrow$ Low

Catalytic rates   Low $\longleftrightarrow$ Maximal

In this scheme, excess water refers to conventional aqueous media. E_1 represents an enzyme in contact with a low amount of water; it is visualized as a rigid enzyme in which its catalytic capacity is severely restricted, and therefore it should have a high thermostability. As water is gradually added, the enzyme progressively (E_2) acquires a larger conformational freedom, but it is still not readily inactivated by high temperatures; the main feature of the E_2 form is that it is capable of catalyzing at high temperatures. As more water is added (excess water), the enzyme acquires maximal flexibility (E_3). It exhibits high catalytic rates, but only in a range of temperature that does not cause extensive alterations of its three-dimensional structure.

It is stressed that the enzyme in the E_2 narrow range of water concentrations exhibits sustained catalysis in the high temperature range. At higher water concentrations (E_3), catalysis at these temperatures diminishes or is abolished. Therefore, at high temperatures, water above a given concentration may be considered as a factor that leads to rapid inactivation and hence abolition of enzyme catalysis. Therefore, in the high temperature range, perhaps it will be necessary to speak of optimal water concentrations for enzyme function.

Enzymes from mesophilic organisms may work at high rates at high temperatures. However, with enzymes that are normally denatured at high temperatures, there is the question as to how fast they can work at such temperatures when placed in a low water environment. Zaks and Klibanov[63] measured the lipase-catalyzed transesterification between tributyrin and primary and secondary alcohols and found that at 100°C rates were about five times faster than at 20°C, their values at 20°C being in the range of 60 to 80 μmol/h per 100 mg of lipase. With the mitochondrial ATPase[31] in a toluene/phospholipid/Triton X-100 system and 1.6% water, the rate at 90°C was faster than at 25°C (Figure 4). However, if the activity is compared to that attained in all water media at 25°C, hydrolysis in the organic solvent system was nearly two orders of magnitude lower than in all water media.

Although increases in temperatures increase reaction rates, the low catalytic rate of the mitochondrial ATPase at 91°C may not be surprising. When dealing with enzymes in the high temperature range, the flexibility of the protein must be diminished to a point in which denaturation is prevented. At that point the enzyme must be a rather rigid structure, and in consequence it would exhibit low catalytic rates. Indeed, in our experiments with the ATPase, the rigidity of the protein had to be adjusted to a point at which it could resist high temperatures, but at which it was still able to catalyze.

The data obtained with the mitochondrial ATPase[31] showed that at amounts of water at which the enzyme has a long half-life at high temperatures, catalysis is low or nonexistent; on the other hand, at water contents that support high catalytic rates, the enzyme exhibits a short half-life at high temperatures. It is worth noting that thermophiles solved the problem of high catalytic rates at high temperatures by adjustments of their protein structures (for data on the ATPase from thermophilic bacteria, see Saishu et al.[70]). Therefore, it is tempting to think that protein engineering in combination with solvent engineering will yield interesting results.

VI. DENATURANTS AND ENZYME CATALYSIS IN ORGANIC SOLVENTS

During catalysis, all enzymes undergo conformational changes; this implies that in a catalytic cycle there are many adjustments between protein groups and the surrounding media. In consequence, the low activity that enzymes express in a low-water environment should be due to restrictions in solvent-protein interactions. Thus, we addressed the question as to whether the activity of enzymes placed in a small water space could be increased by promoting the interplay between the solvent and the enzyme. An important point in the experimental approach was to consider solvent-protein interactions that occur in a catalytic cycle in terms of solubilization of protein groups into the surrounding media. Thus, we searched for agents that could increase the solubilization of protein groups and that at the same time could be introduced into low-water systems.

For a long time it has been known that some agents such as urea and guanidine chloride (GdnHCl) bring about the disruption of the three-dimensional structure of protein.[71] Their precise mechanism of action in protein denaturation is not yet completely defined. However, it is accepted that the action of denaturants involves the solubilization of amino acid residues that are poorly soluble in water.[72] It is also known that denaturants bind to protein groups thereby weakening the forces that maintain the native structure of proteins.[73] Finally, it is very likely that urea and guanidine salts modify the structure of water.[74] Accordingly, the general action of these agents is to promote an increase in the interactions between protein groups and the surrounding solvent.

Urea and guanidine salts have been mostly used in studies of protein unfolding and refolding, but there are many reports that describe their effect on enzyme catalysis. In the majority of the cases, it has been observed that structural changes produced by denaturants run in parallel to decreases in enzyme activity. However, in several enzymes, denaturants bring about activation at concentrations lower than those that cause gross changes of protein structure.[75,76] The inverse has also been described, i.e., inactivation of catalysis without extensive alterations of protein structure.[77,78] An interesting case of activation by denaturants is malate dehydrogenase from chloroplasts; the oxidized form of this enzyme is inactive but gains activity upon reduction of disulfide bonds.[79,80] Scheibe and Fickenscher[79] measured the catalytic activity of the oxidized form of the enzyme and found that at optimal concentrations of GdnHCl the enzyme reaches an activity of about 25% of that of the reduced enzyme.

With this background, experiments were carried out to explore if the low activity of enzymes in a limited water space could be enhanced by denaturants. In these studies, the effect of GdnHCl on the activity of lactate dehydrogenase was explored in a system formed with toluene, phospholipids, and Triton X-100.[56] The phospholipid system had the advantage that in a wide range of water concentrations (up to 6.5% water), enzyme activity is drastically restricted (see Figure 2). With 3.8% water in the system, the activity of lactate dehydrogenase from bovine heart was about 6 μmol/min/mg (in all water mixtures it was around 250 μmol/min/mg). In the presence of increasing concentrations of GdnHCl up to 1.5 to 2.0 M, the

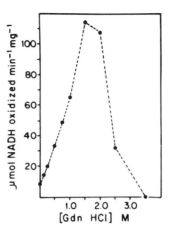

FIGURE 5. Effect of guanidine chloride on the activity of lactate
dehydrogenase from bovine heart. The enzyme was transferred to a system
of toluene/phospholipids/Triton X-100 and 3.8% water (v/v); the water
phase contained 20 mM phosphate buffer (pH 7.4), 30 mM pyruvate,
7.5 mM NADH, and the indicated concentrations of guanidine chloride
(GdnHCl). (From Garza-Ramos, G., Darszon, A., Tuena de Gómez-Puyou,
M., and Gómez-Puyou, A., *Biochem. Biophys. Res. Comm.*, 172, 830,
1990. With permission.)

activity increased; at higher concentrations the activity decreased (Figure 5). With optimal
GdnHCl concentrations, the activity of the enzyme was about 20-fold higher than in its
absence and between 40 and 50% of that detected in all aqueous mixtures. The enhancement
of the activity of lactate dehydrogenase by denaturants has also been observed in systems
formed with octane, hexanol, cetyltrimethylammonium (CTAB), and low amounts of water.[81]
Thus, it would appear that in conditions of limiting water, enzyme activity may be increased
by agents that facilitate protein-solvent interactions, or as indicated above, by facilitating
the solubilization of protein groups that during catalysis have to interact with the solvent.

It is also relevant to point out that the concentration of GdnHCl required for producing
half-maximal inactivation of lactate dehydrogenase is significantly higher in the low water-
organic solvent system than in conventional aqueous mixtures (Table 1). Moreover, it was
observed that the concentration of GdnHCl required to produce half-maximal inactivation
depends on the amount of water in the system, i.e., the lower the amount of water, the
higher the denaturant concentration required.

These results together with those obtained on the thermal stability of enzymes at low
water concentrations suggest that the amount of water or solvent is central for enzyme
inactivation, regardless of whether it is induced by GdnHCl or high temperatures. However,
there is an additional point, and this is illustrated in Figure 6. The effect of GdnHCl on the
activity of bovine heart lactate dehydrogenase was studied in systems formed in organic
solvents with phospholipids or CTAB (+ hexanol) and in all aqueous media. In all conditions
GdnHCl caused activation, but the highest activation was observed in micelles with low
Wo. Moreover, the concentration of GdnHCl needed to produce the highest activation
decreased as the amount of water in contact with the enzyme was increased. This indicates
that perturbations of protein structure, or of the water space in which the enzyme localizes,
or both, release the restrictions in enzyme catalysis imposed by a low-water environment.
According to these results, it is tempting to think that one of the first events in denaturant
action becomes apparent with enzymes placed in low-water systems. This may occur at the
level of the catalytic site or at some other point(s) in the enzyme.

TABLE 1
Inactivation by Guanidine Chloride (GdnHCl) of
Lactate Dehydrogenase Incubated with Various
Amounts of Water

Water % (v/v)	Concentration (M) of GdnHCl required to produce half-maximal inactivation
100	0.9
6	1.1
3.8	2.2
2.0	3.2

Note: The activity of lactate dehydrogenase was measured in conventional water systems (100% water) or in systems formed with toluene, phospholipids, and Triton X-100[31] with the indicated amounts of water. For the case of the phospholipid system, the water phase contained 20 mM phosphate buffer (pH 7.4), 30 mM pyruvate, 7.5 mM NADH, and varying concentrations of GdnHCl. The concentrations of GdnHCl that decreased the activity to 50% are shown; 100% activity was that attained with the optimal concentration of GdnHCl.

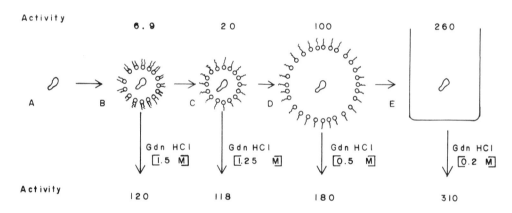

FIGURE 6. Outline of systems formed with organic solvents and different amounts of water: (A) depicts an enzyme dispersed in organic solvents; (B and C) illustrate an enzyme in reverse micelles formed with phospholipids or a synthetic detergent, respectively, at a Wo of 10; (D) is a swollen micelle (Wo = 20) formed with synthetic detergent; and (E) is an enzyme in a test tube with bulk water. The numbers indicate the activity of lactate dehydrogenase in μmol/min/mg. When GdnHCl was introduced at the indicated concentrations, the activities shown were detected; these were the GdnHCl concentrations at which optimal activity was observed.

VII. CONCLUSIONS

It has been described that by using various experimental systems in which the predominant component is an organic solvent, it is possible to study various characteristics of enzymes when placed in contact with increasing amounts of water. Moreover, in organic solvent systems, it seems that it is possible to study enzyme behavior in environments that have water with different properties. In addition, some recent experiments indicate that it is also possible to include denaturants in reverse micelles and thus explore their action in a restricted solvent space. Indeed, the described data suggest that these agents, through enzyme activation, make apparent or amplify one of the primary events in the action of denaturants, and which in conventional water systems is not easily apparent.

Figure 6 depicts a journey through the organic solvent systems in which enzymes may be studied with increasing amounts of water, as well as some of the modifications that can be applied, and how they reflect on enzyme behavior. The starting point is an enzyme dispersed in an organic solvent. Following Klibanov and co-workers, an enzyme in contact with a monolayer of water molecules, or less, is catalytically active and has high thermostability. The following experimental condition is a reverse micelle with a low Wo. This can be a micelle formed with phospholipids or with a synthetic detergent. In both types, catalytic activity is several times lower than in all aqueous media; however, in the phospholipid type activity is lower and with some enzymes (cytochrome oxidase) it is almost nonexistent. Additional features of the phospholipid system are that enzymes exhibit a high resistance to denaturants and high temperatures and that they catalyze at high temperatures. The latter two characteristics are shared by enzymes dispersed in organic solvents. Thus, this state may be considered as the experimental transition between the enzyme dispersed in organic solvents and in the swollen micelle.

The next experimental station is a large micelle with water that has properties that approach those of "bulk" water. In these micelles, enzymes have a high catalytic activity which in many cases is equivalent to that attained in all water media. In this condition, enzymes exhibit a low resistance to high temperatures and denaturants. The end of the journey is an enzyme in the test tube with bulk water. Along the road, it is possible to make side trips. Except when enzymes are dispersed in organic solvents, denaturants may be introduced to induce variations in solvent-protein interactions; as noted, some not easily evident aspects of denaturant action on enzymes may become apparent.

Thus, a trip through low-water territories may prove fruitful for a further understanding of proteins, particularly on the effect of the environment on protein structure and function.

ACKNOWLEDGMENTS

This work was supported by a grant from Dirección General de Asuntos del Personal Académico, Universidad Nacional Autónoma de México, and Consejo Nacional de Ciencia y Technologia (CONACYT), México.

REFERENCES

1. **Kauzmann, W.,** Some factors in the interpretation of protein denaturation, *Adv. Protein Chem.*, 14, 1, 1959.
2. **Brooks, C. L., III and Karplus, M.,** Solvent effects on protein motion and protein effects on solvent motion. Dynamics of the active site region of lysozyme, *J. Mol. Biol.*, 208, 159, 1989.
3. **Williams, R. J. P.,** NMR studies of mobility within protein structures, *Eur. J. Biochem.*, 183, 479, 1989.
4. **Quiocho, F. A., Wilson, D. K., and Vyas, N. K.,** Substrate specificity and affinity of a protein modulated by bound water molecules, *Nature,* 340, 404, 1989.
5. **Holbrook, J. J., Liljas, A., Steindel, S. J., and Rossmann, M. G.,** Lactate dehydrogenase, in *The Enzymes,* Boyer, P. D., Ed., Academic Press, New York, 1975, chap. 4.
6. **Joseph, D., Petsko G. A., and Karplus, M.,** Anatomy of a conformational change: hinged "lid" motion of the triosephosphate isomerase loop, *Science,* 249, 1425, 1990.
7. **Brooks, C. L., III, Karplus, M., and Pettitt, B. M.,** Proteins: a theoretical perspective of dynamics, structure and thermodynamics, in *Advances in Chemical Physics,* Vol. 71, Prigogine, I. and Rice, S. A., Eds., John Wiley & Sons, New York, 1988.
8. **Careri, G., Gratton, E., Yang, P.-H., and Rupley, J. A.,** Correlation of IR spectroscopic, heat capacity, diamagnetic susceptibility, and enzymatic measurements on lysozyme powder, *Nature,* 284, 572, 1980.
9. **Rupley, J. A., Gratton, E., and Careri, G.,** Water and globular proteins, *TIBS,* 8, 18, 1983.
10. **Finney, J. L. and Poole, P. L.,** Protein hydration and enzyme activity: the role of hydration-induced conformation and dynamic changes in the activity of lysozyme, *Comm. Mol. Cell. Biophys.,* 2, 129, 1984.
11. **Poole, P. L. and Finney, J. L.,** Sequential hydration of a dry globular protein, *Biopolymers,* 22, 255, 1983.

12. **Careri, G., Giansanti, A., and Gratton, E.,** Lysozyme films hydration events: an IR and gravimetric study, *Biopolymers,* 18, 1187, 1979.

13. **Bone, S. and Pethig, R.,** Dielectric studies of protein hydration and hydration-induced flexibility, *J. Mol. Biol.,* 157, 571, 1982.

14. **Bone, S.,** Time-domain reflectometry studies of water binding and structural flexibility in chymotripsin, *Biochim. Biophys. Acta,* 916, 128, 1987.

15. **Barzana, E., Karel, M., and Klibanov, A. M.,** Enzymatic oxidation of ethanol in the gaseous phase, *Biotechnol. Bioeng.,* 34, 1178, 1989.

16. **Klibanov, A. M.,** Enzymatic catalysis in anhydrous organic solvents, *TIBS,* 14, 141, 1989.

17. **Dordick, J. S.,** Enzymatic catalysis in monophasic organic solvents, *Enzyme Microb. Technol.,* 11, 194, 1989.

18. **Zaks, A. and Klibanov, A. M.,** The effect of water on enzyme action in organic media, *J. Biol. Chem.,* 263, 8017, 1988.

19. **Zaks, A. and Klibanov, A. M.,** Enzymatic catalysis in nonaqueous solvents, *J. Biol. Chem.,* 263, 3194, 1988.

20. **Zaks, A. and Klibanov, A. M.,** Enzyme-catalyzed processes in organic solvents, *Proc. Natl. Acad. Sci. U.S.A.,* 82, 3192, 1985.

21. **De Meis, L., Behrens, M. I., and Petretski, J. H.,** Contribution of water to free energy of hydrolysis of pyrophosphate, *Biochemistry,* 24, 7783, 1985.

22. **Gómez-Puyou, A., Tuena de Gómez-Puyou, M., and De Meis, L.,** Synthesis of ATP by soluble mitochondrial F1 ATPase and F1-inhibitor-protein complex in the presence of organic solvents, *Eur. J. Biochem.,* 159, 133, 1986.

23. **Wolfenden, R. and Williams, R.,** Solvent water and the biological group-transfer potential of phosphoric and carboxylic anhydrides, *J. Am. Chem. Soc.,* 107, 4345, 1985.

24. **Fletcher, P. D. I., Freedman, R. B., Robinson, B. H., Rees, G. D., and Schomäcker, R.,** Lipase-catalysed ester synthesis in oil-continuous microemulsions, *Biochim. Biophys. Acta,* 912, 278, 1987.

25. **Han, D., Rhee, J. S. and Lee, S. B.,** Lipase reaction in AOT-isooctane reversed micelles: effect of water on equilibria, *Biotechnol. Bioeng.,* 30, 381, 1987.

26. **Luisi, P. L., Giomini, M., Pileni, M. P., and Robinson, B. H.,** Reverse micelles as hosts for proteins and small molecules, *Biochim. Biophys. Acta,* 947, 209, 1988.

27. **Martinek, K., Levashov, A. V., Klyachko, N., Khmelnitski, Y. L., and Berezin, I. V.,** Micellar enzymology, *Eur. J. Biochem.,* 155, 453, 1986.

28. **Escamilla, E., Ayala, G., Tuena de Gómez-Puyou, M., Gómez-Puyou, A., Millán, L., and Darszon, A.,** Catalytic activity of cytochrome oxidase and cytochrome c in apolar solvents containing phospholipids and low amounts of water, *Arch. Biochem. Biophys.,* 272, 332, 1989.

29. **Darszon, A., Philip, M., Zarco, J., and Montal, M.,** Rhodopsin-phospholipid complexes in apolar solvents: formation and properties, *J. Membr. Biol.,* 43, 71, 1978.

30. **Ferreira, S. T. and Verjovski-Almeida, S.,** Fluorescence decay of sarcoplasmic reticulum ATPase. Ligand binding and hydration effects, *J. Biol. Chem.,* 264, 15392, 1989.

31. **Garza-Ramos, G., Darszon, A., Tuena de Gómez-Puyou, M., and Gómez-Puyou, A.,** Enzyme catalysis in organic solvents with low water content at high temperatures. The adenosinetriphosphatase of submitochondrial particles, *Biochemistry,* 29, 751, 1990.

32. **Eromin, A. N. and Metelitsa, D. I.,** Fluorescence of hemoproteins in reversed micelles of surfactants in octane, *Biokhimiya,* 49, 1947, 1984 (translation).

33. **Nicot, C., Vacher, M., Vincent, M., Gallay, J., and Waks, M.,** Membrane proteins in reverse micelles: myelin basic protein in a membrane-mimetic environment, *Biochemistry,* 24, 7024, 1985.

34. **Chatenay, D., Urbach, W., Nicot, C., Vacher, M., and Waks, M.,** Hydrodynamic radii of protein-free and protein-containing reverse micelles as studied by fluorescence recovery after fringe photobleaching. Perturbations induced by myelin basic protein uptake, *J. Phys. Chem.,* 91, 2198, 1987.

35. **Chatenay, D., Urbach, W., Cazabat, A. M., Vacher, M., and Waks, M.,** Proteins in membrane mimetic systems. Insertion of myelin basic protein into microemulsion droplets, *Biophys. J.,* 48, 893, 1985.

36. **Ott, A., Urbach, W., Langevin, D., Ober, R., and Waks, M.,** Light scattering study of surfactant multilayers elasticity. Role of incorporated proteins, *Europhys. Lett.,* 12, 395, 1990.

37. **Ramakrishnan, V. R., Darszon, A., and Montal, M.,** A small angle X-ray scattering study of a rhodopsin-lipid complex in hexane, *J. Biol. Chem.,* 258, 4857, 1983.

38. **Eicke, H. F., Shepherd, J. C. W., and Steinemann, A.,** Exchange of solubilized water and aqueous electrolyte solutions between micelles in apolar media, *J. Colloid Interface Sci.,* 56, 168, 1976.

39. **Tabony, J. and Drifford, M.,** Quasielastic neutron scattering measurements of monomer molecular motions in micellar aggregates, *Colloid Polym. Sci.,* 261, 938, 1983.

40. **Bru, R., Sánchez-Ferrer, A., and Garcia-Carmona, F.,** A theoretical study on the expression of enzymatic activity in reverse micelles, *Biochem. J.,* 259, 355, 1989.

41. **Verhaert, R. M. D., Hilhorst, R., Vermué, M., Schaafsma, T. J., and Veeger, C.,** Description of enzyme kinetics in reversed micelles. I. Theory, *Eur. J. Biochem.,* 187, 59, 1990.
42. **Hilhorst, R., Spruijt, R., Laane, C., and Veeger, C.,** Rules for the regulation of enzyme activity in reversed micelles as illustrated by the conversion of apolar steroids by 20β-hydroxysteroid dehydrogenase, *Eur. J. Biochem.,* 144, 459, 1984.
43. **Tyrakowska, B., Verhaert, R. M. D., Hilhorst, R., and Veeger, C.,** Enzyme kinetics in reversed micelles. III. Behaviour of 20β-hydroxysteroid dehydrogenase, *Eur. J. Biochem.,* 187, 81, 1990.
44. **Martinek, K., Berezin, I. V., Khmelnitski, Y. L., Klyachko, N. L., and Levashov, A. V.,** Enzymes entrapped into reversed micelles of surfactants in organic solvents: key trends in applied enzymology, *Biocatalysis,* 1, 9, 1987.
45. **Laane, C.,** Medium-engineering for bio-organic synthesis, *Biocatalysis,* 1, 17, 1987.
46. **Chopineau, J., Thomas, D., and Legoy, M. D.,** Dynamic interactions between enzyme activity and the microstructured environment, *Eur. J. Biochem.,* 183, 459, 1989.
47. **Menger, F. M., Donohue, J. A., and Williams, R. P.,** Catalysis in water pools, *J. Am. Chem. Soc.,* 95, 286, 1973.
48. **Wong, M., Thomas, J. K., and Nowak, T.,** Structure and state of H_2O in reversed micelles. III, *J. Am. Chem. Soc.,* 99, 4730, 1977.
49. **Tsujii, K., Sunamoto, J., and Fendler, J. H.,** Microscopic viscosity of the interior water pool in dodecylammonium propionate reversed micelles, *Bull. Chem. Soc. Jpn.,* 56, 2889, 1983.
50. **Maitra, A.,** Determination of size parameters of water-aerosol OT-oil reverse micelles from their nuclear magnetic resonance data, *J. Phys. Chem.,* 88, 5122, 1984.
51. **Kumar, C. and Balasubramanian, D.,** Spectroscopic studies on the microemulsions and lamellar phases of the system Triton X-100: hexanol: water in cyclohexane, *J. Colloid Interface Sci.,* 74, 64, 1980.
52. **Ekwall, P., Mandell, L., and Fontell, K.,** Some observations on binary and ternary aerosol OT systems, *J. Colloid Interface Sci.,* 33, 215, 1970.
53. **Walde, P., Giuliani, A. M., Boicelli, C. A., and Luisi, P. L.,** Phospholipid-based reverse micelles, *Chem. Phys. Lipids,* 53, 265, 1990.
54. **Boicelli, C. A., Conti, F., Giomini, M., and Giulliani, A. M.,** Interactions of small molecules with phospholipids in inverted micelles, *Chem. Phys. Lett.,* 89, 490, 1982.
55. **Fung, B. M. and McAdams, J. L.,** The interaction between water and the polar head in inverted phosphatydilcholin micelles, *Biochim. Biophys. Acta,* 451, 313, 1976.
56. **Garza-Ramos, G., Darszon, A., Tuena de Gómez-Puyou, M., and Gómez-Puyou, A.,** High concentrations of guanidine hydrochloride activate lactate dehydrogenase in low water media, *Biochem. Biophys. Res. Comm.,* 172, 830, 1990.
57. **Swezey, R. R. and Epel, D.,** Enzyme stimulation upon fertilization is revealed in electrically permeabilized sea urchin eggs, *Proc. Natl. Acad. Sci. U.S.A.,* 85, 812, 1988.
58. **Grandi, C., Smith, R. E., and Luisi, P. L.,** Micellar solubilization of biopolymers in organic solvents. Activity and conformation of lysozyme in isooctane reverse micelles, *J. Biol. Chem.,* 256, 837, 1981.
59. **Wolf, R. and Luisi, P. L.,** Micellar solubilization of enzymes in hydrocarbon solvents, enzymatic activity and spectroscopic properties of ribonuclease in *n*-octane, *Biochim. Biophys. Res. Comm.,* 89, 209, 1979.
60. **Barbaric, S. and Luisi, P. L.,** Micellar solubilization of biopolymers in organic solvents. V. Activity and conformation of α-chymotrypsin in isooctane-AOT reverse micelles, *J. Am. Chem. Soc.,* 103, 4239, 1981.
61. **Walde, P., Peng, Q., Fadnavis, N. W., Battistel, E., and Luisi, P. L.,** Structure and activity of trypsin in reverse micelles, *Eur. J. Biochem.,* 173, 401, 1988.
62. **Fontana, A.,** Structure and stability of thermophilic enzymes. Studies on thermolysin, *Biophys. Chem.,* 29, 181, 1988.
63. **Zaks, A. and Klibanov, A. M.,** Enzymatic catalysis in organic media at 100°C, *Science,* 224, 1249, 1984.
64. **Wheeler, C. J. and Croteau, R.,** Terpene cyclase catalysis in organic solvent/minimal water media: demonstration and optimization of (+)-α-pinene cyclase activity, *Arch. Biochem. Biophys.,* 248, 429, 1986.
65. **Asther, M. and Meunier, J. C.,** Increased thermal stability of Bacillus licheniformis α-amylase in the presence of various additives, *Enzyme Microb. Technol.,* 12, 902, 1990.
66. **Reslow, M., Adlecreutz, P., and Mattiasson, B.,** Organic solvents for bioorganic synthesis. I. Optimization of parameters for a chymotrypsin catalyzed process, *Appl. Microbiol. Biotechnol.,* 26, 1, 1987.
67. **Ayala, G., Tuena de Gómez-Puyou, M., Gómez-Puyou, A., and Darszon, A.,** Thermostability of membrane enzymes in organic solvents, *FEBS Lett.,* 203, 41, 1986.
68. **Garza-Ramos, G., Darszon, A., Tuena de Gómez-Puyou, M. and Gómez-Puyou, A.,** Catalysis and thermostability of mitochondrial F1-ATPase in toluene phospholipid-low water systems, *Biochemistry,* 28, 3177, 1989.
69. **Luisi, P. L., Häring, G., Maestro, M., and Rialdi, G.,** Proteins solubilized in organic solvents via reverse micelles: thermodynamic studies, *Termochim. Acta,* 162, 1, 1990.

70. **Saishu, T., Nojima, H., and Kagawa, Y.,** Stability of structures of the epsilon subunit and terminator of thermophilic ATPase, *Biochim. Biophys. Acta,* 867, 97, 1986.

71. **Pace, C. N.,** The stability of globular proteins, *CRC Crit. Rev. Biochem.,* 3, 1, 1975.

72. **Tanford, C.,** *The Hydrophobic Effect,* 2nd ed., John Wiley & Sons, New York, 1980.

73. **Lee, J. C. and Timasheff, S. N.,** Partial specific volumes and interactions with solvent components of proteins in guanidine hydrochloride, *Biochemistry,* 13, 257, 1974.

74. **Creighton, T. E.,** *Proteins,* W. H. Freeman, New York, 1983.

75. **Paudel, H. K. and Carlsson, G. M.,** The quaternary structure of phosphorylase kinase as influenced by low concentrations of urea. Evidence suggesting a structural role for calmodulin, *Biochem. J.,* 268, 393, 1990.

76. **Stein, M., Lazaro, J. J., and Wolosivk, R. A.,** Concerted action of cosolvents, chaotropic anions and thioredoxin on chloroplast fructose 1,6-biphosphate reactivity to iodoacetamide, *Eur. J. Biochem.,* 185, 425, 1989.

77. **Tsou, C. L.,** Location of the active site of some enzymes in limited and flexible molecular regions, *TIBS,* 11, 427, 1986.

78. **Strambini, G. B. and Gonnelli, M.,** Effects of urea and guanidine hydrochloride on the activity and dynamical structure of equine liver alcohol dehydrogenase, *Biochemistry,* 25, 2471, 1986.

79. **Scheibe, R. and Fickenscher, K.,** The dark (oxidized) form of the light-activatable NADP-malate dehydrogenase from pea chloroplasts is catalytically active in the presence of guanidine-HCl, *FEBS Lett.,* 180, 317, 1985.

80. **Scheibe, R., Rudolph, R., Reng, W., and Jaenicke, R.,** Structural and catalytic properties of oxidized and reduced chloroplast NADP-malate dehydrogenase upon denaturation and renaturation, *Eur. J. Biochem.,* 189, 581, 1990.

81. **Garza-Ramos, G., Fernández-Velasco, D. A., Ramirez, L., Darszon, A., Shoshani, L., Tuena de Gómez-Puyou, M., and Gómez-Puyou, A.,** submitted.

Chapter 6

THE OPTIMIZATION OF ENZYME CATALYSIS IN ORGANIC MEDIA

Raymond M. D. Verhaert, Riet Hilhorst, Antonie J. W. G. Visser, and Cees Veeger

TABLE OF CONTENTS

I. INTRODUCTION

Although there are some "historic" reports on biomolecules in organic media, the field has developed rapidly only in the last 15 to 20 years, when it was realized that enzymes can be catalytically active in systems that are composed for over 95% of organic solvents.[1-5] The first system of this nature that was investigated more extensively was the reversed micellar medium. Although in some cases enzymes had been shown to be active in virtually dry organic solvents,[1,2] this only attracted the attention of a wider audience after 1985 because of the work by Klibanov and his group.[6-10] This contribution will limit itself to enzymes in reversed micelles.

The existence of reversed micelles (tiny water droplets in a water-immiscible organic solvent, stabilized by a monolayer of surfactant and in some cases a cosurfactant) was first suggested by Hoar and Schulman,[11] who proposed the name "oleopathic hydro-micelles". The first reports on enzyme activity in these aggregates appeared in 1977 to 1978.[12-15] Since that time the number of enzymes that retain their activity in reversed micelles has increased steadily through the years. Furthermore, they have been used to solubilize membrane proteins and to study conformational changes of peptides and proteins upon incorporation into reversed micelles.[16-18] Other potential applications can be found in the area of protein extraction, protein stabilization, protein refolding, and in the fields of analytical chemistry and medicine.

In this chapter, a condensed review of the potential applications will be given, but most attention will be focused on the study of catalytic and structural properties of enzymes in reversed micelles. A good understanding of the factors that affect the catalytic activity is essential for a future use of these systems for the purpose of bioconversions. In some cases, biocatalyst activity was found to be considerably higher than in aqueous solution, but in many cases the substrate concentration was not saturating (e.g., see Katiyar et al.[19]). When the term "superactivity" is limited to those cases where k_{cat} was determined, only a few cases remain.[12,24] Irrespective of whether the substrate concentration was expressed as an overall concentration or as a water-phase concentration, the apparent affinity of the enzyme for its substrate was rarely comparable to that in aqueous solution. If these changes reflect real changes in the properties of the enzyme, the implications for the behavior of enzymes in microheterogeneous systems are large. *In vivo,* many enzymes are associated with membranes and the kinetic results obtained in homogeneous aqueous solution would be of limited use to explain their kinetics in the cellular environment. Structural studies have shown that proteins in reversed micelles can also interact with the surfactant layer and that this can lead to disturbances in their structure. An understanding of this interaction can yield valuable information that can be helpful in understanding kinetic anomalities and can elucidate interactions between proteins and membranes *in vivo.* Section III of this chapter will deal with this subject.

For the application of enzymes in reversed micelles, an understanding of factors influencing their kinetic constants enables engineering of their environment to optimize their performance. In Section IV, enzyme kinetics will be dealt with in detail, and several theories that have been published during the last years will be evaluated.

Finally, some conclusions will be drawn from the information presented in this chapter for the optimization of biocatalysis in reversed micelles.

II. APPLICATIONS OF REVERSED MICELLES

A. ENZYMATIC CONVERSIONS IN REVERSED MICELLES

The first application for enzymes in reversed micelles that was suggested was the conversion of apolar compounds because of the high solubility of such compounds in these media. Optimization of reversed micelles media will be discussed in Section V of this chapter. All classes of enzymes have been incorporated in reversed micelles, including oxidoreductases, that require NAD^+ or NADH as cofactor. Also, systems for cofactor regeneration have been developed.[21-24]

Product recovery from the reversed micellar system is essential for application, but very little attention has been paid to this subject. Larsson et al.[25] used the fact that a reversed micellar phase can coexist with an oil-rich phase for the extraction of apolar product and showed that the enzyme in the micellar phase can be reused. Several extraction steps are necessary using this system to remove all product. A more elegant solution to the problem of product recovery and enzyme recycling is the detergentless microemulsion used by Khmelnitsky et al.[26,27] The system consists of hexane, isopropanol, and aqueous solution. Addition of hexane creates an oil phase that contains the product, whereas the enzyme remains in the aqueous phase.

When naturally occurring surface-active compounds like lecithin are used, the presence of a small amount of surfactant in the product is of minor importance. Enzymes have been shown to retain their activity in systems composed with lecithin.[28,29] Reports on the scaling up of reversed micellar systems are scarce. Two papers have appeared on the use of hollow fiber reactors.[30,31]

B. PROTEIN STABILIZATION

When the freedom of mobility of the protein chains is restricted, the stability increases. This is part of the rationale behind cold storage. In Third World countries, vaccines are often rendered useless because of failure to keep them refrigerated at all times. It might be possible to use storage in reversed micelles as an alternative for cold storage of proteins. The predominantly organic solution prevents bacterial degradation of proteins, whereas the tight enclosure that occurs in micelles of low Wo prevents unfolding. Hydrogenase enclosed in reversed micelles had a residual activity of 68% after 24 d of storage.[32] For lipase, after 12 h of incubation at Wo = 0.65, 90% of activity in AOT (sodium *bis* [2-ethylhexyl] sulfosuccinate) reversed micelles was retained, whereas in aqueous solution 30% was left.[33] A similar stabilization was reported by Schmidli et al.[29] Recovery from the reversed micellar solution could be performed as described in Section II.C. Proteolytic enzymes can also be stored in reversed micelles. As normally only one protein molecule is present per micelle, autodegradation is prevented.

C. PROTEIN EXTRACTION

In the monophasic systems used to measure enzyme activity and stability, enzyme solutions are generally added by injection of a small volume of concentrated enzyme stock solution. It has been shown that transfer from one phase of a biphasic system to another is also possible.[34-36] The biphasic system can be composed of a reversed micellar solution as the upper layer and either a protein-containing solid or a protein-containing aqueous phase below. This field has been reviewed recently.[37,38]

Electrostatic interactions between charged groups on the protein and the surfactant head groups play the most important role in the transfer.[37] Therefore, the pH of the aqueous solution has to be such that the net charge on the protein is opposite to the charge of the surfactant. Increasing the ionic strength shields the interactions, thus causing the transfer profiles to shift to more extreme pH values.[39] Another important factor is the size of the protein to be extracted in relation to the size of the reversed micelles. Wolbert et al.[40] established a relation between the molecular mass and the difference between the pH of optimal transfer and the isoelectric point. The percentage of protein transferred increased when the charge was distributed more asymmetrically over the protein surface. Transfer can also be influenced by the composition of the micellar phase, e.g., surfactant, cosurfactant, or organic solvent. The observed effects can be explained in part by changes in the size of the reversed micelles, since the addition of a nonionic surfactant that increased the amount of water solubilized, and therewith the size of the micelles, broadened the transfer profile and increased the amount of protein transferred.[41] When the Wo was changed by varying the organic solvent used, differences in Wo were not directly related to differences in partitioning.[42] Thermodynamic aspects of protein transfer have been discussed by Fraaije et al.,[43,44] Caselli et al.,[45] and Bratko et al.[46]

Woll et al. have shown that this phase transfer technique can be used to separate a mixture of proteins and to isolate a protein from a fermentor broth.[47] The selectivity can be improved when a surface-active affinity ligand is used.[48] Leser et al. have used the micellar extraction technique to isolate proteins from a dry powder.[49]

For this process to be practically applicable, attention has to be paid to engineering aspects. Our group has shown that two-mixer settler units can be used to perform a continuous extraction with a high yield of active α-amylase.[50] The back extraction into a second aqueous phase can be replaced by a temperature jump that causes expulsion of a part of the solubilized water and of the protein. This step was followed by centrifugal separation of the excess aqueous phase from the micellar phase. The recovery was 73%, and the increase in concentration was 2000-fold with respect to the first aqueous phase.[51,52] Phillips et al.[53] recently developed a method where protein and water are desolubilized by applying pressure on the micellar solution.

D. SOLUBILIZATION OF MICROORGANISMS

Reversed micelles have been used for the solubilization of bacterial cells. In one case, this caused rupture of the cells,[54] followed by selective solubilization of some enzymes; in other cases the cells remained intact and viable.[49,55,56] This apparent contradiction is most likely due to the choice of surfactants and organic solvents. In the first case, CTAB (cetyl-trimethylammonium bromide) and hexanol were used; both compounds are known to have bactericidal effects. In the second case, the system was composed of isopropyl palmitate and TWEEN®. Isopropyl palmitate has a high logP value and is expected not to affect viability.[57,58] In another study, suspensions of bacteria, yeasts, or spores were sonicated in a system composed of soybean phospholipids and toluene.[59] It was shown that bacterial solutions were still capable of NADH oxidation. However, for yeasts and bacilli, viability was lost. The viability of spores was not affected by this treatment. Mitochondria can be solubilized and recovered from AOT/isooctane microemulsion with barely any loss in respiratory activity.[60] Conversions of apolar compounds by whole microorganisms in reversed micelles have not yet been reported.

E. PROTEIN REFOLDING

Since the advent of genetic engineering, many foreign proteins have been cloned in microorganisms. Although in many cases a properly folded active protein is obtained, in several cases the protein is stored (after translation) in a denatured form in inclusion bodies. After isolation, these proteins have to be denatured under reducing conditions and have to be refolded to obtain the functional native protein. The yield of such processes is often low due to the formation of faulty intra- and intermolecular S-S bridges. If the molecules are refolded in a very dilute protein solution, the yield improves. This led Hagen et al.[61] to the idea that enclosing a denatured protein in a reversed micelle and refolding it there would improve yields. For RNase it was shown that it can be transferred in a denatured state into reversed micelles and can be refolded via a series of contacting steps with aqueous solutions and addition of oxidant. Subsequently, the renatured RNase can be reextracted into an aqueous solution of high ionic strength. Unfortunately, for interferon, a more hydrophobic protein, this method was not successful. The authors attributed this to interactions between the protein and the surfactant molecules.[62]

F. ANALYSIS

Enzymatic determination of concentrations of apolar compounds in aqueous solution can be hampered by the low solubility. For example, in aqueous solution bilirubin is assayed in the presence of carrier-molecules like bovine serum albumin and has to dissociate from them before it can be converted. Its solubility in reversed micellar media is high and kinetic analysis of data is more straightforward.[63]

As indicated before, reactions can follow a different course in reversed micelles. This can be advantageous, as in the case of the bioluminescence assay which is 10- to 100-fold more sensitive in reversed micelles. In aqueous solution, the bioluminescence signal fades within a minute, whereas in reversed micelles the intensity remains constant until the ATP is exhausted.[64]

Another example is the coupling of the luminol assay to an enzymic H_2O_2-producing reaction. In aqueous solution, this assay is only operational at pH values above nine or in the presence of a catalyst or co-oxidant, but in reversed micelles, neither of these requirements is needed and the reaction is probably catalyzed by CTAB head groups.[65,66] Hydrogen peroxide can also be detected by coupling its formation to oxidation of dichlorofluorescin to dichlorofluorescein.[67] The reaction rate in reversed micelles is much higher than in aqueous solution, and this reaction has a high sensitivity due to the strong fluorescence of dichlorofluorescein.

G. MEDICAL APPLICATIONS

Intravenous use of proteins as drugs has a large potential. A number of problems are associated with the use of proteins, the main ones being the immunogenicity of the foreign protein and its rapid clearance from the body.[68] These problems can be circumvented by covalent modification of enzymes with polyethylene glycol. Such an enzyme is nonimmunogenic and its residence time in the blood is increased to about 20 d.[69] In aqueous solution the degree of modification is not easily regulated, but carrying out the procedure in reversed micelles allows attachment of a predetermined number of chains, as shown for chymotrypsin with stearoylchloride by Kabanov et al.[70-72]

III. STRUCTURAL STUDIES

Since reversed micelles are thermodynamically stable, optically transparent solutions, optical spectroscopy has proven to be a sensitive technique to probe the behavior of proteins in reversed micelles. The particular type of spectroscopy determines the type of information available from optical experiments. A comprehensive review of spectroscopic methods applied to reversed micellar systems has been given by Vos et al.[73] Light-absorption spectroscopy is usually not very sensitive for changes in the direct environment of the chromophoric group and has, in fact, only practical use in determining enzymatic activity in reversed micellar systems. We will therefore restrict ourselves to discussing three spectroscopic techniques that can be used to investigate protein structure, stability, and dynamics in reversed micellar systems: circular dichroism and fluorescence and triplet-state spectroscopy.

A. CIRCULAR DICHROISM

Far-UV circular dichroism (CD) of the peptide-bond chromophores yields information on the secondary structure of proteins in reversed micelles. The drawback in these measurements is the relatively strong light absorption of components of the surfactant molecules which impedes the registration of high-quality spectra in the range 190 to 240 nm. For instance, CTAB reversed micelles cannot be measured below about 215 nm because of the strong light absorption of bromide. To be able to registrate spectra, one should then use

CTAC instead of CTAB reversed micelles. The most important result of far-UV CD application to small peptides is that the secondary structure changes upon incorporation into reversed micelles.[18] These observations can be explained by the different nature of micellar water as compared to bulk water and by the proximity of the electric field arising from the charged surfactants. The effects are regulated by the hydration level in the micelles. The secondary structural content from far-UV CD is an overall property of the polypeptide chain. If attention is focused on larger polypeptides, there are changes in the far-UV CD induced after incorporation in reversed micelles for proteins like lysozyme and cytochrome *c*.[74,75] There is hardly any change for alcohol dehydrogenase in reversed micelles,[76,77] indicating that the overall protein conformation is not affected upon incorporation into reversed micelles.

In order to demonstrate the effect on secondary structure of a protein upon encapsulation, we wish to present some data on the flavoprotein lipoamide dehydrogenase from *Azotobacter vinelandii*. Figure 1A shows the far-UV CD spectra for this enzyme in aqueous buffer and in CTAC reversed micelles. As can be observed, the two spectra can be superimposed, indicating no change in secondary structure. Analysis of the spectra according to published procedures gives a helix content of 35 ± 5% in both cases. Also shown in the same figure is a spectrum of lipoamide dehydrogenase in CTAC reversed micelles containing 35% of ethanol. It is clear in this example that the helix content decreases and, apparently, the protein starts to unfold.

CD of specific aromatic amino acid residues, coenzymes, or cofactors yields information on the conformation in the immediate environment of the particular molecule. Although the technique is extremely sensitive for slight changes in the vicinity of the chromophoric group, the limited number of chromophores (in contrast to the large number of peptide chromophores in proteins) sets a limit to the sensitivity. The CD spectra of flavoproteins in the light absorption range of the flavin prosthetic group are usually very characteristic for the microenvironment of the flavin and can be considered as fingerprints for the flavoprotein having a particular biocatalytic function. Any change in microenvironment, for instance as a result of a change in water properties or by the influence of an electric field arising from the surfactant interface, would be apparent from a change in optical activity in the visible/near-UV spectral range. Again, lipoamide dehydrogenase can serve as an example. In Figure 1B, visible and near-UV circular dichroism spectra are shown for the enzyme in aqueous solution and in CTAC reversed micelles. Clearly, there is hardly any spectral change when the enzyme is encapsulated in reversed micelles, indicating that the flavin is more or less shielded by amino acid residues from the micellar surface.

B. TIME-RESOLVED FLUORESCENCE SPECTROSCOPY

Fluorescence spectroscopy is a very sensitive technique to obtain information on the direct environment of natural or artificially introduced fluorophores in proteins or peptides since it reports on processes that occur on the same time scale as the lifetime of the fluorescent state, which is typically of nanosecond duration. These processes can vary from dynamic quenching by neighboring amino acids, restricted motion of the fluorophores, and rapid interconversion between protein conformational substates to excited-state reactions like protonation or exciplex formation. These phenomena can play a role in proteins because proteins are not rigid, but are dynamic entities with structural fluctuations covering a time span between subnanoseconds and seconds. Steady-state fluorescence spectroscopy mostly permits us to obtain a global picture of the situation near the fluorescent moiety, for instance, via location and shifts of fluorescence maxima or variation in quantum yield. Detailed information on the dynamics of fluorophores can be retrieved from time-resolved fluorescence spectroscopy.

The most-used fluorescent reporter group in protein research is tryptophan. The photophysics of tryptophan, however, is complex, and time-resolved fluorescence mostly yields

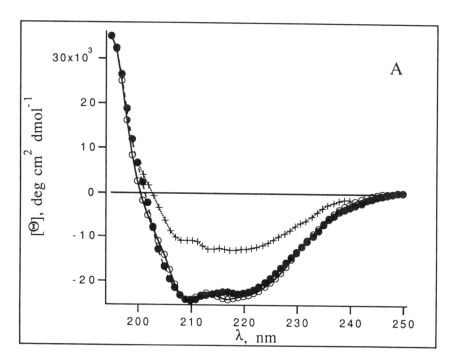

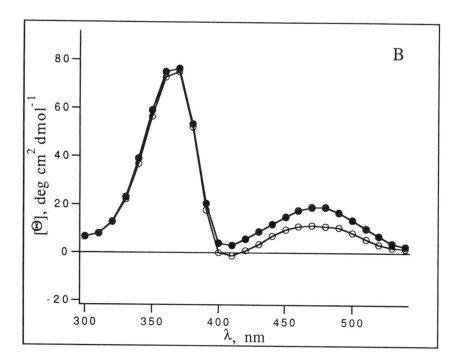

FIGURE 1.   (A) Far-UV CD spectra of lipoamide dehydrogenase (3.3 μ*M*) from *Azotobacter vinelandii* in 50 m*M* HEPES buffer (pH 7.75) in a 0.1-mm path-length cell at 20°C — ○ in buffer, ● in CTAC reversed micelles, and + addition of 35% ethanol to CTAC reversed micelles. (B) Near-UV and visible CD-spectra of lipoamide dehydrogenase (5.0 μ*M*) in a 10-mm path-length cell at 20°C — ○ in buffer and ● in CTAC reversed micelles.

a multitude of lifetime components.[78,79] Recently, the concept of fluorescence lifetime distributions has been introduced as a diagnostic tool to detect the presence of conformational substates and interconversion between them.[80] The theoretical basis behind lifetime distributions — unimodal vs. bimodal distributions, the effect of temperature on the position, and width of the distribution — has greatly enhanced the significance and interpretation of fluorescence decay data. The maximum entropy method to analyze fluorescence decay data in lifetime distributions has distinct advantages above the analysis into discrete lifetimes because the method is *a priori* model-independent.[81] The decay analysis into discrete lifetimes is often hampered by a large correlation among the parameters (amplitude and lifetime) and the interpretation is often ambiguous.

Fluorescence anisotropy decay measurements, on the other hand, are often far easier to explain since these experiments yield information on rotational motions in proteins. These motions can either arise from nanosecond overall motion of the protein or from some subnanosecond internal flexibility of the tryptophan residue. The fluorescence anisotropy decay of tryptophan residues in proteins can be adequately described by the following expression:

$$r(t) = \{\beta_1 \exp(-t/\varphi_{int}) + \beta_2\} \exp(-t/\varphi_{prot})$$

where φ_{int} and φ_{prot} are the correlation times for rapid internal motion and protein rotation, respectively, and β_1 and β_2 are the amplitudes.

The application of time-resolved fluorescence spectroscopy on proteins and peptides in reversed micelles has been rather scarce. Some peptide hormones have been studied in AOT reversed micelles, from which it became evident that the rotation of the single tryptophan residue was restricted to a degree which depends on the localization of the fluorophore either in the inner water core (mobile water) or near the surfactant-water interface (immobilized water).[18] The fluorescence properties of indole derivatives and the proteins lysozyme and azurin were investigated in AOT reversed micelles.[82] The time-resolved fluorescence patterns were described with Gaussian lifetime distributions as a function of the degree of hydration (Wo) in the reversed micelles. A higher degree of water indicated a decrease of heterogeneity of the system as was apparent from a decrease in width of the distribution. Steady-state fluorescence anisotropy of lysozyme revealed that the rotational correlation time of this protein decreases with increasing water content of the micelle.

The viscosity of the micellar interior and of the interface region of AOT reversed micelles has been probed using fluorescence anisotropy decay of rhodamine B and amphiphilic octadecylrhodamine B.[83] At low water content, the anisotropy decay could be appropriately described by a single correlation time reflecting the size of the micelle. As the water core size increases, the experimental correlation time contained a considerable contribution from the internal probe correlation time. Thus, the overall micellar correlation time can no longer be directly measured.

The anisotropy decay of octadecylrhodamine B showed a pattern similar to that of rhodamine B: an important contribution of a rapid internal motion superimposed on the slower decay due to micellar rotation. This experiment indicates that the surfactant boundary region is extremely flexible. Using more viscous glycerol instead of water effectively freezes the internal probe motion and the fluorescence anisotropy decay is then determined by the micellar rotation.[83]

Liver alcohol dehydrogenase (LADH) in reversed micelles is a good example for the use of fluorescence anisotropy decay as a tool to investigate possible changes in the internal structure of the protein when it interacts with the surfactant boundary of reversed micelles.[77] The LADH molecule has a molecular mass of 80,000 Da and consists of two identical subunits, each containing two tightly bound tryptophan residues. Indeed, a relatively small

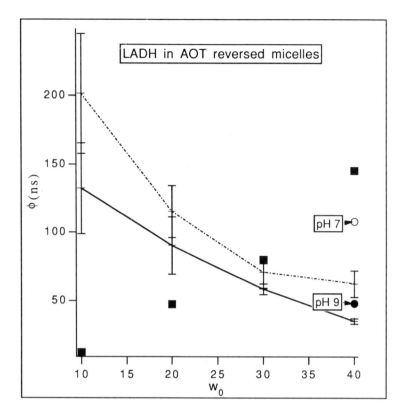

FIGURE 2. Summary of correlation times as function of Wo from fluorescence anisotropy decay of liver alcohol dehydrogenase in AOT reversed micelles at 25°C (adapted from Vos et al.[77]) — solid line, pH 8.8; broken line, pH 7.0; ■, correlation times of empty reversed micelles; ○, calculated φ_{inside} at pH 7; and ●, calculated φ_{inside} at pH 8.8.

contribution of a rapid component in AOT reversed micelles was observed which can be attributed to a rapid internal motion of tryptophan (φ_{int}) due to the interaction of the micellar interface with the enzyme.[77] The predominant slower-decaying component of the anisotropy represents the overall rotation of the protein, which we will further discuss. In order to interpret the fluorescence anisotropy decay of LADH in reversed micelles, one has to assume two independent rotational motions: first, the rotation of the micelle filled with protein as a whole (φ_{mic}) and, second, the rotation of the protein within the micelle (φ_{inside}). The observed rotational correlation time (φ_{obs}) is composed of φ_{mic} and φ_{inside} in the following way:

$$\frac{1}{\varphi_{obs}} = \frac{1}{\varphi_{mic}} + \frac{1}{\varphi_{inside}}$$

Luminescence quenching is one of the methods that can be used to measure the size of empty micelles.[75] At low Wo values (Wo = 10), relatively long correlation times are observed both for AOT and CTAB reversed micelles.[77] At this Wo, the longest protein axis is much larger than the micellar diameter and several micelles must aggregate to solubilize the enzyme. With increasing Wo, a gradual decrease of correlation times is observed. The results are summarized for the LADH-AOT reversed micellar system in Figure 2. Since the micellar correlation time of φ_{mic} increases with increasing Wo, the size of the micelles becomes so large that the protein acquires rotational freedom within the micelle, thereby increasing the contribution of φ_{inside}. In Figure 2, the results at two pH values are incorporated. The isoelectric point of LADH is at pH 8.7, which implies that at pH 7.0 the overall charge is positive, while at pH 8.8 the overall charge is zero. At Wo = 40 (which corresponds to micelles sufficiently large to accommodate the whole protein), there is a distinct effect of association of LADH in pH 7.0 buffer with the negatively charged AOT surfactant leading to an effectively longer correlation time than for LADH in pH 8.8 buffer.

C. TRIPLET-STATE KINETICS

Fluorescence methods are not appropriate for measurements of exchange rates between reversed micelles because of the widely different time scales of the two processes. A convenient spectroscopic relaxation method for measuring exchange rates is based upon the relatively long lifetime of the excited triplet state of certain residues in proteins. Strambini and co-workers have made use of tryptophan phosphorescence of alcohol dehydrogenase and alkaline phosphatase in reversed micelles.[84,85] Changes in the dynamical structure in the vicinity of the phosphorescent tryptophan residues were manifested by a change in water content from very low to possible maximum values. Our group has used cytochrome *c* in which the porphyrin iron was replaced by zinc.[86] These results will be briefly summarized. The incorporation of zinc in the heme moiety results in very efficient triplet formation enabling the photoexcited triplet state to be utilized as a probe for specific interactions between the protein and the surfactant layer of reversed micelles.

Furthermore, because the unperturbed lifetime of the triplet state (τ_o) is in the order of 10 ms under anaerobic conditions, intermicellar exchange rates (k_{ex}) can be determined in reversed micelles from quenching of the triplet state by electron transfer, bearing in mind that triplet zinc porphyrin is a powerful reducing agent. The reversed micellar system then consists of Zn-porphyrin cytochrome *c* and ionic electron acceptors which are initially hosted in separate micelles. The quencher concentration in the medium ([Q]) must be chosen in such a way that the decay of the triplet state (T) is only limited by the exchange of molecules between different reversed micelles. Thus, only a fraction of all reversed micelles contains quencher. The decay is then described by:

$$T(t) = T(o)\exp\{-(1/\tau_o + k_{ex}[Q])t\}$$

From this equation, one can see that the exchange rate constant can be determined from the exponential decay by plotting $1/\tau_{obs}$ vs. [Q]. The triplet state of Zn-porphyrin cytochrome *c* in AOT reversed micelles decays in a nonexponential fashion, illustrating that the interaction with AOT perturbs the microenvironment of the heme. Therefore, measurements of exchange rates were restricted to CTAB reversed micelles where the photoexcited triplet Zn-porphyrin in cytochrome *c* was found to decay exponentially.

The results with the two different quenchers methyl viologen and potassium ferricyanide yielded very accurate k_{ex} values for relatively small (Wo = 10) and relatively large (Wo = 40) CTAB reversed micelles (Table 1). The higher exchange rate with methyl viologen as compared to potassium ferricyanide was explained by electrostatic interaction between negatively charged ferricyanide and the positively charged surfactant head groups thereby retarding the electron transfer process as compared to that between donor Zn-porphyrin and acceptor methyl viologen. With both quenchers, a higher exchange rate was found for smaller micelles which was tentatively ascribed to stronger curvature of the surfactant interface at Wo = 10. Using this technique, it would be interesting to compare the exchange between two proteins in separate micelles. An obvious example may be the quenching of triplet Zn-porphyrin cytochrome *c* by ferricytochrome *c*. Another promising system may consist of Zn-porphyrin cytochrome *c* and some ruthenated protein as acceptor.[87] In these acceptor

TABLE 1
Rate Constants of Exchange Between
Zn-Porphyrin Cytochrome *c* and Quenchers
in CTAB-Reversed Micelles at 25°C

Quencher	Wo	$k_{ex} \times 10^{-7}$ ($M^{-1}s^{-1}$)
Potassium ferricyanide	10	3.3
	40	1.0
Methyl viologen	10	4.7
	40	1.5

proteins, ruthenium amines are attached to surface histidines. One can then compare electron transfer reactions in normal aqueous solution and in reversed micelles and study the role of the nature of surfactant and micellar size on these reactions.

IV. PERFORMING KINETICS IN REVERSED MICELLES

A. DEFINING A CONCENTRATION
1. The Concentration of Hydrophilic Compounds

As soon as enzyme kinetics in microemulsions started, the controversy on defining the concentration of compounds in these media was initiated. It was suggested that in all cases the concentration of both polar and apolar solutes had to be expressed with respect to the total volume of the medium.[88] The authors reasoned that similarities existed between a substrate solubilized in an aqueous solution and a substrate dispersed by means of a detergent. In both cases, the volume of importance is the total space available to the molecules, thus the total volume of the medium, regardless of whether these molecules are surrounded by a surfactant shell or not (Figure 3A).

Other groups reasoned that, in the case of hydrophilic substrates, the only volume of interest for the reactions occurring in the aqueous phase of the solution is that of the aqueous phase.[89] Thus, the concentration should be expressed with respect to the aqueous volume only (Figure 3B). Since the flux of substrates is considered to be fast (see below) compared to most chemical and enzymatic reactions (i.e., the turnover rate of the enzyme), it was stated that, kinetically, the system could be regarded as a two-phase system: one is a kinetically continuous aqueous phase containing the enzyme and the substrate and the other is a non-relevant, organic phase. Thus, the volume of the latter would not affect the catalytic process which occurs in the aqueous phase. This second approach is called the pseudophase approach. The substrate concentration used in the second approach can easily be converted into that of the first approach by multiplying its value by the volume fraction of water in the microemulsion. Experimental data backing both approaches were reported in the literature.[90-92]

A problem even more confusing than the one above was raised at the same time. Related to the difficulty of defining a substrate concentration is that of defining the proton concentration. However, for two reasons the problem of pH is essentially different from that of other solutes. The first reason is trivial: a proper definition of pH exists only in aqueous solution and at most a concentration of protons can be regarded. Thus, the term acidity (basicity) of the water has to be used. The second reason is that the presence of protons in the aqueous phase depends on the characteristics of that aqueous phase. Unlike other solutes, the total concentration of protons and hydroxylic ions can and will be modified in the microemulsion. Experiments indicate that in the small volume of interest the dissociation constant of water is decreased.[93,94] Thus, considering a single micelle filled with a limited amount of water molecules, it is not useful to study the possibility of observing a proton in

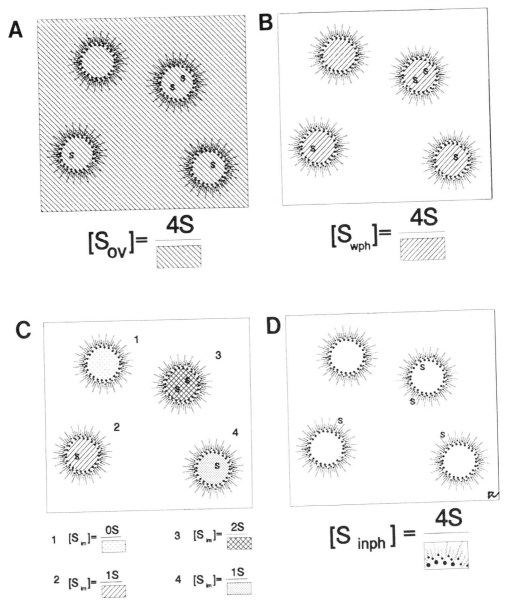

FIGURE 3. Four ways of defining a concentration in reversed micelles. (A) The overall approach, (B) the waterphase approach, (C) the intramicellar or intradroplet approach, and (D) the interphase approach. Abbreviations: ov: overall; wph: waterphase; im: intramicellar; and inph: interphase.

the droplet, but measurements need to be aimed at revealing the acid-base behavior of the water.

At the moment, the pH phenomenon gains less attention in the literature on enzyme reactions in these media. Besides some clarifying studies,[95] important reasons for this were that only small pH effects are measured and that experiments are routinely performed in microemulsions with a well-buffered aqueous phase.

As mentioned above, a dependence of the rate of the enzyme reaction in reversed micelles on both the overall and the water phase concentration was found. Changing the point of view from that of the experimentalist to that of the enzyme, it becomes clear that the concentration experienced by the enzyme is most important. Thus, for instance, the concentration

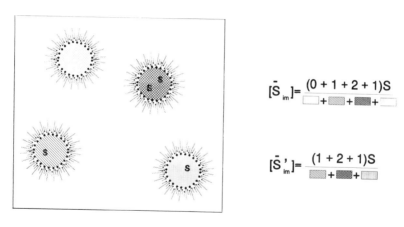

FIGURE 4. The relation between the microscopic definition and the macroscopic expression for the average intramicellar concentration in reversed micelles and for the average intramicellar concentration in filled reversed micelles.

of the substrate in the droplet in which the reaction actually occurs also has to be considered. In every droplet, a local substrate concentration can be defined (Figure 3C). This local concentration depends on the total overall concentration of substrate, on the number of droplets, on the volume of the droplet, and on the way the substrate molecules are distributed among the droplets. This local concentration can be converted to a macroscopic definition (Figure 4). This approach can be called the intradroplet or intramicellar approach.[96]

2. The Concentration of Apolar Compounds

In principle, the same three definitions of concentration can be used for apolar compounds. The second (pseudophase approach) and the third (intradroplet approach) definitions do not seem very useful at first sight since these compounds are hardly solubilized in the aqueous phase. However, the observations that enzymes which are active in reversed micelles are surrounded by water suggest that the apolar compound has to migrate through the aqueous layer around the protein.[97] At the moment, the partition coefficient is routinely used to calculate the amount of substrate available to the protein.[98] This ratio between the concentration of the solute in the organic phase and in the aqueous phase does not allow for any accumulation of substrate in the interfacial region. Indeed, the ability of the interface to host polar, apolar, and amphipolar compounds has to be taken into account also. The interfacial concentration would be defined as the molar amount of substrate per volume unit of interface (Figure 3D). This fourth way of defining a concentration in these media has not been related to the enzymology yet due to the complexity of both defining and measuring the volume of the interface and due to the difficulty of measuring the amount of substrate in the interface. Only one report is known in which the exact partitioning behavior of a series of alcohols was measured.[99] The results were used to explain that the altered substrate specificity of alcohol dehydrogenase in reversed micelles presented in earlier reports was only an apparent change. Indeed, for explaining and predicting enzyme performance in these media, all four concentrations have to be kept in mind. The importance of each depends on its conjugant rate constant (see below) and the intrinsic rate constants of the enzyme.

3. The Substrate Concentration in Enzyme-Containing Reversed Micelles

No consensus on the structure of a protein containing reversed micelle exists. Already for proteins which are only confined to the aqueous phase of micelles, a variety of possible enzyme micelle structures can be distinguished. When a protein is incorporated in the water core of the micelle, redistribution of detergent and water molecules can occur. This phenomenon is important for measuring or calculating the substrate concentration in the droplet with the enzyme. For simplicity, we will concentrate only on the effect of enzyme incorporation on the volume of the water pool in that micelle and assume that the number of detergent molecules of that micelle does not change. If the volume of the protein is negligible compared to that of the droplet, we can assume that the concentration of substrate with respect to the volume of the droplet is not changed. Also, the water volume and therefore the substrate concentration in that droplet is unaltered. If a larger protein is incorporated in a reversed micelle, three possibilities exist: (1) the micelle increases in size since both protein and water are retained in the micelle,[100] (2) the micellar size does not change, but the water is retained in the micelle by redistribution (e.g., between the surfactant head groups),[101,102] and (3) the micelle increases in size, but part of the water is redistributed over all other micelles.

Only in the last case (3) is the concentration of substrate in enzyme-filled micelles different from that in other substrate-containing micelles. Thus, except for this case, the local concentration of substrate in the vicinity of the enzyme can be measured by considering the local concentration in "normal" reversed micelles.

B. THE FLUX OF REACTANTS IN THE MEDIUM

Solute transport rates have only barely been addressed in studies of enzymes in reversed micelles. As mentioned above, the enzyme reaction was in most cases assumed to be rate limiting, and the water phase was regarded as a continuous phase (pseudophase) on the time scale of the enzyme reaction. Indeed, in such a system, the concentration of reactants is the average concentration of all droplets which equals the water phase concentration. In the case of diffusion limitation, the kinetic picture changes. Now the rate of substrate transport from the bulk to the enzyme environment has to be considered. This case is similar to a macroscopic two-phase system. In the latter, the limitation is generally the transport of the substrate from the organic phase to the aqueous enzyme-containing phase. For reversed micelles, besides transport through an interface, a second pathway of substrate flux is present. Transport can occur directly from one droplet to another. This process is accompanied by formation of transient droplet dimers.[103] Four steps can be distinguished in this pathway: migration of the two reversed micelles resulting in a collision, the actual fusion of the micelles into a dimer, the mixing of contents, and the fission of the fused droplet. This process is followed by transport of the solute in the droplet to the active site of the enzyme.

1. Diffusion of the Micelle

The diffusion rate of a reversed micelle is dependent on the viscosity of the surrounding organic solvent, the micellar size, and temperature. The average distance traveled by a reversed micelle in a microsecond can be calculated to be approximately 30 nm. The rate constant for collision in reversed micelles is in the order of 10^{10} ($M^{-1}s^{-1}$). The diffusion coefficient of a reversed micelle that is enlarged due to the incorporation of a protein will be decreased. For instance, the diffusion coefficient of a reversed micelle containing enoate reductase (r_E = 6.5 nm) in an AOT reversed micellar solution of Wo = 10 (r = 2.9 nm) is a factor of three lower compared to that of an "empty" micelle.[75,104] Similar to kinetics in aqueous media, the diffusion of the small reactant (i.e., the substrate-containing reversed micelles) and the size of the enzyme may mainly determine the collision frequency.

2. Exchange of the Contents of Reversed Micelles

The process of the merging of two reversed micelles and mixing of their contents followed by micellar division was first suggested by Eicke and co-workers.[103] The rate constants of this process have been measured by fluorescence spectroscopy and stopped-flow methods. For the extensively studied systems of AOT/water/alkane, the exchange rate constants (k_{ex})

vary from 10^6 to 10^8 (M^{-1} s^{-1}). The value of k_{ex} was found to be slightly dependent on the method used.[105] The presence of small molecules was considered to be of less importance because the rate-determining step is the fusion of the interfaces of the two droplets.

The values that are measured with micelles containing small solutes are also used to describe enzyme reactions in reversed micelles, but it is not known to which extent the exchange-rate constant is affected by the presence of an enzyme in the water pool. The increased size of the enzyme-containing micelle may also decrease the rate constant for interface merging. Unfortunately, no data are yet available on the effect of enzyme-detergent interactions in the process of fusion of the two interfaces of colliding micelles.

3. Transport Within the Droplet

The self-diffusion coefficient of water in AOT reversed micelles was measured by SANS.[106] Its value was, independent of the Wo of the system, lower than in bulk water (8.0×10^{-10} and 2.3×10^{-10} m^2s^{-1}, respectively). This observation is consistent with the theory that the water in the vicinity of the interphase is bound. It also suggests that the rate of diffusion of solutes in the water pool is decreased. Despite this, no difference in the rate of substrate binding between enzymes in reversed micelles and homogeneous solvents was measured.[107,108] Apparently, the intramicellar transport rate of the substrate to the active site of the enzyme does not affect the overall rate of substrate transport and binding under the experimental conditions that were used.

4. Transport Across the Interface

In two-phase systems, the diffusion of substrates to the interface is considered to be fast compared to the actual transport. It has been shown that the diffusion of compounds in the organic phase can be estimated to be in the order of 5×10^{-10} to 10^{-9} (m^2s^{-1}).[109-111] Thus, the rate constant of collision between a solute in the organic phase and a reversed micelle in a 0.2 M AOT solution in octane (Wo = 10) can be calculated to be 6×10^{10} ($M^{-1}s^{-1}$) and the rate constant for the reversal process becomes 8×10^8 (s^{-1}). If we, for simplicity, consider the transport through the surfactant layer as a diffusion-controlled migration in a homogeneous, highly viscous environment, a rate constant of 7×10^5 (s^{-1}) is calculated. In this approximation, an effective layer depth of 1.2 nm, the length of an AOT molecule, and a viscosity similar to that of glycerol were taken. The second-order rate constant for the transport process from the continuous oil phase to the inside wall of the microdroplet becomes approximately 10^7 ($M^{-1}s^{-1}$). Indeed, in most cases, the enzyme turnover, but not the enzyme-substrate complexation in aqueous systems, is slower than the process of interfacial transport.

C. THE RATE OF THE ENZYME REACTION

The enzymic conversion cycle of one substrate molecule to a product is schematized (Figure 5). Supply and removal of substrate and product to and from the enzyme-containing reversed micelle are represented irrespective of their mechanism by the first-order rate constants k_f^S, k_r^S, k_f^P, k_r^P. In the scheme, the concentration of most importance seems to be that of the substrate and product in the droplet in which the reaction occurs. However, concentration factors are, implicitly, also present in the rate constants for mass transfer. As can be envisaged from the scheme (Table 2 and Figures 5 and 6A), it is difficult to relate the various parameters directly with macroscopically observable parameters. This is caused by the dependence of the rate on concentrations and by rate constants which act both locally and globally. The expression for the general case (Figure 6A) was demonstrated to explain the experimental data on enoate reductase and hydroxysteroid dehydrogenase.[96,112-114] This expression can be simplified for three different limiting cases. In these simplifications, a distinction is made between the mechanism of mass transfer and the molar ratio of substrate

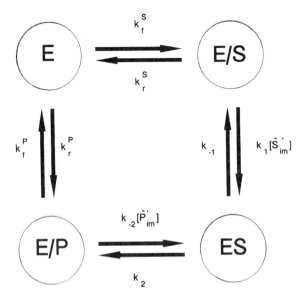

FIGURE 5. Model for the enzymatic conversion of substrate
S to product *P* in reversed micelles.

to reversed micelle. No assumptions have to be made on the rate constants of the various steps of the reaction. The three cases are depicted in Figures 6B to 6D with their concomitant rate expressions. In these equations, all concentrations are represented with respect to the total volume.

The first two limiting cases (B and C) consider the reaction kinetics for the conversion of a hydrophilic substrate. These compounds are assumed to be restricted to the water pool of the reversed micelle. In the first one, the overall concentration of substrate is lower than the droplet concentration. In the second, the overall substrate concentration exceeds that of the droplets. The third limiting case (D) represents the case of the conversion of a highly apolar compound. As already mentioned, the apparent K_m and the k_{cat} differ in these media from the values in water, but the "true" K_m is assumed to have the aqueous value. Both Meier and Luisi[107] and Jobe et al.[108] have reported binding constants that are unaffected by enclosure in reversed micelles and justify such an assumption. Further evidence can be found in the fact that $\Delta H_{cat}^{\neq}$ and ΔH_M^0 in reversed micelles are the same as in bulk water.[20] The relation between the values that are observed in reversed micelles (K_m^{app} and k_{cat}^{app}) and those in aqueous solvent (K_m and k_{cat}) are given in Figures 6B to D. Unfortunately, classification of an enzymatic conversion in only one limiting case is prone to error. Even a small partitioning of the substrate between the aqueous phase and the remainder of the solution can result in a considerable decrease of the reaction rate.[115]

The main features of the model are summarized:

1. The rate equation for the conversion of apolar and polar compounds is very different. This is caused by the mass transfer mechanism. The conversion of compounds with an intermediate polarity has to be described with the full (expanded) rate equation.
2. The two rate equations for the hydrophilic substrate show only a small difference. This difference, however, can result in a considerable change in the apparent K_m. Performing enzyme kinetics by changing the substrate concentration from the first extreme ($[S_{ov}]/[M] < 0.2$) to the second ($[S_{ov}]/[M] > 5$) yields a variable K_m.
3. The K_m value for hydrophilic substrates decreases (φ is always smaller than unity) when the mass transfer is fast. In principle, the apparent affinity of the enzyme for

TABLE 2
Relation Between the Symbols Used in the Rate Equations and the Macroscopic Parameters (Initial Conditions)

Symbol	Unit	Meaning	All cases	Limiting cases		Apolar substrate
				Polar substrate		
				$[S_{ov}]/[M] < 0.2$	$[S_{ov}]/[M] > 5$	
$\bar{S}'_{im}$	(M)	Intramicellar substrate concentration in substrate-filled reversed micelles	$\dfrac{1}{1-\exp(-\bar{n})}\dfrac{[S_{ov}]}{\varphi}$	$[M]/\varphi$	$[S_{ov}]/\varphi$	$[M]/\varphi$
k_f^S	(s⁻¹)	Substrate supply to enzyme-containing reversed micelles	$k_{ex}[M]\{1 - \exp\left(-\dfrac{\bar{n}}{2}\right)\} + k_{in}^S[S_{org}]$	$\tfrac{1}{2}k_{ex}[S_{ov}]$	$k_{ex}[M]$	$k_{in}^S[S_{ov}]$
k_r^S	(s⁻¹)	Substrate depletion from the enzyme-containing reversed micelles	$k_{ex}[M]\{2^{-\bar{m}/(1-\exp(-\bar{m}))}\exp\left(-\dfrac{\bar{n}}{2}\right)\} + k_{out}^S[\bar{S}_{im}]$	$\tfrac{1}{2}k_{ex}[M]$	0	$k_{out}^S[M]/\varphi$
k_f^P	(s⁻¹)	Product removal from enzyme-containing reversed micelles	Analogous to k_r^S	$\tfrac{1}{2}k_{ex}[M]$	$\tfrac{1}{2}k_{ex}[M]$	$k_{out}^P[M]/\varphi$
k_r^P	(s⁻¹)	Product import into the enzyme-containing reversed micelles	Analogous to k_f^S	0	0	0

Note: In the expressions, the following symbols were used: $[M]$ is the droplet concentration; $[S_{ov}]$ is the molar amount of substrate expressed with respect to the total volume; $[S_{wph,ov}]$ is the molar amount of substrate in the water phase expressed with respect to the total volume;[115] $\bar{n}$ represents the ratio of $[S_{wph,ov}]$ to $[M]$; $\bar{m}$ is the same ratio in reversed micelles containing an enzyme molecule (for simplicity, it can be assumed that $\bar{m} = \bar{n}$); φ is the volume fraction of water in the medium; k_{ex} is the second-order exchange rate constant; and $k_{in/out}^{S/P}$ represent the second-order rate constant of the transfer of the substrate S or product P from the bulk organic phase into (*in*) and out of (*out*) the enzyme-containing reversed micelle.

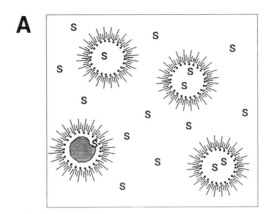

A

$$\frac{v}{[E_0]} = \frac{k_2}{1 + k_2\left(\dfrac{1}{k_f^s} + \dfrac{1}{k_f^P}\right) + \dfrac{[M]}{\varphi}\dfrac{k_{-2}}{k_f^P} + \dfrac{1}{[\overline{S}_{im}']}\left(K_m + \dfrac{k_{-1}k_{-2}[M]}{k_1k_f^P\varphi}\right)\left(1 + \dfrac{k_r^S}{k_f^S}\right)}$$

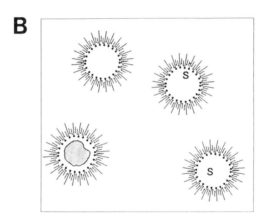

B

$$\frac{v}{[E_0]} = \frac{k_2}{1 + \dfrac{2k_2}{k_{ex}[S_{ov}]} + \dfrac{2k_{-2}}{\varphi k_{ex}} + \dfrac{\varphi K_m}{[S_{ov}]} + \dfrac{2k_{-1}k_{-2}}{k_{ex}k_1[S_{ov}]}}$$

$$k_{cat}^{app} = \frac{k_2}{1 + \dfrac{2k_{-2}}{\varphi k_{ex}}}$$

$$K_m^{app} = \frac{\varphi K_m + \dfrac{2k_2}{k_{ex}} + \dfrac{2k_{-1}k_{-2}}{k_{ex}k_1}}{1 + \dfrac{2k_{-2}}{\varphi k_{ex}}}$$

FIGURE 6. The model and the equations for enzymic conversion of polar and apolar compounds. (A) General model; (B) to (D) limiting cases. For the explanation of the symbols, see Table 2.

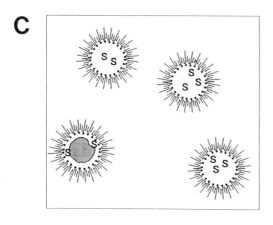

$$\frac{v}{[E_0]} = \frac{k_2}{1 + \dfrac{3k_2}{k_{ex}[M]} + \dfrac{k_{-2}}{\varphi k_{ex}} + \dfrac{\varphi K_m}{[S_{ov}]} + \dfrac{2k_{-1}k_{-2}}{k_{ex}k_1[S_{ov}]}}$$

$$k_{cat}^{app} = \frac{k_2}{1 + \dfrac{3k_2}{k_{ex}[M]} + \dfrac{k_{-2}}{\varphi k_{ex}}}$$

$$K_m^{app} = \frac{\varphi K_m + \dfrac{2k_{-1}k_{-2}}{k_{ex}k_1}}{1 + \dfrac{3k_2}{k_{ex}[M]} + \dfrac{k_{-2}}{\varphi k_{ex}}}$$

FIGURE 6 (continued).

the substrate can be increased 100-fold. In contrast to hydrophilic substrates, the affinity for apolar compounds seems to be decreased. This affinity drop can range up to 10^4 to 10^5.

4. The maximum turnover number in reversed micelles (k_{cat}^{app}) is the same as in water, only under well-defined conditions. Those conditions are determined by the droplet concentration and the rate of the backward reaction.
5. The rate constant for the backward reaction (k_{-2}) is present in the rate equation for initial conditions. Thus, in contrast to initial velocity measurements in homogeneous media, the backward reaction can always compete with the conversion in forward direction. This effect is caused by the non-negligible concentration of the product in the reversed micelle after just one cycle (Figure 5).
6. The rate equations for hydrophilic substrates can yield a bell-shaped dependence of the activity on the water content of the system (Wo). Theoretically, this observation is further elaborated for δ-chymotrypsin by Verhaert and Hilhorst.[115]

D. COMPARISON OF THIS THEORY WITH OTHER MODELS

Recently, a number of studies have been undertaken to explain the anomalous kinetics of enzymes in microemulsions.[116-120] Most of these models are put forward to explain the Wo dependence of the activity and the increased activity of some enzymes compared to aqueous solution. In this section, we will try to compare the kinetic theory of the previous

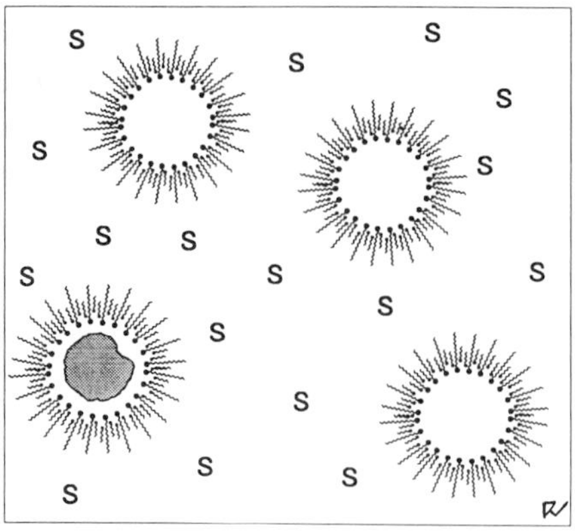

$$\frac{v}{[E_0]} = \frac{k_2}{1 + \dfrac{\varphi K_m}{[M]} + \dfrac{\varphi k_2}{k^P_{out}[M]} + \dfrac{k_{-2}}{k^P_{out}} + \dfrac{\varphi k_{-1}k_{-2}}{k^P_{out}k_1[M]} + \dfrac{K_m P^S}{[S_{ov}]} + \dfrac{k_{-1}k_{-2}P^S}{k_1 k^P_{out}[S_{ov}]} + \dfrac{k_2}{k^S_{in}[S_{ov}]}}$$

$$k^{app}_{cat} = \frac{k_2}{1 + \dfrac{\varphi K_m}{[M]} + \dfrac{\varphi k_2}{k^P_{out}[M]} + \dfrac{k_{-2}}{k^P_{out}} + \dfrac{\varphi k_{-1}k_{-2}}{k^P_{out}k_1[M]}}$$

$$K^{app}_m = \frac{P^J K_m + \dfrac{k_{-1}k_{-2}P^S}{k_2 k^P_{out}} + \dfrac{k_2}{k^S_{in}}}{1 + \dfrac{\varphi K_m}{[M]} + \dfrac{\varphi k_2}{k^P_{out}[M]} + \dfrac{k_{-2}}{k^P_{out}} + \dfrac{\varphi k_{-1}k_{-2}}{k^P_{out}k_1[M]}}$$

FIGURE 6 (continued).

section with those presented in the literature. The pseudophase approach presented by Khmelnitski and co-workers dealing with the partitioning of substrates in the reversed micelles is not considered.[99]

The model presented by Maestro et al. was first described in Ref. 116 and, more extensively, in Ref. 121. In this elegant model, an imaginary box containing reversed micelles is considered. This box contains one reversed micelle with an enzyme and several other reversed micelles with exactly one substrate molecule. Starting from this state, the processes of collision, fusion, and intramicellar enzyme reaction are described. The model is based on the laws of diffusion and the second-order rate constant for the enzyme-substrate reaction in aqueous solvents. Similar to the previous model, it assumes no changes in the enzyme itself.

First, the process of forming one fused micelle with an enzyme and a substrate molecule is subdivided in a diffusive term and a term for the fusion of the collided micelles. The latter term is estimated from literature values. The rate constant for the complete process can be calculated this way. The calculated value ranges from 10^8 to 10^{10} ($M^{-1}s^{-1}$) for a substrate micelle and an enzyme micelle with radii of 2 and 3 nm, respectively. This value is much higher than the second-order rate constant for micellar exchange in the literature.

Subsequently, the rate of the enzyme reaction is related to the diffusion of the substrate from the interface of the micelle to the surface of the enzyme. Maestro et al. find values for this rate in the order of 10^{10} $(M^{-1}s^{-1})$. This number can be obtained using an enzyme $(r = 1.5$ nm) with a surrounding water layer radius of 1.0 nm, a substrate radius of 0.3 nm, and a diffusion coefficient of 8×10^{-10} (m^2s^{-1}).[122]

The last step of their model is the actual binding of the substrate *and* its conversion. The value of the combined rate constant is that of the same reaction steps in aqueous solutions. They chose a rate constant for this process of 5×10^8 $M^{-1}s^{-1}$. In aqueous solutions, the diffusion-complexation process is typically 10^7 $M^{-1}s^{-1}$ and the actual conversion is in the order of 10^{-3} to 10^5 s^{-1}.[123]

Unfortunately the last, but important, step of the reaction cycle — product removal from the enzyme-containing micelle — is not a part of their model.

Despite the similarities of this model with the theory above (the enzyme is assumed to be unchanged in the reversed micelle and the rate of the reaction is considered from a microheterogeneic point of view), the main differences need to be stressed:

1. The model starts from a purely diffusional approach.
2. An initial reaction rate is derived.
3. The rate constant of enzyme-substrate association is increased by using a limited space of diffusion and a very high value for the second-order enzyme reaction. Combination of these factors yields the bell-shaped dependence of enzyme activity on the water content of the microemulsion. In our model, the increased rate of enzyme-substrate complexation in the micelle is solely brought about by an increased local substrate concentration.

The model described by Bru et al. is essentially different from the previous one. This model assumes different locations of the enzyme in the micellar environment.[117] A distinction is made between the turnover number of an enzyme located in the bound water layer, the surfactant tail layer, and the bulk-like water part of the reversed micelle. In the model, it is assumed that the enzyme reaction is much slower than the solubilisate exchange and that the enzyme is operating under substrate-saturating conditions. Any exchange of the enzyme between the various environments is considered to be much slower than the reaction. The volume of each environment is calculated using the equilibrium between monomeric and dimeric micelles, a shape factor for the dimeric micelle, and a penetration factor for the surfactant tails to stick into the apolar solvent. The activity profile of α-chymotrypsin measured by Martinek and co-workers can be explained by this model:[89] In bulk water, a k_2 value of 1200 (s^{-1}) is used; in the boundary layer of the micelle, the k_2 is 13,500 (s^{-1}); and in the apolar region, the k_2 value is set to zero. The constant, governing the equilibrium between micelle dimers and monomers, was set to 2.5×10^6 (M^{-1}), indicating that at Wo = 10, nearly all micelles (99%) are aggregated to dimeric micelles.

Indeed, the presence of the enzyme in the boundary layer where the enzyme has its highest turnover number results in the observation of superactivity. The model predicts strong inhibition of the reaction if the enzyme has a high turnover number in either free water or in the apolar region, but is inactive in the boundary layer. Unfortunately, experimental proof for the partitioning of enzymes between the various phases and for an environment-dependent turnover number is difficult to obtain.

The effect of substrate partitioning on enzyme activity was considered in a subsequent paper.[97] The dependence of the (double reciprocal extrapolated) maximum activity of tyrosinase on Wo (at constant φ) was found to be consistent with a partitioning ratio of the enzyme over the three phases (free water, bound water, and surfactant tails) of 1:0.45:0.21. Only the free-water fraction was active. The activity and the K_m in the free-water phase were calculated (540 s^{-1} and 13.6 mM), but no buffer values are available. It was concluded that, at constant Wo, the partitioning behavior of the enzyme was not dependent on the volume fraction of the water phase present. It has to be stressed that the enzyme activities which were measured are also in accordance with other theories.

The substrate was calculated to be mainly located in the boundary water — an observation which is in accordance with the prediction of the substrate polarity. The decreasing rate of the reaction upon increasing the volume fraction of water was explained by a dilution effect of the substrate in the neighborhood of the enzyme.

The results presented in this paper were also explained by assuming a linear dependence of the rate of the reaction on the water phase concentration of the substrate, indicating that the enzyme reaction only occurs in the aqueous phase and that other phases in the medium only act as reservoirs for the substrate. This agreement of the new experimental results with the model presented earlier by Levashov and co-workers[124] demonstrates that a biphasic model can be considered as well and these authors correctly separated the system in a purely free-water phase and a residual phase containing bound water, detergent, and organic solvent. Once again, the importance of knowledge of an exact localization of the substrates is stressed.

The model presented in 1988 by Kabanov and co-workers is based on the earlier observations that the enzyme is most active when its size is equal to that of an empty reversed micelle.[119] By assigning different turnover numbers to enzymes in tightly fitting micelles, oversized micelles, and micelles that are too small, the activity is calculated using the mean radius of the droplet and the variation in this micellar size as the variable to calculate the rate of reaction. The concentration of substrate is supposed to be saturating. In this model, it is assumed that the turnover number in tightly fitting micelles is higher than in water. The turnover in oversized micelles is smaller than or equal to that in optimal reversed micelles, and in micelles that are actually too small, it is set to zero. Thus, the activity of enzymes in reversed micelles depends on the average size of the reversed micelles and the degree of fluctuations of the micellar radius around that average value. For the simplest case of an enzyme not being active in large micelles (acid phosphatase, cytochrome c), this model predicts that no activity at all can be observed in a medium with very large micelles.

An optimum activity is centered around the Wo which yields the best-sized micelles. If the activity in larger micelles is not zero, the optimum can be shifted to slightly increased Wo values since the contribution of adverse fluctuations leading to small micelles is decreased. A hyperbolic curve will be observed when the turnover number does not change upon increasing the size of the micelle. Thus, by applying the proper micelle-dependent turnover numbers, superactivity and the Wo-activity dependence can be explained and simulated. Until now, no experimental evidence has been reported for the existence of size-dependent enzyme turnover numbers.

The model presented by Karpe and Ruckenstein is based on the intramicellar distribution of substrates.[125] The effect of the charged interface of the reversed micelle is calculated using the electrical potential function. These calculations show good agreement with intradroplet H^+ distribution measured by El Seoud[126] and Bardez.[95] In a subsequent paper,[118] this study was used to explain the Wo behavior of enzymes. It was assumed that transport and diffusion factors are fast compared to the enzyme reactions, but that the distribution within the micelle can be regarded as at equilibrium. It was demonstrated that a Wo optimum curve could be predicted by using the hydrolysis of N glutaryl-L-phenylalanine-p-nitroaniline (GPNA) by α-chymotrypsin in anionogenic AOT reversed micelles as an example.[93] Despite the original way of calculating the intramicellar substrates' availability, the model rules out a similar Wo dependence of an enzyme converting a substrate with a charge opposite to that of the detergent used to prepare the microemulsion. Moreover, no Wo dependence is expected in microemulsions with uncharged detergentia or in microemulsions with an aqueous phase of a high salinity. Yet, the conversion of charged substrates in reversed micelles with uncharged or oppositely charged detergentia have been reported to depend on the Wo.[29,121,127]

The model presented by Oldfield can best be compared with the limiting case of a low substrate concentration of our theory (Figure 6B).[96,120] In deriving the equations, Oldfield used the steady-state approximation and a time-averaged view on the substrate occupation of micelle. An important difference of the Oldfield model with the theory presented in Section IV.C is that in the latter case, the behavior of the substrate in the enzyme-containing micelle is assumed to be the same as in bulk water (k_1), while in Oldfield's theory, no physical interpretation on this rate constant is given. In our theory, a K_m with the same value as in water can be (and was) used to explain the experimental results available.[112-114]

V. THE OPTIMIZATION OF ENZYME ACTIVITY IN REVERSED MICELLES

If one wants to use enzymes in reversed micelles for bioconversion, this application can be optimized for activity and stability. When looking at activity, one has to distinguish between maximal catalytic potential, i.e., under substrate saturation, or maximal activity at a given substrate concentration. Not only can the composition, pH, ionic strength, or volume fraction of the aqueous phase be adjusted, but also the concentration and type of (co)surfactant and the organic solvent can be manipulated.

In the past, the predictions were mainly based on the partitioning of the substrate and the product between the different (micro)phases. They were demonstrated to be valid for a variety of applications.[58,128]

At the moment, the attempts to describe and to understand the enzyme kinetics in these media gain more attention than generalizing the complex variety of processes to rules of thumb. The recent attempts already show that the efficient application of enzymes in these media is bound to limits which are not only restricted to the medium itself, as proposed earlier, but also to the characteristics of the enzyme. Yet the models presented in the literature have not led to general predictions of optimal enzyme activity and stability in organic solvents. The cause for this is that many factors influencing the enzyme activity in the models, like intramicellar substrate distribution, and environment or micellar size-dependent turnover numbers, cannot be determined experimentally. Furthermore, the effect of interactions with surfactant head groups (see Section III) on catalytic activity has not been studied in detail. In the next sections, the most important components which have to be optimized will be discussed in short. Also, the prerequisites for the various components to be applied will be summarized.

A. THE ENZYME: STABILITY AND ACTIVITY

The enzyme has to be stable in reversed micelles, both under operational and nonoperational (storage) conditions. Unfortunately, stability of enzymes towards their environment is, also in aqueous solutions, difficult to predict. In aqueous environments, small amounts of detergent can already denature the enzyme, while in reversed micelles enzymes are incorporated without detrimental effects in the presence of hundred-fold-increased concentrations of detergent. In fact, for many enzymes, the stability of the enzyme under nonoperational conditions is increased compared to that in water.[22,29,33,89] This fact has been explained by an enzyme-structural point of view. Due to the decreased mobility of the environment of the protein (e.g., the water around the protein), fluctuations of enzyme domains are limited in reversed micelles. This decreased mobility is expected to decrease the probability of irreversible denaturing fluctuations in the protein backbone. This explanation is the basis for the model that an enzyme which is located in the interface of a reversed micelle[117] or in a small reversed micelle[119] is more active than an enzyme in bulk water. This explanation cannot generally be used to understand the stability of enzymes in reversed micelles. Redox enzymes are more stable in reversed micelles than in water when oxidized, but in their reduced form they are more labile in reversed micelles.[129] So the backbone motion is not the only factor determining the stability in reversed micelles. Apparently, also factors concerning (reaction-dependent) charge or polarity changes may play a role.

The stable incorporation of an enzyme in a reversed micelle does not necessarily lead to an easy detection of any activity. The three main factors responsible for this are obvious from the variety of theories and experiments reported: (1) nondenaturing but inhibiting solvent-enzyme interactions occur, (2) the reaction is limited by the amount of substrate available, or (3) the reaction is inhibited by excess substrate or product. Substrate limitation easily occurs when the substrate mainly partitions into the organic phase or the interphase (see below). Product inhibition occurs if the enzyme has a relatively high rate constant for the formation of enzyme-substrate complex starting from an enzyme and a product molecule, i.e., if the backward reaction is faster than the product-removal reaction. Product inhibition in reversed micelles is not, as in aqueous solutions, occurring gradually when the concentration of product in the medium increases, but starts immediately after the formation of one product molecule in the reversed micelle. Thus, enzymes with a relatively high affinity for the product and enzymes with an insufficient substrate affinity compared to the partitioning behavior of the substrate molecule cannot conveniently be applied in reversed micelles.

B. SUBSTRATES AND PRODUCTS

In most applications, the conversion of apolar substrates is stressed. This is because a high substrate concentration can be used and because high conversion yields can be reached. Until now, the results in this field are limited. Conversion yields of 100% have been reported,[23,27] but, to our knowledge, none of these processes are used on a commercial scale. Both the product-recovery problem and the substrate-partitioning behavior can be held responsible for this: despite the decreased diffusion limitation, this partitioning causes the rate of the reaction to be very low since only a small amount of substrate is located in the aqueous phase and readily available to the protein. Indeed, the reaction rate will already be increased if the substrate is preferentially located in the interphase where it is more accessible to the protein. A drawback of this interfacial location is that high substrate concentrations can affect the properties of the micelle and therewith enzyme performance. Obviously this approach, based on the availability of the substrate, forms the basis for the practical optimalization rules by Hilhorst, Laane, and co-workers.[58,128] Another general feature of the theories on partitioning and enzyme activity is that reversed micelles are most fit as a medium for the production of compounds which are less polar than the original compounds. In this way, the equilibrium position is shifted due to the selectively increased extraction of the product into the organic phase. This phenomenon has been discussed at length in the literature.[89]

With respect to the reaction of polar substrates, several methods for the optimization can be mentioned. From the theory of Karpe and Ruckenstein,[125] it can be deduced that conversion of a charged substrate can best be carried out in a microemulsion with a detergent which has the same charge. Other theories suggest that the concentration that is necessary for an optimal conversion of a polar substrate in aqueous medium can be reduced in reversed micelles by a factor inversely proportional to the volume fraction of water present in that medium.

The amount of substrate that can be added maximally to the system is not easily determined. Substrate inhibition seems to occur in reversed micelles at concentrations which are much lower than in aqueous media. The observation that this occurred for a charged polar substrate when its concentration exceeded that of the droplets demonstrates that care has to be taken about the substrate-droplet concentration ratio when an enzyme reaction is optimized.[113]

Substrates that contain both polar sites and apolar sites are most abundant. In the literature, most of such molecules (e.g., GNPA) are considered to be located exclusively in the waterpool of the reversed micelle. A view on all theories and suggestions presented on partitioning and enzyme kinetics teaches that, already a very low degree of partitioning leads to a serious fall of the enzyme activity because the enzyme is not saturated with substrate anymore. This is caused by the large volume of the organic- and interphase compared to the volume of the aqueous phase.

Thus, in many cases that are reported in the literature, the substrate available to the enzyme in the aqueous phase is not the total amount of solute added. Already a low affinity of the substrate for the interphase or the organic phase can induce a dramatic change in the dependence of the rate of the reaction on the (overall) substrate concentration. As an example of this effect, a partition coefficient of 0.2 (= $[S_{org}]/[S_{wph}]$) was calculated to increase the apparent K_m by a factor of four.[115]

A last important substrate-dependent parameter for the optimization of enzyme activity is related to this one. The concentration of substrates of intermediate polarity in the aqueous phase is also dependent on the volumes of the organic phase and the interphase. By adding detergent to a reversed micellar solution (e.g., to study the Wo dependence), any substrate with interface affinity can be extracted out of the aqueous phase into the interphase. This results in a lower amount of substrate directly available to the enzyme although the volume fraction of water is barely changed.

C. OPTIMIZING THE REVERSED MICELLAR MEDIUM

Implicitly, some parameters that can be optimized to use the reversed micellar solution for an efficient quantitative conversion of polar and apolar substrates have been mentioned. The most important ones are the rate constants, which are related to mass transfer in the system, and the ability of the solution to keep the enzyme in the active state and its role in the substrate location.

One possibility of increasing the mass transfer in reversed micelles is the addition of cosurfactants. This method acts in two directions. The first is that substrates with a polarity intermediate between the organic phase and the aqueous phase are more readily extracted into the interphase, thereby minimizing the diffusive distance between preferential location and the active site of the enzyme. The second is that the addition of cosurfactants also enhances the intermicellar transfer of polar substrates. This process is induced by an increased flexibility of the interface of the reversed micelles[130] due to the addition of cosurfactants.

A second way of increasing the mass transfer seems to be an increase of the number of droplets. In our theory, typically a minimum droplet concentration of 10 μM is required for the conversion of apolar substrates. In the conversion of polar substrates, the droplet concentration only becomes limiting if the turnover number of the enzyme exceeds the rate of substrate supply ($k_{ex}[M]$).

A general method for increasing a reaction rate is increasing the temperature. In this case, all enzymic and micellar reactions are increased.

Based on the results of storage stability measurements (see above), one can expect that proteins are also more stable in reversed micelles at higher temperatures, but three enzymes were shown to denature at lower temperatures in reversed micelles under acidic conditions.[131] Performing enzyme reactions at slightly elevated temperatures may be promising, but systematic studies of temperature on the whole process of enzyme action in reversed micelles are lacking.

Optimization of the reversed micellar media with respect to their enzyme stability is even more difficult. To date, most studies are performed on a few reversed micellar media and common parameters (pH, Wo). The composition of a medium optimal for stability has to be tested experimentally since this requirement conflicts with those for an optimal enzyme activity:[22,132] for stability, the environment has to be rather rigid,[98] but still a certain amount of water has to be available to the protein.[131]

ACKNOWLEDGMENTS

This work was supported by the Netherlands Technology Foundation (STW), the Netherlands Foundation for Chemical Research (SON), the Netherlands Organisation for Scientific Research (NWO), and by a fellowship of the Royal Netherlands Academy of Arts and Sciences (R. V.). We thank Ms. Sarah Fulton and Ms. Laura Ausma for typing and editing the manuscript.

REFERENCES

1. **Dastoli, F. R. and Price, S.,** Catalysis by xanthine oxidase suspended in organic media, *Arch. Biochem. Biophys.,* 118, 103, 1967.
2. **Dastoli, F. R. and Price, S.,** Further studies on xanthine oxidase in nonpolar media, *Arch. Biochem. Biophys.,* 122, 289, 1967.
3. **Hanahan, D. H.,** The enzymatic degradation of phosphatidyl choline in diethyl ether, *J. Biol. Chem.,* 155, 199, 1953.
4. **Poon, P. H. and Wells, M. A.,** Physical studies of egg phosphatidylcholine in diethyl ether-water solutions, *Biochemistry,* 13, 4928, 1974.
5. **Gitler, C. and Montal, M.,** Formation of decane-soluble proteolipids: influence of monovalent and divalent cations, *FEBS Lett.,* 28, 329, 1972.
6. **Klibanov, A. M.,** Enzymes work in organic solvents, *Chemtech,* 16, 354, 1986.
7. **Grunwald, J., Wirz, B., Scollar, M. P., and Klibanov, A. M.,** Asymmetric oxidoreductions catalyzed by alcohol dehydrogenase in organic solvents, *J. Am. Chem. Soc.,* 108, 6732, 1986.
8. **Zaks, A. and Klibanov, A. M.,** Enzyme-catalyzed processes in organic solvents, *Proc. Natl. Acad. Sci. U.S.A.,* 82, 3192, 1985.
9. **Zaks, A. and Klibanov, A. M.,** Enzymatic catalysis in organic media at 100°C, *Science,* 124, 1249, 1984.
10. **Klibanov, A. M.,** Enzymic catalysis in anhydrous organic solvents, *Trends Biochem. Sci.,* 14, 141, 1989.
11. **Hoar, T. P. and Schulman, J. H.,** Transparent water-in-oil dispersions: the oleopathic hydro-micelle, *Nature,* 152, 102, 1943.
12. **Martinek, K., Levashov, A. V., Khmelnitsky, Yu. L., Klyachko, N. L., and Berezin, I. V.,** Colloidal solution of water in organic solvents: a microheterogeneous medium for enzymatic reactions, *Science,* 218, 889, 1982.
13. **Martinek, K., Levashov, A. V., Klyachko, N. L., and Berezin, I. V.,** Catalysis by water-soluble enzymes in organic solvents. Stabilization of enzymes against denaturation (inactivation) through their inclusion in reversed micelles, *Dokl. Akad. Nauk SSSR,* 236, 951, 1977.
14. **Bonner, F. J., Wolf, R., and Luisi, P. L.,** Micellar solubilization of biopolymers in hydrocarbon solvents. I. A structural model for protein-containing reverse micelles, *J. Solid-Phase Biochem.,* 5, 255, 1980.
15. **Han, D. and Rhee, J. S.,** Batchwise hydrolysis of olive oil by lipase in AOT-isooctane reverse micelles, *Biotechnol. Lett.,* 7, 651, 1985.
16. **Chatenay, D., Urbach, W., Cazabat, A. M., Vacher, M., and Waks, M.,** Proteins in membrane mimetic systems. Insertion of myelin basic protein into microemulsion droplets, *Biophys. J.,* 48, 893, 1985.
17. **Nicot, C., Vacher, M., Vincent, M., Gallay, J., and Waks, M.,** Membrane proteins in reverse micelles: myelin basic protein in a membrane-mimetic environment, *Biochemistry,* 24, 7024, 1985.
18. **Gallay, J., Vincent, M., Nicot, C., and Waks, M.,** Conformational aspects and rotational dynamics of synthetic adrenocorticotropin-(1-24) and glucagon in reverse micelles, *Biochemistry,* 26, 5738, 1987.
19. **Katiyar, S. S., Kumar, A., and Kumar, A.,** The phenomenon of super activity in dihydrofolate reductase entrapped inside reversed micelles in apolar solvents, *Biochem. Int.,* 19, 547, 1989.
20. **Fletcher, P. D. I., Freedman, R. B., Mead, J., Oldfield, C., and Robinson, B. H.,** Reactivity of α-chymotrypsin in water-in-oil microemulsions, *Colloids Surf.,* 10, 193, 1984.
21. **Hilhorst, R., Laane, C., and Veeger, C.,** Enzymatic conversion of apolar compounds in organic media using an NADH-regenerating system and dihydrogen as reductant, *FEBS Lett.,* 159, 225, 1983.
22. **Verhaert, R. M. D., Schaafsma, T. J., Laane, C., Hilhorst, R., and Veeger, C.,** Optimization of the photo-enzymatic reduction of the carbon-carbon double bond of α-β unsaturated carboxylates in reversed micelles, *Photochem. Photobiol.,* 49, 209, 1989.
23. **Laane, C. and Verhaert, R. M. D.,** Photochemical, electrochemical, and hydrogen-driven enzymatic reductions in reversed micelles, *Isr. J. Chem.,* 28, 17, 1987/88.

24. **Larsson, K. M., Adlercreutz, P., and Mattiasson, B.,** Activity and stability of horse-liver alcohol dehydrogenase in sodium dioctylsulfosuccinate/cyclohexane reverse micelles, *Eur. J. Biochem.*, 166, 157, 1987.

25. **Larsson, K. M., Olsson, U., Adlercreutz, P., and Mattiasson, B.,** Enzymatic catalysis in microemulsions. Enzyme reuse and product recovery, *Biotechnol. Bioeng.*, 35, 135, 1990.

26. **Khmelnitsky, Yu. L., Zharinova, I. N., Berezin, I. V., Levashov, A. V., and Martinek, K.,** Detergentless microemulsions: a new microheterogeneous medium for enzymatic reactions, *Ann. N.Y. Acad. Sci.*, 501, 161, 1987.

27. **Khmelnitsky, Y. L., Hilhorst, R., and Veeger, C.,** Detergentless microemulsions as media for enzymatic reactions. Cholesterol oxidation catalyzed by cholesterol oxidase, *Eur. J. Biochem.*, 176, 265, 1988.

28. **Peng, Q. and Luisi, P. L.,** The behavior of proteases in lecithin reverse micelles, *Eur. J. Biochem.*, 188, 471, 1990.

29. **Schmidli, P. K. and Luisi, P. L.,** Lipase-catalyzed reactions in reverse micelles formed by soybean lecithin, *Biocatalysis*, 3, 367, 1990.

30. **Lüthi, P. and Luisi, P. L.,** Enzymatic synthesis of hydrocarbon-soluble peptides with reverse micelles, *J. Am. Chem. Soc.*, 106, 7285, 1984.

31. **Doddema, H. J., van der Lugt, J. P., Lambers, A., Liou, J. K., Grande, H. J., and Laane, C.,** Enzymatic oxidation of steroids in organic solvent using a STR-Plug flow reactor and continuous product separation, *Ann. N.Y. Acad. Sci.*, 501, 178, 1987.

32. **Hilhorst, R., Laane, C., and Veeger, C.,** Photosensitized production of hydrogen by hydrogenase in reversed micelles, *Proc. Natl. Acad. Sci. U.S.A.*, 79, 3927, 1982.

33. **Chang, P. S. and Rhee, J. S.,** Characteristics of lipase-catalyzed glycerolysis of triglyceride in AOT-isooctane reversed micelles, *Biocatalysis*, 3, 343, 1990.

34. **Leser, M. E., Wei, G., Luisi, P. L., and Maestro, M.,** Application of reverse micelles for the extraction of proteins, *Biochem. Biophys. Res. Commun.*, 135, 629, 1986.

35. **Luisi, P. L., Henniger, F., Joppich, M., Dossena, A., and Casnatti, G.,** Solubilization and spectroscopic properties of α-chymotrypsin in cyclohexane, *Biochem. Biophys. Res. Commun.*, 74, 1384, 1977.

36. **Luisi, P. L., Bonner, F. J., Pellegrini, A., Wiget, P., and Wolf, R.,** Micellar solubilization of proteins in aprotic solvents and their spectroscopic characterisation, *Helv. Chim. Acta*, 62, 740, 1979.

37. **Dekker, M., Hilhorst, R., and Laane, C.,** Isolating enzymes by reversed micelles, *Anal. Biochem.*, 178, 217, 1989.

38. **Jolivalt, C., Minier, M., and Renon, H.,** Separation of proteins by using reversed micelles, *ACS Symp. Ser.*, 419, 87, 1990.

39. **Dekker, M., van't Riet, K., Baltussen, J. W. A., Bijsterbosch, B. H., Hilhorst, R., and Laane, C.,** Reversed micellar extraction of enzymes: investigations on the distribution behaviour and extraction efficiency of α-amylase, in *Proc. 4th Eur. Congr. Biotechnol.*, Vol. II, Elsevier, Amsterdam, 1987, 507.

40. **Wolbert, R. B. G., Hilhorst, R., Voskuilen, G., Nachtegaal, H., Dekker, M., and van't Riet, K.,** Protein transfer from an aqueous phase into reversed micelles. The effect of protein size and charge distribution, *Eur. J. Biochem.*, 184, 627, 1989.

41. **Dekker, M., Baltussen, J. W. A., van't Riet, K., Bijsterbosch, B. H., Laane, C., and Hilhorst, R.,** Reversed micellar extraction of enzymes; effect of nonionic surfactants on the distribution and extraction efficiency of α-amylase, in *Biocatalysis in Organic Media* (Studies in Organic Chemistry, no. 29), Laane, C., Tramper, J., and Lilly, M. D., Eds., Elsevier, Amsterdam, 1987, 29.

42. **Jolivalt, C., Minier, M., and Renon, H.,** Extraction of α-chymotrypsin using reversed micelles, *J. Colloid Interface Sci.*, 135, 85, 1990.

43. **Fraaije, J. G. E. M.,** Interfacial Thermodynamics and Electrochemistry of Protein Partitioning in Two-Phase Systems, Ph.D. thesis, Agricultural University, Wageningen, The Netherlands, 1987.

44. **Fraaije, J. G. E. M., Rijnierse, E. J., Hilhorst, R., and Lyklema, J.,** Protein partitioning and ion copartitioning in liquid two-phase systems, *Colloid Polym. Sci.*, 268, 855, 1990.

45. **Caselli, M., Luisi, P. L., Maestro, M., and Roselli, R.,** Thermodynamics of the uptake of proteins by reverse micelles: a first approximation model, *J. Phys. Chem.*, 92, 3899, 1988.

46. **Bratko, D., Luzar, A., and Chen, S. H.,** Electrostatic model for protein/reverse micelle complexation, *J. Chem. Phys.*, 89, 545, 1988.

47. **Woll, J. M., Dillon, A. S., Rahaman, R. S., and Hatton, T. A.,** Protein separations using reversed micelles, in *Protein Purification: Micro to Macro*, Burgess, R., Ed., Alan R. Liss, New York, 1987, 117.

48. **Woll, J. M., Hatton, T. A., and Yarmush, M. L.,** Bioaffinity separations using reversed micellar extraction, *Biotechnol. Prog.*, 5, 57, 1989.

49. **Leser, M., Wei, G., Lüthi, P., Haering, G., Hochkoeppler, A., Blöchliger, E., and Luisi, P. L.,** Applications of enzyme-containing reverse micelles, *J. Chim. Phys.*, 84, 1113, 1987.

50. **Dekker, M., van't Riet, K., Weijers, S. R., Baltussen, J. W. A., Laane, C., and Bijsterbosch, B. H.,** Enzyme recovery by liquid/liquid extraction using reversed micelles, *Chem. Eng. J.*, 33, B27, 1986.

51. **Dekker, M.,** Enzyme Recovery Using Reversed Micelles, Ph.D. thesis, Agricultural University, Wageningen, The Netherlands, 1990.

52. **Dekker, M., van't Riet, K., van der Pol, J. J., Baltussen, J. W. A., Hilhorst, R., and Bijsterbosch, B. H.,** Temperature effect on the reversed micellar extraction of enzymes, *Chem. Eng. J.,* 46, B69, 1991.

53. **Phillips, J. B., Nguyen, H., and John, V. T.,** Protein recovery from reversed micellar solutions through contact with a pressurized gas phase, *Biotechnol. Prog.,* 7, 43, 1991.

54. **Giovenco, S., Verheggen, F., and Laane, C.,** Purification of intracellular enzymes from whole bacterial cells using reversed micelles, *Enz. Microb. Technol.,* 9, 470, 1987.

55. **Häring, G., Luisi, P. L., and Meussdoerffer, F.,** Solubilization of bacterial cells in organic solvents via reverse micelles, *Biochem. Biophys. Res. Commun.,* 127, 911, 1985.

56. **Pfamatter, A., Guadalupe, A. A., and Luisi, P. L.,** Solubilization and activity of yeast cells in water-in-oil microemulsion, *Biochem. Biophys. Res. Commun.,* 161, 1244, 1989.

57. **Laane, C., Boeren, S., and Vos, K.,** On optimizing organic solvents in multi-liquid-phase biocatalysis, *Trends Biotechnol.,* 3, 251, 1985.

58. **Laane, C., Boeren, S., Vos, K., and Veeger, C.,** Rules for optimization of biocatalysis in organic solvents, *Biotechnol. Bioeng.,* 30, 81, 1987.

59. **Darszon, A., Escamilla, E., Gómez-Puyou, A., and Tuena de Gómez-Puyou, M.,** Transfer of spores, bacteria and yeast into toluene containing phospholipids and low amounts of water: preservation of the bacterial respiratory chain, *Biochem. Biophys. Res. Commun.,* 151, 1074, 1988.

60. **Hochkoeppler, A. and Luisi, P. L.,** Solubilization of soybean mitochondria in AOT/isooctane water-in-oil microemulsions, *Biotechnol. Bioeng.,* 33, 1477, 1989.

61. **Hagen, A. J., Hatton, T. A., and Wang, D. I. C.,** Protein refolding in reversed micelles, *Biotechnol. Bioeng.,* 35, 955, 1990.

62. **Hagen, A. J., Hatton, T. A., and Wang, D. I. C.,** Protein refolding in reversed micelles: interactions of the protein with micelle components, *Biotechnol. Bioeng.,* 35, 966, 1990.

63. **Oldfield, C. and Freedman, R. B.,** Kinetics of bilirubin oxidation catalysed by bilirubin oxidase in a water-in-oil microemulsion system, *Eur. J. Biochem.,* 183, 347, 1989.

64. **Klyachko, N. L., Rubtsova, M. Yu., Levashov, A. V., Gavrilova, E. M., Egorov, A. M., Martinek, K., and Berezin, I. V.,** Bioluminescent analysis in a biomembrane-like medium consisting of surfactant, hydrocarbon solvent, and water: ATP determination using firefly luciferase, *Ann. N.Y. Acad. Sci.,* 501, 267, 1987.

65. **Visser, A. J. W. G. and Santema, J. S.,** A luminol-mediated assay for oxidase reactions in reversed micellar systems, in *Analytical Applications of Bioluminescence and Chemiluminescence,* Krika, L. J., Torpe, G. H. G., and Whitebread, T. P., Eds., Academic Press, London, 1984, 559.

66. **Hoshino, H. and Hinze, W. L.,** Exploitation of reversed micelles as a medium in analytical chemiluminescence measurements with application to the determination of hydrogen peroxide using luminol, *Anal. Chem.,* 59, 487, 1987.

67. **Sánchez Ferrer, A., Santema, J. S., Hilhorst, R., and Visser, Q. J. W. G.,** Fluorescence detection of enzymatically formed hydrogen peroxide in aqueous solution and in reversed micelles, *Anal. Biochem.,* 187, 129, 1990.

68. **Abuchowski, A., McCoy, J. R., Palczuk, N. C., Van Es, T., and Davis, F. F.,** Effect of covalent attachment of polyethylene glycol on immunogenicity and circulating life of bovine liver catalase, *J. Biol. Chem.,* 252, 3582, 1977.

69. **Pool, R.,** ''Hairy enzymes'' stay in the blood, *Science,* 248, 305, 1990.

70. **Kabanov, A. V., Levashov, A. V., and Martinek, K.,** The imparting of membrane-active properties to water-soluble enzymes by their artificial hydrophobization — a new approach to the enzymatic reactions in SAA-water-organic solvent systems, *Vestn. Mosk. Univ. Khim.,* 41, 591, 1986.

71. **Kabanov, A. V., Levashov, A. V., and Martinek, K.,** Transformation of water-soluble enzymes into membrane active form by chemical modification, *Ann. N.Y. Acad. Sci.,* 501, 63, 1987.

72. **Kabanov, A. V., Levashov, A. V., Alakhov, V. Y., Kravtsova, T. N., and Martinek, K.,** Hydrophobized proteins penetrating lipid membranes, *Collect. Czech. Chem. Commun.,* 54, 835, 1989.

73. **Vos, K., Laane, C., and Visser, A. J. W. G.,** Spectroscopy of reversed micelles, *Photochem. Photobiol.,* 45, 863, 1987.

74. **Steinmann, B., Jackle, H., and Luisi, P. L.,** A comparative study of lysozyme conformation in various reverse micellar systems, *Biopolymers,* 25, 1133, 1987.

75. **Vos, K., Laane, C., Weijers, S. R., van Hoek, A., Veeger, C., and Visser, A. J. W. G.,** Time-resolved fluorescence and circular dichroism of porphyrin cytochrome c and Zn-porphyrin cytochrome c incorporated in reversed micelles, *Eur. J. Biochem.,* 169, 259, 1987.

76. **Meier, P. and Luisi, P. L.,** Micellar solubilization of biopolymers in hydrocarbon solvents. II. The case of horse liver alcohol dehydrogenase, *J. Solid-Phase Biochem.,* 5, 269, 1980.

77. **Vos, K., Laane, C., van Hoek, A., Veeger, C., and Visser, A. J. W. G.,** Spectroscopic properties of horse liver alcohol dehydrogenase in reversed micellar solutions, *Eur. J. Biochem.,* 169, 275, 1987.

78. **Creed, D.,** The photophysics and photochemistry of the near-UV absorbing amino acids. I. Tryptophan and its simple derivatives, *Photochem. Photobiol.,* 39, 537, 1984.

79. **Beechem, J. M. and Brand, L.,** Time-resolved fluorescence of proteins, *Annu. Rev. Biochem.,* 54, 43, 1985.

80. **Alcala, J. R., Gratton, E., and Prendergast, F. R.,** Fluorescence lifetime distribution in proteins, *Biophys. J.,* 51, 597, 1987.

81. **Livesey, A. K. and Bronchon, J. C.,** Analyzing the distribution of decay constants in pulse-fluorimetry using the maximum entropy method, *Biophys. J.,* 52, 693, 1987.

82. **Ferreira, S. and Gratton, E.,** Hydration and protein substates: fluorescence of proteins in reverse micelles, *J. Mol. Liq.,* 45, 253, 1990.

83. **Visser, A. J. W. G., Vos, K., van Hoek, A., and Santema, J. S.,** Time-resolved fluorescence depolarization of rhodamine B and octadecylrhodamine B in Triton X-100 micelles and Aerosol OT reversed micelles, *J. Phys. Chem.,* 92, 759, 1988.

84. **Strambini, G. B. and Gonnelli, M.,** Protein dynamical structure of tryptophan phosphorescence and enzymatic activity in reverse micelles. I. Liver alcohol dehydrogenase, *J. Phys. Chem.,* 92, 2850, 1988.

85. **Gonnelli, M. and Strambini, G. B.,** Protein dynamical structure by tryptophan phosphorescence and enzymatic activity in reverse micelles. II. Alkaline phosphatase, *J. Phys. Chem.,* 92, 2854, 1988.

86. **Vos, K., Lavalette, D., and Visser, A. J. W. G.,** Triplet-state kinetics of Zn-porphyrin cytochrome c in micellar media. Measurement of intermicellar exchange rates, *Eur. J. Biochem.,* 169, 269, 1987.

87. **Gray, H. B. and Malström, B. G.,** Long-range electron transfer in multisite metalloproteins, *Biochemistry,* 28, 7499, 1989.

88. **Laane, C., Hilhorst, R., and Veeger, C.,** Design of reversed micellar media for the enzymatic synthesis of apolar compounds, *Methods Enzymol.,* 136, 216, 1987.

89. **Martinek, K., Levashov, A. V., Klyachko, N., Khmelnitski, Y. L., and Berezin, I. V.,** Micellar enzymology, *Eur. J. Biochem.,* 155, 453, 1986.

90. **Fletcher, P. D. I., Rees, G. D., Robinson, B. H., and Freedman, R. B.,** Kinetic properties of α-chymotrypsin in water-in-oil microemulsions: studies with a variety of substrates and microemulsion systems, *Biochim. Biophys. Acta,* 832, 204, 1985.

91. **Barbaric, S. and Luisi, P. L.,** Micellar solubilization of biopolymers in organic solvents. Activity and conformation of α-chymotrypsin in isooctane AOT reverse micelles, *J. Am. Chem. Soc.,* 103, 4239, 1981.

92. **Fletcher, P. D. I. and Parrott, D.,** The partitioning of proteins between water-in-oil microemulsions and conjugate aqueous phases, *J. Chem. Soc., Faraday Trans. I,* 84, 1131, 1988.

93. **Seoud, O. A. and Chinelatto, A. M.,** Acid-base indicator equilibria in Aerosol OT reversed micelles in heptane. The use of buffers, *J. Colloid Interface Sci.,* 95, 163, 1983.

94. **Bardez, E., Monnier, E., and Valeur, B.,** Absorption and fluorescence probing of the interface of Aerosol OT reversed micelles and microemulsions, *J. Colloid Interface Sci.,* 112, 200, 1986.

95. **Valeur, B. and Bardez, E.,** Proton transfer in reverse micelles and characterization of the acidity in the water pools, in *Structure and Reactivity in Reverse Micelles,* Pileni, M. P., Ed., Elsevier/North-Holland, Amsterdam, 1989, 103.

96. **Verhaert, R. M. D.,** Photoinduced Charge Separation and Enzyme Reactions in Reversed Micelles, Ph.D. thesis, Agricultural University, Wageningen, The Netherlands, 1989.

97. **Bru, R., Sánchez-Ferrer, A., and García-Carmona, F.,** The effect of substrate partitioning on the kinetics of enzymes acting in reverse micelles, *Biochem. J.,* 268, 679, 1990.

98. **Martinek, K., Klyachko, N. L., Kabanov, A. V., Khmelnitski, Yu. L., and Levashov, A. V.,** Micellar enzymology: its relation to membranology, *Biochim. Biophys. Acta,* 981, 161, 1989.

99. **Khmel'nitskii, Yu. L., Neverova, I. N., Polyakov, V. I., Grinberg, V. Ya., Levashov, A. V., and Martinek, K.,** Kinetic theory of enzymic reactions in reversed micellar systems. Application of the pseudophase approach for partitioning substrates, *Eur. J. Biochem.,* 190, 155, 1990.

100. **Luisi, P. L.,** Enzymes hosted in reverse micelles in hydrocarbon solution, *Angew. Chem.,* 24, 439, 1985.

101. **Levashov, A. V., Khmelnitski, Yu. L., Klyachko, N. L., Chernyak, V. Ya., and Martinek, K.,** Enzymes entrapped into reversed micelles in organic solvents. Sedimentation analysis of the protein-Aerosol OT-H$_2$O-octane system, *J. Colloid Interface Sci.,* 88, 444, 1982.

102. **Shapiro, Yu. E., Budanov, N. A., Levashov, A. V., Klyachko, N. L., Khmelnitski, Yu. L., and Martinek, K.,** Carbon-13 NMR study of entrapping proteins (α-chymotrypsin) into reversed micelles of surfactants (Aerosol OT) in organic solvents (*n*-octane), *Collect. Czech. Chem. Commun.,* 54, 1126, 1989.

103. **Eicke, H. F., Shepherd, J. C. W., and Steinemann, A.,** Exchange of solubilized water and aqueous electrolyte solutions between micelles in apolar media, *J. Colloid Interface Sci.,* 56, 168, 1976.

104. **Kuno, S., Bacher, A., and Simon, H.,** Structure of enoate reductase from a *Clostridium tyrobutyricum,* *Biol. Chem. Hoppe-Seyler,* 366, 463, 1985.

105. **Fletcher, P. D. I., Howe, A. M., and Robinson, B. H.,** The kinetics of solubilisate exchange between water droplets of a water-in-oil microemulsion, *J. Chem. Soc. Faraday Trans. I,* 83, 985, 1987.

106. **Tabony, J. and Drifford, M.,** Quasielastic neutron scattering measurements of monomer molecular motions in micellar aggregates, *Colloid Polym. Sci.,* 261, 938, 1983.

107. **Meier, P. and Luisi, P. L.,** Micellar solubilization of biopolymers in hydrocarbon solvents. II. The case of horse liver alcohol dehydrogenase, *J. Solid-Phase Biochem.,* 5, 269, 1980.

108. **Jobe, D. J., Dunford, H. B., Pickard, M., and Holzwarth, J. F.,** Kinetics of azide binding to chloroperoxidase in water and reversed micelles of SDS, hexanol and water, in *Reactions in Compartmentalized Liquids,* Knoche, W. and Schomacher, R., Eds., Springer-Verlag, Berlin, 1989, 41.

109. **Lindman, B., Stilbs, P., and Moseley, M. E.,** Fourier transform NMR self-diffusion and microemulsion structure, *J. Colloid Interface Sci.,* 83, 569, 1981.

110. **Stilbs, P. and Lindman, B.,** Aerosol OT aggregation in water and hydrocarbon solution from NMR self-diffusion measurements, *J. Colloid Interface Sci.,* 99, 290, 1984.

111. **Langevin, D.,** Structure of reversed micelles, in *Structure and Reactivity in Reverse Micelles,* Pileni, M. P., Ed., Elsevier/North-Holland, Amsterdam, 1989, 13.

112. **Verhaert, R. M. D., Hilhorst, R., Vermuë, M., Schaafsma, T. J., and Veeger, C.,** Description of enzyme kinetics in reversed micelles. I. Theory, *Eur. J. Biochem.,* 187, 59, 1990.

113. **Verhaert, R. M. D., Tyrakowska, B., Hilhorst, R., Schaafsma, T. J., and Veeger, C.,** Enzyme kinetics in reversed micelles. II. Behavior of enoate reductase, *Eur. J. Biochem.,* 187, 73, 1990.

114. **Tyrakowska, B., Verhaert, R. M. D., Hilhorst, R., and Veeger, C.,** Enzyme kinetics in reversed micelles. III. Behavior of 20 β-hydroxysteroid dehydrogenase, *Eur. J. Biochem.,* 187, 81, 1990.

115. **Verhaert, R. M. D. and Hilhorst, R.,** Enzymes in reversed micelles: IV. Theoretical analysis of a one-substrate one-product conversion and suggestions for an efficient application, *Rec. Trav. Chim. Pays-Bas,* 110, 236, 1991.

116. **Bianucci, M., Maestro, M., and Walde, P.,** Bell-shaped curves of the enzyme activity in reverse micelles: a simplified model for hydrolytic reactions, *Chem. Phys.,* 141, 273, 1990.

117. **Bru, R., Sánchez-Ferrer, A., and García-Carmona, F.,** A theoretical study on the expression of enzymic activity in reverse micelles, *Biochem. J.,* 259, 355, 1989.

118. **Ruckenstein, E. and Karpe, P.,** Enhanced enzymic activity in reverse micelles, *Biotechnol. Lett.,* 12, 241, 1990.

119. **Kabanov, A. V., Levashov, A. V., Klyachko, N. L., Nametkin, S. N., Pshezhetskii, A. V., and Martinek, K.,** Enzymes entrapped in reversed micelles of surfactants in organic solvents: a theoretical treatment of the catalytic activity regulation, *J. Theor. Biol.,* 133, 327, 1988.

120. **Oldfield, C.,** Evaluation of steady-state kinetic parameters for enzymes solubilized in water-in-oil microemulsion systems, *Biochem. J.,* 272, 15, 1990.

121. **Maestro, M.,** Enzymatic activity in reversed micelles: some modelistic considerations on bell shaped curves, *J. Mol. Liq.,* 42, 71, 1989.

122. **Tabony, J.,** Quasi-elastic neutron scattering measurements of molecular motions in micelles and microemulsions, *Chem. Phys. Lett.,* 113, 75, 1985.

123. **Hiromi, K.,** *Kinetics of Fast Enzyme Reactions. Theory and Practice,* John Wiley & Sons, New York, 1979.

124. **Levashov, A. V., Klyachko, N. L., Pantin, V. I., Khmelnitski, Y. L., and Martinek, K.,** *Bioorg. Khim.,* 6, 929, 1980.

125. **Karpe, P. and Ruckenstein, E.,** Effect of hydration ratio on the degree of counterion binding and pH distribution in reverse micelles with aqueous core, *J. Colloid Interface Sci.,* 137, 408, 1990.

126. **El Seoud, O. A.,** Acidities and basicities in reversed micellar systems, in *Reverse Micelles,* Luisi, P. L. and Straub, B. E., Eds., Plenum Press, New York, 1981.

127. **Kumar, A., Kumar, A., and Katiyar, S. S.,** Activity and kinetic characteristics of glutathione reductase in vitro in reverse micellar waterpool, *Biochim. Biophys. Acta,* 996, 1, 1989.

128. **Hilhorst, R., Spruijt, R., Laane, C., and Veeger, C.,** Rules for the regulation of enzyme activity in reversed micelles as illustrated by the conversion of apolar steroids by 20 β-hydroxysteroid dehydrogenase, *Eur. J. Biochem.,* 144, 459, 1984.

129. **Hilhorst, R.,** Enzymatic Reactions in Reversed Micelles, Ph.D. thesis, Agricultural University, Wageningen, The Netherlands, 1984.

130. **Zana, R. and Lang, J.,** Dynamics of microemulsions, in *Solution Behaviour of Surfactants. Theoretical and Applied Aspects,* Mittal, K. L. and Fendler, E. J., Eds., Plenum Press, New York, 1982, 1195.

131. **Battistel, E., Luisi, P. L., and Rialdi, G.,** Thermodynamic study of globular protein stability in microemulsions, *J. Phys. Chem.,* 92, 6680, 1988.

132. **Larsson, K.,** Enzyme Catalysis in Microemulsions, Ph.D. thesis, Lund University, Sweden, 1990.

Chapter 7

EXPRESSION OF ENZYME ACTIVITY IN REVERSE MICELLES

Francisco García-Carmona, Roque Bru, and Alvaro Sánchez-Ferrer

TABLE OF CONTENTS

I. INTRODUCTION

Micellar enzymology is concerned with enzyme catalysis in reverse micelles. These are microdroplets of water dispersed in a hydrophobic medium from which they are shielded by a monolayer of surfactant molecules. The hydrophilic part (the head) of these molecules, which is often charged, is turned toward the water core of the droplet, while its hydrophobic part (the tail) is exposed to the apolar medium.

Since the first report in 1977 on the catalytic activity of a protease,[1] more than 40 enzymes have been found to be active in this system (for review, see References 2 to 4). The reasons for which the new field of reverse micelles has attracted so much attention are the following:

1. They can simulate *in vitro* conditions of the *in vivo* action of enzymes due to compartmentalization.
2. They can be used to study conformational changes in the structure of nonenzymatic proteins (myelin basic protein) and the role of highly structured water in these changes.[5]
3. They open up the kinetic study of important water-insoluble substrates[6-8] and also provoke changes in the kinetic parameters expressed by some enzymes ("superactivity"[9-12] and substrate specificity[13,14]).
4. They represent a simple system which can be studied by different techniques (spectrophotometry,[15] fluorometry,[16,17] CD,[18] NMR,[19] and FTIR[20]) because they are an optically transparent pseudohomogeneous medium which can be prepared experimentally without any sophisticated equipment and with very cheap commercially available surfactants.
5. Reverse micelles have a great potential application in biotechnology because they provide the appropriate environment for bioconversion of polar and apolar compounds used in organic synthesis,[21] microparticle synthesis,[22,23] and luminescence.[24] They also permit the extraction of biologically active proteins[25] and, finally, can be used in analytical[26] and medical applications.[2]

This versatility has resulted in a great quantity of different experimental data which have led to the formulation of diverse theoretical models for their explanation. This chapter is basically concerned with the discussion of the most representative models published until now. However, before dealing with these models it is convenient to give a brief overview of the basic concepts studied in micellar enzymology.

II. BASIC CONCEPTS IN REVERSE MICELLES

A. EXPERIMENTAL CONDITIONS

Entrapment of enzymes in reverse micelles is a very simple procedure, and different approaches have been used. The enzyme can be solubilized by spontaneous transfer of the protein in a two-phase system consisting of approximately equal volumes of the aqueous protein solution and the organic solvent containing the surfactant[27] (Figure 1a). A second procedure consists of the solubilization of lyophilized enzyme in previously formed hydrated reverse micelles[9] (Figure 1b). The third and, recently, most widely used method consists of microinjection of enzyme dissolved in a buffered aqueous medium into a stirred solution of surfactant dissolved in organic solvent[1] (Figure 1c). This last method avoids the lengthy time required for the solubilization of the proteins and the problem of determining the actual amount of water in the reverse micelles, both of which are drawbacks of the first two methods.

Once the enzyme is inside the reverse micelles, the substrate has to be prepared under

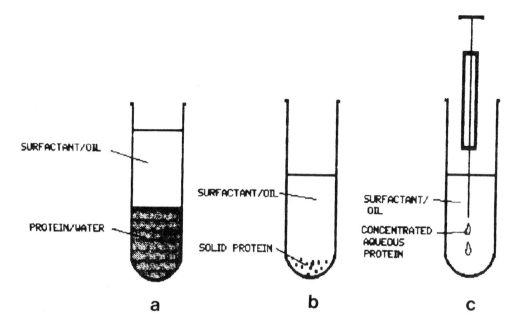

FIGURE 1.   Methods for entrapment of enzymes in reverse micelles: (a) solubilization by phase transfer; (b) solubilization of dry enzyme; and (c) solubilization by microinjection of buffered aqueous solution with enzyme.

the same micellar conditions. These are governed by two different factors: the size and the concentration of micelles. The size is termed Wo in micellar enzymology, and this represents the molar ratio [H$_2$O]/[SURF]. The concentration of micelles under a fixed Wo is represented as θ, the water volume fraction in the system.

Only one of these parameters can be changed at a time and this must be kept in mind in order to avoid difficulties in understanding the results.

Due to the transparency of the system, experiments can be followed by the different techniques outlined above.

Kinetically speaking, the velocity of the reaction is estimated in steady state and is obtained from the linear part of the product accumulation curve. The kinetic obtained in reverse micelles until now obeys the classical Michaelis-Menten equation when the substrate concentration is maintained higher than enzyme concentration.

B. EFFECT OF MICELLE SIZE

Micelle size has a particular effect on enzyme activity in reverse micelles. The three most widely shown profiles are presented in Figure 2. The saturation curve (a) may be interpreted by the need of the enzyme for free water in order to reach its maximal activity; above this, the size of the micelle has no effect on activity. The bell shape (b) is explained by the existence of an optimal inner cavity for catalytic activity which is supposed to correspond to the size of the enzyme.

The third case (c), where enzyme activity decreases as Wo increases, is more rare and might represent a decrease in the conformational flexibility of the protein, which permits a greater catalytic efficiency of the enzyme at low Wo.[28]

In addition, there are some enzymes that present in reverse micelles a k$_{cat}$ bigger than the one expressed in water, giving rise to the concept of ''superactivity''. This effect can be explained by a higher reactivity of the structured water in the micelle and/or the relatively high rigidity of the enzyme molecule caused by the surfactant layer.

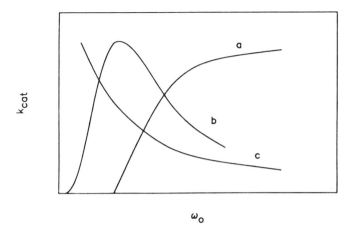

FIGURE 2.   Effect of micelle size (Wo) on enzyme activity: (a) saturation
curve; (b) bell-shape profile; and (c) continuous decrease in activity when
Wo increases.

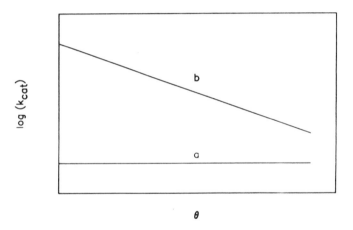

FIGURE 3.   Effect of micelle concentration (θ) on enzyme activity: (a)
enzyme which does not depend on θ, and (b) enzyme whose change in
activity can be as high as one or two orders of magnitude when θ changes.

C. EFFECT OF MICELLE CONCENTRATION

The alteration of the concentration of identically sized micelles (θ) at a constant Wo
has two different responses in the k_{cat} of the enzymes in reverse micelles (Figure 3). The
first (a) is displayed by a group of enzymes whose activity does not depend on θ. The second
(b) is presented by enzymes whose catalytic activity can be increased by as much as 10- or
100-fold when θ is changed.

The difference between these two groups stems from the presence of anchoring groups
of a different apolar nature in the enzyme belonging to the second group, which permits it
to interact with the micellar monolayer. This assumption has been tested by the conversion
of an enzyme of the first group into an enzyme of the second by covalently attaching apolar
residues to it.[28]

D. THEORETICAL MODELS

To consolidate a branch of science, it is necessary to mathematize its results. In enzymology, this consists of the design of kinetic-structural models to explain the experimen-

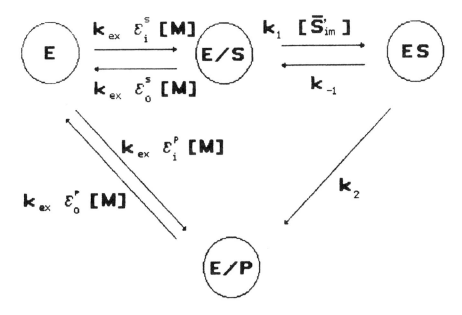

SCHEME 1.   Scheme of a reaction obeying classical Michaelis-Menten kinetics in reverse micelles. E/S denotes the presence of E and S in one reverse micelles without a complex being formed. [M] is the total concentration of reverse micelles in the medium. The rate constant governing the intermicellar exchange reaction is $k_{ex}(M^{-1}s^{-1})$. ϵ_i^S and ϵ_o^S are the efficiencies of substrate transport into and out of the enzyme-filled reverse micelle, respectively. The average concentration of substrate in substrate-filled reverse micelles containing substrate is given by $[\bar{S}'_{im}]$. (From Verhaert, R. M. D., Hilhorst, R., Vermue, M., Schaafsma, T. J., and Veeger, C., *Eur. J. Biochem.*, 187, 59, 1990. With permission.)

tally obtained results. Recently, the more active groups in micellar enzymology have gone to great lengths to explain their results in terms of different kinetic theories.

The whole set of models described can be classified into two groups:

1. Diffusional models — Diffusion between micelles is a determinant factor in the expression of catalytic activity.
2. Nondiffusional models — The diffusion is not a limiting factor for the catalytic process because the experimental exchange rate measured (in the order of microseconds) is much faster than the maximum turnover (in the range of seconds). For this reason, the diffusional factors are not included in the kinetic equations.

III. DIFFUSIONAL MODELS

The above models assume that the intrinsic kinetic parameters of the enzyme are not affected by the incorporation of the enzyme in reverse micelles. Thus, changes in the enzymatic reaction rate can only be explained as a consequence of the intermicellar exchange reaction (diffusion).

The whole reaction rate has to be considered in two steps. The first is the collision of a micelle containing substrate with a micelle containing enzyme, followed by exchange and concentration of reactants in the water pool. In the second step the enzymatic reaction takes place.

The first diffusional model was proposed by Veeger's group[29] in Holland for a highly hydrophilic substrate which only partitions in the water phase.

The reaction proposed for enzymatic catalysis in this model[29] is shown in Scheme 1.

Taking into account the exchange rates into and out of, $\epsilon_i^p = 0$, and by using the King and Altman method, the initial rate can be derived as:

$$v = \frac{k_2 \cdot [E_o]}{1 + \dfrac{([\epsilon_o^s/\epsilon_i^s] + 1) \cdot K_m}{[\overline{S}_{im}']} + \dfrac{k_2}{k_{ex} \cdot [M]} \cdot \left(\dfrac{1}{\epsilon_i^s} + \dfrac{1}{\epsilon_o^p}\right)} \tag{1}$$

where $[\overline{S}_{im}']$ is the average intramicellar substrate concentration, $[M]$ is the micelle concentration, k_{ex} is the exchange rate between reverse micelles, and ϵ is the efficiency of intramicellar exchange of the solute and of the product into and out of the enzyme-filled micelles (ϵ_i^s, ϵ_i^p, ϵ_o^s, and ϵ_o^p, respectively).

The equation rate thus obtained becomes a non-Michaelis equation because it includes an extra term which represents the interexchange between reverse micelles containing substrate. It also includes a K_m multiplied by substrate concentration, because the efficiencies (ϵ) include the overall substrate concentration.

In order to transform Equation 1 into a Michaelis equation, two limits must be assumed. The first, when the substrate concentration is low, i.e., $[S_{ov}] << [M]$, gives:

$$v = \frac{k_2 \cdot [E_o]}{1 + \dfrac{(\theta \cdot K_m + [2k_2/k_{ex}])}{[S_{ov}]} + \dfrac{\theta \cdot K_m}{[M]} + \dfrac{2k_2}{k_{ex} \cdot [M]}} \tag{2}$$

The above equation means that the apparent K_m is not only modified by a term θ, but also by the ratio between the turnover of the enzyme and the exchange rate of the reversed micellar medium.

The second limit is when $[S_{ov}] >> [M]$. In this case, all reverse micelles contain one or more substrate molecules. Rearranging the terms in Equation 1 results in:

$$v = \frac{k_2 \cdot [E_o]}{1 + \dfrac{\theta \cdot K_m}{[S_{ov}]} + \dfrac{3k_2}{k_{ex} \cdot [M]}} \tag{3}$$

The observed K_m will be equal to the K_m observed in an aqueous solution multiplied by a constant factor and θ. The observed V_{max} is also modified by this factor, which depends on the ratio of the turnover rate of the enzyme and the intermicellar exchange.

This model has been extended to include bisubstrate enzymes and shows analog dependences as in Equation 1. Verhaert et al.[30] fitted their experimental results to the model (Figure 4), achieving good agreement only when there were a very high exchange rate in the reverse micellar solution in comparison with the turnover rate of the enzyme and a relatively high concentration of substrate-filled micelles. The lack of adjustment at lower substrate concentration is due to the fact that the condition $[S_{ov}] >> [M]$ is very difficult to maintain experimentally, even under the most favorable conditions, i.e., at high Wo. This can be better understood by looking at the changes in $[M]$ when Wo is varied (Figure 5). At high Wo, the concentration in micelles is in the range of mM, which is the range for the K_m for many enzymes (0.1 to 10 mM).

At the same time that Verhaert proposed his model, Oldfield[31] in England proposed another diffusional model — which is to all intents a simplified version of Verhaert's — where the initial condition is $[S_{ov}] << [M]$. The problems with this model are once again the restrictive conditions which are difficult to fulfill experimentally.

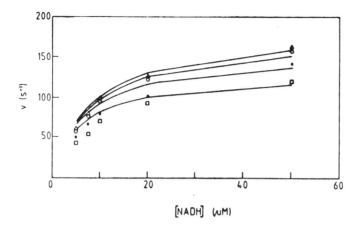

FIGURE 4. Activity of enoate reductase in AOT reverse micelles. The symbols represent experimental values with 2-methylbutenoic acid at (□) 0.075 mM, (*) 0.15 mM, (○) 0.3 mM, and (△) 0.6 mM, and the solid line represents the simulated curves. (From Verhaert, R. M. D., Tyrakowska, B., Hilhorst, R., Schaafsma, T. J., and Veeger, C., *Eur. J. Biochem.*, 187, 73, 1990. With permission.)

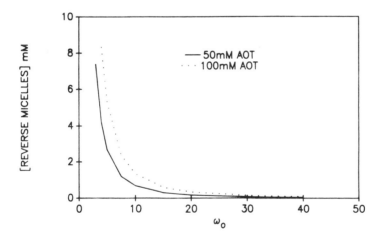

FIGURE 5. Changes in micelle concentration with Wo. The data are calculated using Equation 17.

IV. NONDIFFUSIONAL MODELS

The nondiffusional models assume that the changes in the kinetic parameters k_{cat} and K_m are a consequence of the changes in the protein structure. These changes are due to the differences between the physicochemical characteristics of the reverse micelles and of the buffered aqueous medium. Two models have been proposed: (1) the polydispersed model, as proposed by Kabanov et al. in U.S.S.R.,[32] and (2) the multiphasic model, as proposed by Bru et al. in Spain.[33,34]

A. POLYDISPERSED MODEL

This model assumes the existence of three different micelles, M_{opt}, M_s, and M_1, which corresponds to micelles with optimal, suboptimal, and superoptimal radii to express enzymatic activity. The enzymatic activity can be explained as a result of this polydispersion of micelles. Enzyme molecules can be exchanged between micelles of these three types according to the following:

$$\overline{M}_{opt} + M_s \underset{\longleftarrow}{\overset{K_1}{\rightleftharpoons}} M_{opt} + \overline{M}_s$$

$$\overline{M}_{opt} + M_l \underset{\longleftarrow}{\overset{K_2}{\rightleftharpoons}} M_{opt} + \overline{M}_l$$

where $\overline{M}_l$, $\overline{M}_{opt}$, and $\overline{M}_s$ are enzyme-containing micelles, and K_1 and K_2 are their corresponding equilibrium constants.

The dynamic interaction between micelles permits the formation of dimers that subsequently give monomers. This intermicellar interaction can be described by the following set of equilibria:

$$\overline{M}_s + M \underset{\longleftarrow}{\overset{K_3}{\rightleftharpoons}} \overline{M}_s^*$$

$$\overline{M}_{opt} + M \underset{\longleftarrow}{\overset{K_4}{\rightleftharpoons}} \overline{M}_{opt}^*$$

$$\overline{M}_l + M \underset{\longleftarrow}{\overset{K_5}{\rightleftharpoons}} \overline{M}_l^*$$

where M denotes empty micelles of all three types ($[M] = [M_s] + [M_{opt}] + [M_l]$); $\overline{M}_s^*$, $\overline{M}_{opt}^*$, $\overline{M}_l^*$ are deformed micelles with no enzymatic activity, and K_3, K_4, and K_5 are the equilibrium constants.

If we consider the k_{cat} for each micellar form ($k_{cat}^{(opt)}$, $k_{cat}^{(l)}$, $k_{cat}^{(s)} = 0$), the observed catalytic constant can be expressed by means of the maximal reaction rate (V_{max}) and the overall enzyme concentration $[E]_T$ as follows:

$$k_{cat} = \frac{V_{max}}{[E]_T} = \frac{1}{[E]_T} (k_{cat}^{(opt)}[\overline{M}_{opt}] + k_{cat}^{(l)}[\overline{M}_l]) \tag{4}$$

Taking into account the mass balance for micelles containing enzyme, the equilibrium constants ($K_1 \ldots K_5$), and micellar sizes at given Wo, Equation 4 becomes

$$k_{cat} = \frac{k_{cat}^{(opt)} + k_{cat}^{(l)} K_2 \chi}{1 + K_1\phi + K_2\chi + \psi\left(\dfrac{1}{K_4} + \dfrac{K_1}{K_3}\phi + \dfrac{K_2}{K_5}\chi\right)[SURF]} \tag{5}$$

where [SURF] is the concentration of surfactant, and ϕ, χ and ψ are related to the gaussian distribution function of micellar sizes with Wo. Using this equation, it is possible to simulate the different shapes of k_{cat} vs. Wo and k_{cat} vs. θ. However, no experimental data have been fitted. In addition, it has to be noted that this model can only be used for amphiphilic compounds and not for highly hydrophilic forms.

When the hydrophilic/lipophilic balance of the substrate permits partition in different phases, the substrate can reach the water pool with no diffusional impediment.

The first model to take into account this kind of amphiphilic substrate was proposed by Levashov et al.[35] This model considers the micellar system as two phases — organic (bulk) phase and micellar phase — related by a partition coefficient $P_s = [S]_{mic}/[S]_b$. The enzymatic reaction under this condition is shown in Scheme 2.

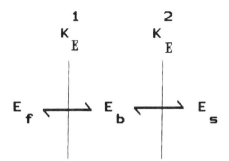

SCHEME 2. Distribution of the enzyme among the micellar microenvironments. The phases can be considered as continuous when the time scale of the process under study is higher than that of exchange between micelles.[17] (From Bru, R., Sánchez-Ferrer, A., and García-Carmona, F., *Biochem. J.*, 259, 355, 1989. With permission.)

The initial rate in the micellar phase related to θ can be expressed as:

$$v = \frac{k_{cat,mic}[S]_{mic}[E]_{mic}}{K_m^{mic} + [S]_{mic}} \cdot \theta \tag{6}$$

and as a function of overall volume:

$$v = \frac{k_{cat,app}[E]_T[S]_T}{K_m^{app} + [S]_T} \tag{7}$$

where $k_{cat,app} = k_{cat,mic}$, $[E]_T = [E]_{mic}\,\theta$ and:

$$K_m^{app} = K_m^{mic}\,\frac{1 + \theta(P - 1)}{P} \tag{8}$$

P_s can be calculated from Equation 8 by plotting K_m^{app} vs. θ.

In the same way, Equation 7 can be linearized in two forms:

$$\frac{1}{v} = \frac{K_m^{mic}}{V_{max}}\left[\frac{1 + \theta(P - 1)}{P}\right]\frac{1}{[S]_T} + \frac{1}{V_{max}} \tag{9}$$

and

$$\frac{1}{v} = \left[\frac{P - 1}{P} \cdot \frac{K_m^{mic}}{V_{max} \cdot [S]_T}\right] \cdot \theta + \frac{1}{V_{max}} + \frac{K_m^{mic}}{V_{max}[S]_T P} \tag{10}$$

B. MULTIPHASIC MODEL
1. Theory

This theory is basic on the physicochemical existence of different phases in reverse micelles. To understand this model, the four following aspects must be taken into account:

1.   Physicochemical microenvironments of reverse micelles
2.   Fusion of the reverse micelles
3.   Enzyme behavior in the different microenvironments of reverse micelles
4.   Substrate distribution in the reverse micellar system

a. Physicochemical Microenvironments of Reverse Micelles

Reverse micelles are spherical aggregates consisting of a water core separated from a continuous apolar phase by a surfactant shell. It is well known that in the absence of water some surfactants such as AOT can form these aggregates, while others such as CTAB or SDS need a cosurfactant (short-chain alcohol or cholesterol) to generate such structures.

The presence of water increases reverse micelle size, which is determined by Wo. There is an empiric relationship between Wo and reverse micelle radius[36] when AOT is used as surfactant.

$$r \text{ (hydrodynamic radius)/nm} = 0.175 \text{ Wo} + 1.5 \qquad (11)$$

If we assume that the AOT molecule length is *circa* 1.5 nm, the water droplet radius (ranging from 0 to 20 nm) can be expressed as:

$$r_w \text{ (water droplet)/nm} = 0.175 \text{ Wo} \qquad (12)$$

The presence of a polar head (which may be charged) in the surfactant molecule leads us to assume that water molecules which hydrate the polar head are subject to different forces to those of bulk water. The former can form hydrogen bonds with polar heads and can undergo electrostatic attractions when the polar head is charged. As a result, the hydration water becomes more structured than bulk water.

El Seoud[37] outlined an AOT reverse micelle structure when the amount of water exceeds the hydration requirements of the surfactant. It consists of three different domains:

1.   Surfactant apolar tails — Their penetration in the apolar solvent depends on the type of surfactant and apolar solvent.
2.   Bound water — Its properties (hydrogen bonding, effective dielectric constant, viscosity, etc.) are qualitatively different from those of bulk water[38] because of the hydrophilic interactions with polar heads of surfactant.
3.   Free water — Its properties and structure become closer to those of bulk water as Wo increases. Measurements of water activity vs. Wo display a hyperbolic shape,[39] reaching a value higher than 0.98 at Wo = 10

In accordance with this structural model, reverse micelles can possess up to three different microenvironments when water in the system exceeds the hydration requirements of surfactant polar heads (three-domain reverse micelles). If we either decrease the amount of water or increase the amount of surfactant, reverse micelles will only consist of two microenvironments, without free water (two-domain reverse micelles), and will consist of only one microenvironment with a further decrease of water (dry reverse micelles). In practice, it is usual to start from dry reverse micelles, and then water is added to obtain a desired Wo value.

b. Fusion of the Reverse Micelles

Amphiphilic molecules of surfactant in water/oil mixtures self-assemble spontaneously to form not only spherical or ellipsoidal aggregates, but also continuous liquid-crystal structures with cylindrical or lamellar arrangements, depending on the composition of the medium.

For example, starting from a composition that yields a continuous liquid-crystal aggregate, reverse micelles can be obtained by increasing the concentration of the appropriate component, in this case the apolar solvent.

These interconvertible modes of aggregation suggest that such stable aggregates are ruled by a dynamic equilibrium. In the case of reverse micellar aggregates, the dynamic equilibrium involves a fusion and dissociation process[36] that can be represented through the following equation:

$$M + M \underset{}{\overset{K_{eq}}{\rightleftarrows}} D$$

Thus,

$$K_{eq} = \frac{[D]}{[M]^2} \tag{13}$$

where [D] is the dimeric micelle concentration and [M] is the monomeric micelle concentration. The degree of association (ξ) can be defined as follows:

$$\xi = \frac{2[D]}{[M]_i} \tag{14}$$

where $[M]_i$ is the micellar concentration.

In the dimerization process, the water droplet volume is twice that of a monomeric micelle and, if it is considered that dimeric micelles tend to have a spherical shape, there is a loss of surface area (Wo < 2Wo) defined by a factor f:

$$Wo_{(d)} = Wo_{(m)} \cdot f \tag{15}$$

This factor is estimated at *circa* 1.33 for a sphere.

In consequence, the dimerization process involves a microdispersion of micelle size with a standard deviation which depends on micelle concentration. Thus, water droplet radius of monomeric micelles (r_w) is estimated by extrapolation at infinite micellar dilution.

The number of micelles can be evaluated as:

$$N_m = \frac{3V_{H_2O}}{4\pi r_w^3} \tag{16}$$

where V_{H_2O} is the overall volume of water in nanometers. The r_w is correlated with Wo through Equation 11. If we assume that all the micelles are in monomeric state, their overall concentration can be expressed by Equations 12 and 16 as:

$$[M]_i = \frac{N_m}{N_A \cdot V} = 1.33 \frac{[SURF]}{Wo^2} \tag{17}$$

where N_A is Avogadro's number, [SURF] is the surfactant concentration, and V is the overall volume of micellar solution. Taking into account the micelle balance:

$$[M]_i = [M] + 2[D] \tag{18}$$

Equation 14 can be expressed as a function of Equations 13, 17, and 18:

$$\xi = 1 + \frac{1 - (8K_{eq}[M]_i + 1)^{1/2}}{4K_{eq}[M]_i} \qquad (19)$$

c. Enzyme Behavior in the Different Microenvironments of Reverse Micelles

Enzyme molecules, like any other molecule in a heterogeneous medium, tend to be distributed among the different phases that make up the system. Reverse micelles can be regarded as a microheterogeneous medium where solubilized molecules are subject to a partition towards the different phases.

By relating the diffusion-controlled rate constant (*circa* 10^{10} M^{-1} s^{-1}) with the exchange rate constant (*circa* 10^7 M^{-1} s^{-1}), it has been shown[36] that 1 in 1000 to 10,000 encounters between micelles results in solute exchange. When the rate of the chemical or enzymatic reaction in question is much slower than the time scale for micelle exchange, the dispersed phase can be regarded as a pseudo-continuous phase, since transport of solutes between micelles is not rate limiting.[4]

On this basis, the distribution of enzymes can be described through some simple relationships. The volume filled by every microenvironment in the micellar solution is given by:

$$V_s = V_S^M \cdot \tau \cdot mol\ S \left(\frac{1 - \xi}{f}\right) \qquad (20)$$

$$V_b = V_{H_2O}\left[\frac{(1 - \xi)n}{Wo_{(m)}} + \frac{\xi n}{Wo_{(d)}}\right] \qquad (21)$$

$$V_b = V_{H_2O} - V_b \qquad (22)$$

$$V_{os} = V - (V_s + V_b + V_f) \qquad (23)$$

where V_s, V_b, V_f, and V_{os} are the volumes of surfactant apolar tails, bound water, free water, and organic solvent, respectively; *mol S* is the number of surfactant moles; V_S^M is the volume occupied by 1 mol of surfactant; τ is the penetration factor of surfactant apolar tails in the oil; and n is the number of water molecules bound per surfactant polar head. The surfactant fraction lost because of the dimerization process is the term ξ/f.

Assuming that the active enzyme is only present in the reverse micelles (Scheme 2), enzyme distribution among the three micellar microenvironments is

$$K_E^1 = \frac{[E]_b}{[E]_f} \qquad (24)$$

$$K_E^2 = \frac{[E]_s}{[E]_b} \qquad (25)$$

If the reaction medium contains a definite number of enzyme micromoles, its distribution will be

$$\mu mol\ of\ E = [E]_f V_f + [E]_b V_b + [E]_s V_s \qquad (26)$$

Applying Equations 24, 25, and 26 gives:

$$[E]_f = \frac{\mu mol\ of\ E}{V_f + K_E^1 V_b + K_E^1 K_E^2 V_s} \qquad (27)$$

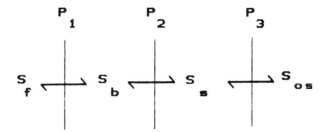

SCHEME 3. Substrate distribution in the reverse micelle system. Schematic representation of reverse micelle as a multiphasic system. The phases can be considered as continuous when the time scale of the process under study is higher than that of exchange between micelles.[17] (From Bru, R., Sánchez-Ferrer, A., and García-Carmona, F., *Biochem. J.*, 268, 679, 1990. With permission.)

$$[E]_b = K_E^1 [E]_f \tag{28}$$

$$[E]_s = K_E^1 K_E^2 [E]_f \tag{29}$$

When the enzyme works in bulk water, it expresses a catalytic constant k, but if it works in a different environment it will express a different catalytic constant depending on the conformation acquired in the new medium. In this way, k_f, k_b, and k_s can be defined as catalytic constants expressed by the enzyme in free water, bound water, and surfactant apolar tails, respectively. Because free water is similar to bulk water, it can be assumed that $k_f = k$.

The enzymatic activity (V_{max}) must be expressed as a function of the overall volume of the cuvette (V) since this is the activity which can be measured experimentally. This activity can be considered as the sum of the enzymatic activities in each microenvironment with respect to the overall volume and expressed as:

$$V_{max} = \sum_{i=f,b,s} k_i [E]_i j \tag{30}$$

where j represents the volume fraction (V_i/V_T) of each phase.

d. Substrate Distribution in the Reverse Micellar System

The distribution of substrate among phases that make up reverse micelles and the continuous oil phase (Scheme 3) is not limited by diffusion, as already mentioned, because exchange between solutes in the reverse micelles is faster than the enzymic reactions,[4] and so, there is nothing to hinder the diffusion of reactants. The equilibrium distribution of the substrate at zero time[40] is ruled by the respective partition coefficients:

$$P_1 = \frac{[S]_b}{[S]_f} \tag{31}$$

$$P_2 = \frac{[S]_s}{[S]_b} \tag{32}$$

$$P_3 = \frac{[S]_{os}}{[S]_s} \tag{33}$$

The mass balance will be

$$[S]_T = [S]_f \cdot \alpha + [S]_b \cdot \beta + [S]_s \cdot \gamma + [S]_{os} \cdot (1 - \alpha - \beta - \gamma) \qquad (34)$$

$$[E]_T = [E]_f \cdot \alpha + [E]_b \cdot \beta + [E]_s \cdot \gamma \qquad (35)$$

where α is the volume fraction (V_f/V) of free water, β is the volume fraction of bound water, and γ is the volume of surfactant tails.

Each particular substrate concentration can be expressed as a function of the partition coefficients, of the relative volumes, and of the overall substrate concentration by substituting and rearranging Equations 31 to 33 in Equation 34:

$$[S]_f = \frac{[S]_T}{\alpha + P_1\beta + P_1P_2\gamma + P_1P_2P_3(1 - \alpha - \beta - \gamma)]} \qquad (36)$$

$$[S]_b = P_1[S]_f \qquad (37)$$

$$[S]_s = P_1P_2[S]_f \qquad (38)$$

$$[S_{os}] = P_1P_2P_3[S]_f \qquad (39)$$

Assuming a different K_m for each phase, since the enzyme in one phase can only catalyze the substrate in the same phase, the initial velocity expressed in each phase is described by:

$$v_i = \frac{V_{max_i}[S]_T}{K_{m_i}^{app} + [S]_T} \qquad (40)$$

where $K_{m_i}^{app}$ in each phase is

$$K_{m_f}^{app} = K_{m_f}[\alpha + P_1\beta + P_1P_2\gamma + P_1P_2P_3(1 - \alpha - \beta - \gamma)] \qquad (41)$$

$$K_{m_b}^{app} = K_{m_b} \frac{\alpha + P_1\beta + P_1P_2\gamma + P_1P_2P_3(1 - \alpha - \beta - \gamma)}{P_1} \qquad (42)$$

$$K_{m_s}^{app} = K_{m_s} \frac{\alpha + P_1\beta + P_1P_2\gamma + P_1P_2P_3(1 - \alpha - \beta - \gamma)}{P_1P_2} \qquad (43)$$

Thus, the overall activity is

$$v = \sum_{i=f,b,s} v_i \qquad (44)$$

2. Simulation and Adjustment of Experimental Data

In order to check the validity of the multiphasic model, it is advantageous to describe first the theoretical curves obtained from it and then to fit the experimental data using tyrosinase (also called polyphenoloxidase) as model enzyme because it can oxidize *o*-diphenols of different hydrophobicity to quinones.

The steps to be followed are

1.   Simulation of the dependence of V_{max} on Wo at a constant θ, then the fitting of the experimental results obtained with tyrosinase by nonlinear regression to the model —

This gives the partition coefficients of the enzyme among the micellar domains and its catalytic constants (one each per micellar phase). These numerical values are used subsequently in the equations.

2. Simulation of the dependence of the reaction rate on θ at a constant Wo — Again, the experimental results are fitted by nonlinear regression to the global model. As a result, P_1, P_2, P_3, $K_{m,f}$, $K_{m,b}$, and $K_{m,s}$ are obtained.
3. Introduction into the model of the dependence of the reaction rate on the new parameter ρ, which represents the ratio $[S]_T/\theta$

a. Dependence of Enzymatic Activity on Wo

Figures 6a to 6c show the simulation plots of V_{max} vs. Wo when the enzyme expresses the highest activity in only one of the three microenvironments. Three basic patterns of behavior can be seen with a common critical point about Wo = 8 to 10. This point represents (1) a start in activity when the enzyme is more active in the free water, (2) maximal activity when the enzyme is more active in bound water, or (3) a sudden fall in the activity when the enzyme works mostly in surfactant apolar tails. In the three cases, the changes in enzyme behavior are related to the appearance of free water.

The most frequently described pattern is that of Figure 6b, where the enzyme expresses its maximal catalytic activity in bound water, while it is markedly less active in the other two domains. It is not uncommon to talk about an optimal Wo.

Acid phosphatase and peroxidase have been shown to be superactive enzymes.[2] This phenomenon has also been shown in N-*trans*-cinnamoyl α-chymotrypsin deacylation.[11] Figure 7 shows that enzyme activity in reverse micelles rises to a maximum higher than that in bulk water, this activity decreasing to the same level expressed in bulk water as micelle size increases. Although no experimental explanation for this behavior has been found, the model proposed here fits the data (Figure 7, solid line).

The versatility of this model not only explains the superactivation but also the superinhibition (Figure 8), even though this behavior has not yet been shown experimentally. Such behavior would require the enzyme to catalyze the substrate efficiently in surfactant apolar tails and free water, but less efficiently in bound water.

This model was used to fit the experimental data obtained from tyrosinase. Figure 9(a) shows the dependence of V_{max} on Wo by using TBC as substrate. V_{max} was determined graphically by means of double reciprocal plots and the continuous line is the theoretical adjustment of the data. From the pattern observed, it can be deduced that tyrosinase mostly works in the free water of the water pool of the reverse micelles since no activity is detected before Wo = 8. When the maximal reaction rate is considered, the model predicts that the Wo pattern will not depend on the kind of substrate or its concentration. Thus, if a hydrophilic substrate such as 4MC is used instead of TBC, the Wo profile is only affected by the rate of the ES complex breakdown. Figure 9(b) shows that this dependence is clearly similar to that obtained with TBC. Thus, it is reasonable to assume that mushroom tyrosinase solubilizes mainly in the free water of the reverse micelles. It has been shown that such a profile is bell-shaped for grape polyphenoloxidase when using 4MC as substrate in Brij 96/cyclohexane reverse micelles.[41] These results seem to indicate that the type of Wo pattern may be a consequence of the external groups of the enzyme and their interaction with the reverse micellar medium. Indeed, grape polyphenoloxidase is a thylakoid-bound enzyme[42] which is extracted by detergents,[43] whereas mushroom tyrosinase is a soluble enzyme.[44]

To fit the results accurately to the model, it is assumed that only three water molecules are tightly bound per surfactant head because no physical studies on the hydration requirements of Brij 96 have been reported. However, it is clear that the hydration number of this surfactant has to be smaller than that of ionic surfactants, mainly because of the absence of the counterion.[3]

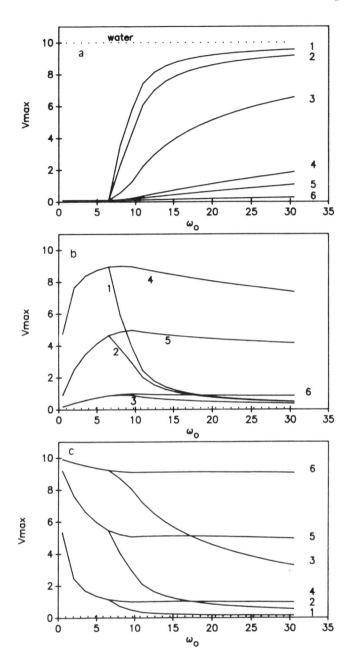

FIGURE 6.   Dependence of catalytic activity V_{max} vs. micelle size (Wo). The number of bound water molecules has been taken as 9, and the system contains 1.1% (v/v) of water. Other conditions are $K_{eq} = 2.5 \times 10^8$ M^{-1}, f = 1.33, $V_S^M = 400$ ml mol^{-1}, $\tau = 0.5$, and the amount of E = 10 μmol; (a) $k_f = 1000$, $k_b = 10$, and $k_s = 10$; (b) $k_f = 10$, $k_b = 1000$, and $k_s = 10$; (c) $k_f = 10$, $k_b = 10$, and $k_s = 1000$. For K_E^1 and K_E^2, see Table 1. The broken line represents the activity in water. (From Bru, R., Sánchez-Ferrer, A., and García-Carmona, F., *Biochem. J.*, 259, 355, 1989. With permission.)

TABLE 1
Key to Curve
Numbers

K_E^1	K_E^2	Number
0.1	0.1	1
0.1	1	2
0.1	10	3
10	0.1	4
10	1	5
10	10	6

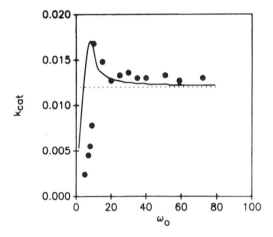

FIGURE 7.   Superactivity of *N-trans*-cynnamoil α-chymotrypsin deacylation. Dark circles (●) represent experimental data taken from Martinek et al.[2] These values have been simulated and the results are represented as a solid line. The dotted line represents the activity in bulk water. Conditions are $K_E^1 = 0.05$, $K_E^2 = 5$, $k_f = 1200$, $k_b = 13,500$, $k_s = 0$, and $n = 9$. Other conditions are the same as those in Figure 6. (From Bru, R., Sánchez-Ferrer, A., and García-Carmona, F., *Biochem. J.*, 259, 355, 1989. With permission.)

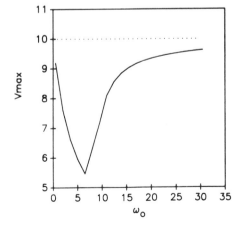

FIGURE 8.   Simulation of a superinhibition case. The dotted line is the activity in water. Conditions are $K_E^1 = 0.1$, $K_E^2 = 1$, $k_f = 1000$, $k_b = 10$, $k_s = 1000$, and $n = 9$. Other conditions are the same as in Figure 6. (From Bru, R., Sánchez-Ferrer, A., and García-Carmona, F., *Biochem. J.*, 259, 355, 1989. With permission.)

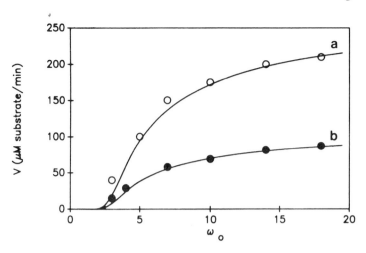

FIGURE 9.   Effect of Wo on V_{max}. This effect was checked for TBC ($\circ$) and 4MC ($\bullet$) at $\theta = 0.002$. The experimental points were fitted by using the multiphasic model (35,36) with the following sets of parameters ($\pm$ SE): curve a — $K_E^1 = 0.437 \pm 0.018$, $K_E^2 = 0.478 \pm 0.032$, $k_f = 32{,}202 \pm 474$ μM TBC/(min.μM — E), $k_b = 0$, and $k_s = 0$; curve b — $K_E^1 = 0.441 \pm 0.017$, $K_E^2 = 0.484 \pm 0.030$, $k_f = 13{,}052 \pm 177$ μM 4MC/(min.μM — E), $k_b = 0$, and $k_s = 0$; and in both curves, $n = 3$. Solid lines are the theoretical curves. (From Bru, R., Sánchez-Ferrer, A., and García-Carmona, F., *Biochem. J.*, 268, 679, 1990. With permission.)

b. Dependence of Enzymatic Activity on θ

The effect produced by θ when no substrate is considered on enzymatic activity can be understood as a relative variation in phase volumes where the enzyme can solubilize. At low Wo, micelles are of the two-domain reverse micelle type and dynamic equilibrium does not significantly modify the volumes of the two microenvironments nor, consequently, the enzyme partition. Likewise, at high Wo, the increase in free water by dimerization does not cause a noticeable variation in volume with respect to overall free water, and so the effect on enzyme distribution is negligible. This parameter (θ) becomes important only when free water begins to appear (in our simulation at Wo = 8), because variations in the volume of each phase are very significant for enzyme distribution (Figure 10).

Quite the reverse seems to occur with θ when the substrate is introduced and is consequently distributed among the reverse micelle domains and the oil phase. The effect of substrate must be studied in a θ where no changes in V_{max} are observed in order to avoid any collateral effect of enzyme distribution.

When the effect of θ is studied with tyrosinase (Figure 11), a decrease in the initial reaction rate is shown at different substrate concentrations. The experimental data were fitted to the model (solid lines) and the partition coefficients of the substrate calculated. These indicate that the substrate is mainly solubilized in the surfactant tails and to a less extent in other reverse micelle phases and oil.

This fact suggests that the decrease observed in the activity might be due to a dilution of the substrate in the interphase. If this is so, θ must only affect K_m but not V_{max}. To test this hypothesis, the double reciprocal values of the data in Figure 11 were plotted in Figure 12a, showing that the maximal rate remains constant and that the K_m varies proportionally with θ (Figure 12b).

The set of parameters that best fitted the experimental results yielded a K_m value of 13.5 mM instead of the 7.8 M obtained when using the model of Levashov et al.[35] (with a value of P = 8000, calculated in Equation 10 and substituted in Equation 8). The K_m calculated by the multiphasic model seems to be more realistic, while it is hard to explain

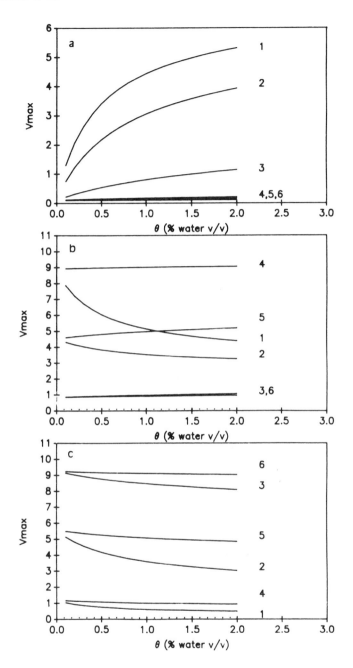

FIGURE 10.   Evolution of catalytic activity vs. water percentage (v/v) at Wo = 8. All the other conditions are the same as in Figure 6, including the key to symbols. In (a), the activity level in water is 10. (From Bru, R., Sánchez-Ferrer, A., and García-Carmona, F., *Biochem. J.*, 259, 355, 1989. With permission.)

the high K_m obtained by the polydispersed model, since the substrate is practically all dissolved in the same micellar reduced volume as the enzyme.

c. Dependence of Activity on ρ

In the case of TBC, which clearly partitions in the surfactant tails, there seems to be a definite dependence between the reaction rate and the ratio $\rho = [S]_T/\theta$. In the expression of K_m (see Equations 41 to 43), two kinds of contribution can be distinguished: on the one

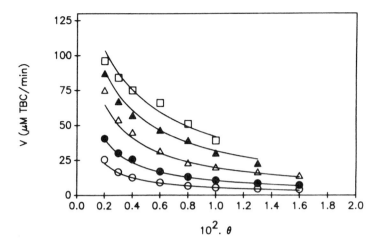

FIGURE 11.   Effect of θ on the reaction rate. Several TBC concentrations were checked: 2.5 mM (○), 5 mM (●), 10 mM (△), 17.5 mM (▲), and 25 mM (□) at Wo = 10. The experimental points were fitted by using the model with the following set of parameters ($\pm$SE): $P_1 = 47.30 \pm 1.04$, $P_2 = 9.67 \pm 0.23$, $P_3 = 0.00017 \pm 0.00004$, $K_{M,f} = 13.56 \pm 0.29$ mM, and $K_{M,b}$ and $K_{M,s}$ are not considered because the enzyme is not active in those domains. Solids lines are the theoretical curves. Other conditions are as in Figure 9. (From Bru, R., Sánchez-Ferrer, A., and García-Carmona, F., *Biochem. J.*, 268, 679, 1990. With permission.)

hand, the contribution of the micellar phases, expressed by the term $\alpha + P_1\beta + P_1P_2\gamma$, and on the other hand, the contribution of the continuous oil, expressed by the term $P_1P_2P_3(1 - \alpha - \beta - \gamma)$. The former can be considered as a homogeneous block when Wo is far from the critical values; in this case the ratio α:β:γ remains practically constant. In such conditions, the relationship between K_m and θ will depend only on the partition coefficients of the substrate. The distinction between micellar and nonmicellar substrates will be governed mainly by P_3. If P_3 is low enough, as in the present case (with TBC), the contribution of the continuous oil compared with that of the micellar domains should be neglected and thus K_m/θ and θ are constants. As shown in Figure 13a, the reaction rate depends on ρ in a hyperbolic shape, in accordance with the equation:

$$v = \sum_{i=f,b,s} \frac{V_{max_i}\rho}{\dfrac{K_{m_i}^{app}}{\theta + \rho}} \tag{45}$$

which comes from Equation 40 by introducing the parameter ρ.

Equation 45 describes the sum of three hyperbolas. Thus, a double reciprocal plot of $1/v$ vs. $1/\rho$ should give a nonlinear dependence. A deviation from linearity is to be expected as $1/\rho$ decreases. Replotting the data from Figure 13a, a linear dependence is observed (Figure 13b). This result implies that the enzyme works in one phase only, which is in accordance with our hypothesis that the enzyme is only active in the free water phase. Thus, it is only necessary to use one catalytic constant and one K_m to fit the experimental data from Figures 9 and 11 to the model.

In short, the kinetic behavior of tyrosinase in Brij 96/cyclohexane reverse micelles can be simulated with the multiphasic model by using Equation 40. The enzyme seems to be active only in free water; thus $i = f$ and K_m^{app} follows Equation 41.

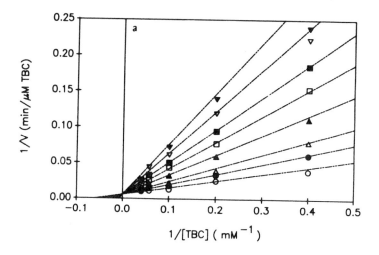

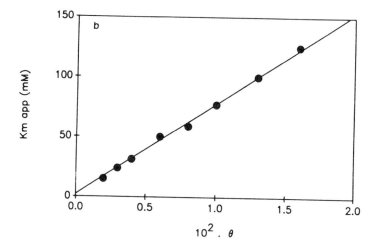

FIGURE 12. (a) Double reciprocal plot of the experimental and theoretical data. $10 \cdot \theta$ values are 0.2 (○); 0.3 (●); 0.4 (△); 0.6 (▲); 0.8 □; 1 (■); 1.3 (▽), and 1.6 (▼). The straight lines are the theoretical values. (b) Effect of θ on the apparent Michaelis constant. $K_{m,app}$ is the value calculated by means of the double-reciprocal plot in (a). The solid line represents the theoretical dependence calculated by using Equation 41. (From Bru, R., Sánchez-Ferrer, A., and García-Carmona, F., *Biochem. J.*, 268, 679, 1990. With permission.)

V. CONCLUSIONS

The theoretical models developed to explain the behavior of enzyme entrapped in reverse micelles have focused on the kinetic parameters expressed by the enzyme in the different microenvironments of reverse micelles and on the relation between substrate partition in these microenvironments and enzyme activity.

The diffusional models, like these of Verhaert et al.[39] and Oldfield,[31] are based on the fact that no changes occur in the kinetic characteristics of the enzymes (K_m and V_{max}) during their entrapment in reverse micelles. Thus, the apparent changes in the above parameters are due to the restriction in substrate accessibility (diffusional factors). In the case of highly hydrophilic substrates (with no partition), these models assume that the only way the enzyme

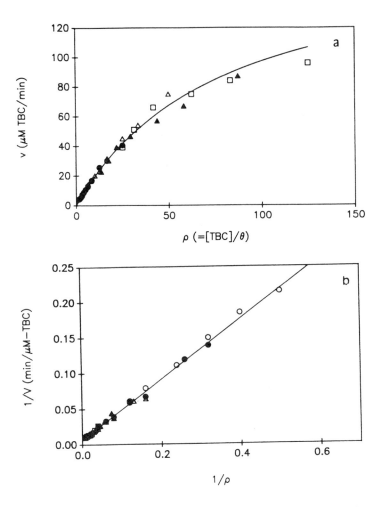

FIGURE 13.   (a) Effect of ρ (= [substrate]/θ ratio) on the reaction rate; (b) double-reciprocal plot. Symbols represent the same concentrations as in Figure 11. See text for explanation. (From Bru, R., Sánchez-Ferrer, A., and García-Carmona, F., *Biochem. J.*, 268, 679, 1990. With permission.)

can obtain substrate is by fusion. However, no experimental data have been found to prove that this fusion is the limiting step in the reaction rate since the measurement of the content exchange between micelles is in microseconds,[17] which is far from the time scale of normal enzymatic reactions (seconds). In addition, for this type of substrate, Verhaert et al.[29] proposed a velocity equation (Equation 1) which does not obey the Michaelis-Menten equation, even though practically all the experimental data obtained until now for reverse micelles follow a Michaelis-Menten kinetic. When the substrate can partition, Verhaert et al.[29] and Oldfield[31] used the basic idea of Levashov et al.[35] Even with this type of substrate, Verhaert et al.[29] also admit the diffusional limitations between organic and aqueous phases. This assumption seems very difficult to maintain since the interface is extremely great in reverse micelles.

In addition to the above diffusional models, two other models have been developed. These consider that the kinetic parameters K_m and V_{max} change when the enzymes are entrapped in reverse micelles since conformational changes have been detected in the proteins when entrapped in reverse micelles.[16,18,45-47]

The polydispersion model proposed by Kabanov et al.[32] bases its theory on the existence of an optimal radius for maximal enzyme activity, and this polydispersion of reverse micelles is the factor used to explain the behavior of the enzyme as regards θ and Wo. For this polydispersion model, it is not necessary to assume the interchange of micelle contents as a limiting factor to explain the changes in enzymatic activity, because it assumes that the enzyme works with infinite substrate. This is also true with the diffusional models when they use partitional substrates. However, it is difficult to apply this assumption to the diffusional models in the case of highly hydrophobic substrates.

The multiphasic model, as proposed by Bru et al.[33,34] is based on the experimental physicochemical data of the three different phases described in reverse micelles. These phases are considered as pseudocontinuous because of the high exchange rate between micelles.[17] The kinetic parameters of the enzyme depend on the microenvironment (surfactant tail, free and bound water, and oil) where the enzyme and the substrate can be distributed. The equations obtained by this multiphase model explain the dependence of the enzymatic activity on Wo and θ, the not-yet-described superinhibition and the superactivity phenomenon. The latter phenomenon can be explained only by the polydispersed model and not by the diffusional models. It should be noted that in the multiphasic model the substrate can be in one phase only (e.g., highly hydrophilic substrate). This does not invalidate the equations proposed, because the activity is expressed as the sum of activities in the different microenvironments of the micelle. However, this multiphasic model presents some simplifications. First, the water pool is considered as two discrete phases (free and bulk water) when, in fact, there is a continuous change in the characteristics of the water from the center to the surfactant heads. Second, it does not consider the molecular size of the enzyme in relation to micelle size.

In conclusion, the models here discussed serve to open up the way for new and more accurate models to express enzymatic activity in reverse micelles.

ABBREVIATIONS

AOT	Dioctyl sodium sulphosuccinate
Brij 96	10 Oleyl ether
CTAB	Hexadecyltrimethylamonium bromide
4MC	4-Methylcatechol
SDS	Sodium dodecyl sulfate
TBC	4-*tert*-Butylcatechol
FTIR	Fourier transform-infrared spectroscopy

LIST OF PARAMETERS

α	Volume fraction of free water
β	Volume fraction of bound water
γ	Volume fraction of surfactant tails
ϵ	The efficiency of intramicellar exchange of the solute and of the product into and out of the enzyme-filled micelles (ϵ_i^s, ϵ_i^p ϵ_o^s and ϵ_o^p, respectively)
θ	Fraction of water (Far from critical Wo values, it is equivalent to the micellar fraction.)
ξ	Degree of association
ξ/f	Surfactant fraction lost because of the dimerization process
ρ	Represents the ratio $[S]_T/\theta$
τ	Penetration factor of surfactant apolar tails in the oil

ϕ, χ and ψ	Coefficients related to the gaussian distribution function of micellar sizes with Wo
[D]	Dimeric micelle concentration
j	α, β, or γ
k_{ex}	Exchange rate between micelles
k_i	Catalytic constant
K_E^1, K_E^2	Enzyme partition coefficients
$K_1 \ldots K_5$	Equilibrium constants
M	Empty micelles of all three types
[M]	Monomeric micelle concentration
$[M_i]$	Micellar concentration
M_{opt}, M_s and M_1	Micelles with optimal, suboptimal, and superoptimal radii to express enzymatic activity
$\overline{M}_1$, $\overline{M}_{opt}$, and $\overline{M}_s$	Enzyme-containing micelles
$\overline{M}_s^*$, $\overline{M}_{opt}^*$, $\overline{M}_1^*$	Deformed micelles with no enzymatic activity
mol S	Number of surfactant moles
n	Number of water molecules bound per surfactant polar head
N_A	Avogadro's number
N_m	Number of micelles
$P_{1,2,3}$	Partition coefficients of the substrate
r_w	Water droplet radius of monomeric micelles
$[S_{ov}]$	Overall substrate concentration
$[\overline{S}_{im}]$	Average intramicellar substrate concentration
[SURF]	Concentration of surfactant
V	Overall volume of micellar solution
V_s, V_b, V_f, and V_{os}	Volumes of surfactant apolar tails, bound water, free water, and organic solvent, respectively
V_s^M	Volume occupied by 1 mol of surfactant
Wo	Micelle size

SUBSCRIPTS

app	Apparent
b	Bound water
f	Free water
i	f, b, or s
mic	Micellar
n	Number of water molecules bound per surfactant polar head
os	Organic solvent
s	Surfactant tails
T	Total

ACKNOWLEDGMENTS

The authors thank Ms. Manuela Pérez Gilabert for her kind help in preparing this chapter. This work was partially supported by CICYT (Proyecto BIO91-0790). Dr. Bru is a holder of a grant from Instituto de Fomento, Murcia (Spain), and Dr. Sánchez-Ferrer is a holder of a reincorporacion grant from Ministerio de Educación y Ciencia (Spain).

REFERENCES

1. **Martinek, K., Levashov, A. V., Klyachko, N. L., and Berezin, I. V.**, Catalysis by water-soluble enzymes in organic solvents. Stabilization of enzymes against denaturation through their inclusion into reversed micelles of surfactants, *Dokl. Akad. Nauk SSSR*, 236, 920, 1977.
2. **Martinek, K., Levashov, A. V., Klyachko, N. L., Kmelnitski, Y. L., and Berezin, I. V.**, Micellar enzymology, *Eur. J. Biochem.*, 155, 453, 1986.
3. **Luisi, P. L. and Magid, L. J.**, Solubilization of enzymes and nucleic acids in hydrocarbon micellar solutions, *CRC Crit. Rev. Biochem.*, 20, 409, 1987.
4. **Luisi, P. L., Giomini, M., Pileni, M. P., and Robinson, B. H.**, Reverse micelles as host for proteins and small molecules, *Biochim. Biophys. Acta*, 947, 209, 1988.
5. **Waks, M.**, Proteins and peptides in water-restricted environments, *Prot. Struc. Func. Gen.*, 1, 4, 1986.
6. **Bru, R., Sánchez-Ferrer, A., and García-Carmona, F.**, Characterization of cholesterol oxidase activity in AOT-isooctane reverse micelles and its dependence on micelle size, *Biotechnol. Lett.*, 11, 237, 1989.
7. **Hilhorst, R., Spruijt, R., Laane, C., and Veeger, C.**, Rules for the regulation of enzyme activity in reversed micelles as illustrated by the conversion of apolar steroids by 20β-hydroxy-steroid dehydrogenase, *Eur. J. Biochem.*, 144, 459, 1984.
8. **Lee, K. M. and Biellmann, J. F.**, Cholesterol oxidase in microemulsion: enzymatic activity on a substrate of low water solubility and inactivation by hydrogen peroxide, *Bioorg. Chem.*, 14, 262, 1986.
9. **Menger, F. M. and Yamada, K.**, Enzyme catalysis in water pools, *J. Am. Chem. Soc.*, 101, 6731, 1979.
10. **Barbaric, S. and Luisi, P. L.**, Micellar solubilization of biopolymers in organic solvents. V. Activity and conformation of α-chymotrypsin in isooctane-AOT reverse micelles, *J. Am. Chem. Soc.*, 103, 4239, 1981.
11. **Belonogova, O. V., Likhtenshtein, G. I., Levashov, A. V., Kmelnitski, Y. L., Klyachko, N. L., and Martinek, K.**, Use of the spin label method to study the state of the active site and microsurroundings of α-quimotrypsin, solubilized in octane, using surfactant Aerosol OT, *Biokimiya*, 48, 379, 1983.
12. **Fletcher, P. D. I., Freedman, R. B., Mead, J., Olfield, C., and Robinson, B. H.**, Reactivity of α-quimotrypsin in water-in-oil microemulsions, *Colloids Surf.*, 10, 193, 1984.
13. **Martinek, K., Khmelnitski, Y. L., Levashov, A. V., and Berezin, I. V.**, Substrate specificity of alcohol dehydrogenase in colloidal solution of water in an organic solvent, *Dokl. Akad. Nauk SSSR*, 263, 737, 1982.
14. **Malakhova, E. A., Kurganov, B. I., Levashov, A. V., Berezin, I. V., and Martinek, K.**, A new approach to the study of enzymatic reactions with water-insoluble substrates. Pancreatic lipase enclosed in reversed micelles of a surfactant in organic solvent, *Dokl. Akad. Nauk SSSR*, 270, 474, 1983.
15. **Wolf, R. and Luisi, P. L.**, Micellar solubilization of enzymes in hydrocarbon solvents. Enzymatic activity and spectroscopic properties of ribonuclease in *n*-octane, *Biochem. Biophys. Res. Commun.*, 89, 209, 1979.
16. **Nicot, C., Vacher, M., Vincent, M., Gallay, J., and Waks, M.**, Membrane proteins in reverse micelles: myelin basic protein in a membrane-mimetic environment, *Biochemistry*, 24, 7024, 1985.
17. **Vos, K., Lavalette, D., and Visser, J. W. G.**, Triplet-state kinetics of Zn-porphyrin cytochrome c in micellar media. Measurement of intermicellar exchange rates, *Eur. J. Biochem.*, 169, 269, 1987.
18. **Steinmann, B., Jäckle, H., and Luisi, P. L.**, A comparative study of lysozyme conformation in various reverse micellar systems, *Biopolymers*, 25, 1133, 1986.
19. **Thompson, K. F. and Gierasch, L. M.**, Conformation of a polypeptide solubilizate in a reverse micelle water pool, *J. Am. Chem. Soc.*, 106, 3648, 1984.
20. **Walde, P. and Luisi, P. L.**, A continuous assay for lipases in reverse micelles based on Fourier transform infrared spectroscopy, *Biochemistry*, 28, 3353, 1989.
21. **Verhaert, R. M. D., Schaafsma, T. J., Laane, C., Hilhorst, R., and Veeger, C.**, Optimization of the photo-enzymatic reduction of the carbon-carbon double bond of α-β unsaturated carboxylates in reverse micelles, *Photochem. Photobiol.*, 49, 209, 1989.
22. **Boutonnet, M., Kizling, J., Stenius, P., and Maire, G.**, The preparation of monodisperse colloidal metalparticles from microemulsions, *Colloid Surf.*, 5, 209, 1982.
23. **Kurihara, K., Kizling, J., Stenius, P., and Fendler, J. H.**, Laser and pulse radiolytically induced colloidal gold formation in water and in water-in-oil microemulsion, *J. Am. Chem. Soc.*, 105, 2574, 1983.
24. **Belyaeva, E. I., Brovko, L. Y., Ugarova, N. N., Klyachko, N. L., Levashov, A. V., Martinek, K., and Berezin, I. V.**, Regulation of the catalytic activity of firefly luciferase in the system of Brij 96/water/octane, *Dokl. Akad. Nauk SSSR*, 273, 494, 1983.
25. **Dekker, M., Hilhorst, R., and Laane, C.**, Isolating enzymes by reversed micelles, *Anal. Biochem.*, 178, 217, 1989.
26. **Sánchez-Ferrer, A., Santema, J. S., Hilhorst, R., and Visser, A. J. W. G.**, Fluorescence detection of enzymatically formed hydrogen peroxide in aqueous solution and in reversed micelles, *Anal. Biochem.*, 187, 129, 1990.
27. **Luisi, P. L., Henninger, F., Joppich, M., Dossena, A., and Casnat, G.**, Solubilization and spectroscopic properties of α-quimotrypsin in cyclohexane, *Biochem. Biophys. Res. Commun.*, 74, 1384, 1977.

28. **Kabanov, A. V., Namyotkin, S. N., Levashov, A. V., and Martinek, K.,** Transmembrane transport of artificial hydrophobized proteins (enzymes), *Biol. Membr. (Moscow),* 2, 985, 1985.
29. **Verhaert, R. M. D., Hilhorst, R., Vermue, M., Schaafsma, T. J., and Veeger, C.,** Description of enzyme kinetics in reversed micelles. I. Theory, *Eur. J. Biochem.,* 187, 59, 1990.
30. **Verhaert, R. M. D., Tyrakowska, B., Hilhorst, R., Schaafsma, T. J., and Veeger, C.,** Enzyme kinetics in reversed micelles. II. Behaviour of enoate reductase, *Eur. J. Biochem.,* 187, 73, 1990.
31. **Oldfield, C.,** Evaluation of steady-state kinetic parameters for enzymes solubilized in water-in-oil microemulsion systems, *Biochem. J.,* 272, 15, 1990.
32. **Kabanov, A. V., Levashov, A. V., Klyachko, N. L., Namyotkin, S. N., and Pshezhetsky, A. N.,** Enzymes entrapped in reversed micelles of surfactants in organic solvents: a theoretical treatment of the catalytic activity regulation, *J. Theor. Biol.,* 133, 327, 1988.
33. **Bru, R., Sánchez-Ferrer, A., and García-Carmona, F.,** A theoretical study on the expression of enzymic activity in reverse micelles, *Biochem. J.,* 259, 355, 1989.
34. **Bru, R., Sánchez-Ferrer, A., and García-Carmona, F.,** The effect of substrate partitioning on the kinetics of enzymes acting in reverse micelles, *Biochem. J.,* 268, 679, 1990.
35. **Levashov, A. V., Klyachko, N. L., Pantin, V. I., Khmelnitski, Y. L., and Martinek, K.,** Catalysis by water-soluble enzymes included in reverse micelles of surface-active agents in non-aqueous solvents, *Bioorg. Khim.,* 6, 929, 1980.
36. **Fletcher, P. D. I., Howe, A. M., and Robinson, B. H.,** The kinetics of solubilizate exchange between water droplets of a water-in-oil microemulsion, *J. Chem. Soc. Faraday Trans. I,* 83, 985, 1987.
37. **El Seoud, O. A.,** Acidities and basicities in reversed micellar systems, in *Reverse Micelles,* Luisi, P. L. and Straub, B. E., Eds., Plenum Press, New York, 1984, 81.
38. **Kuntz, I. D. and Kauzmann, W.,** Hydratation of proteins and polypeptides, *Adv. Protein Chem.,* 28, 239, 1974.
39. **Higuchi, W. I. and Misra, J.,** Physical degradation of emulsions via the molecular diffusion route and its possible prevention, *J. Pharm. Sci.,* 51, 459, 1962.
40. **Herries, D. G., Bishop, W., and Richards, F. M.,** The partitioning of solutes between micellar and aqueous phases: measurements by gel filtration and effect on the kinetics of some bimolecular reactions, *J. Phys. Chem.,* 68, 1842, 1964.
41. **Sánchez-Ferrer, A., Bru, R., and García-Carmona, F.,** Kinetic properties of polyphenoloxidase in organic solvents: a study in Brij 96-cyclohexane reverse micelles, *FEBS Lett.,* 233, 363, 1988.
42. **Harel, E. and Mayer, A. M.,** Partial purification and properties of catechol oxidases in grapes, *Phytochemistry,* 10, 17, 1971.
43. **Sánchez-Ferrer, A., Bru, R., and García-Carmona, F.,** Novel procedure for extraction of a latent grape polyphenoloxidase using temperature-induced phase separation in Triton X-114, *Plant Physiol.,* 91, 1481, 1989.
44. **Robb, D. A.,** Tyrosinase, in *Copper Proteins and Copper Enzymes,* Vol. 2, Lontie, R., Ed., CRC Press, Boca Raton, 1984, 207.
45. **Grandi, C., Smith, R. E., and Luisi, P. L.,** Micellar solubilization of biopolymers in organic solvents. Activity and conformation of lysozyme in isooctane reverse micelles, *J. Biol. Chem.,* 256, 837, 1981.
46. **Walde, P., Peng, Q., Fadnavis, N. W., Battistel, E., and Luisi, P. L.,** Structure and activity of trypsin in reverse micelles, *Eur. J. Biochem.,* 173, 401, 1988.
47. **Delahodde, A., Vacher, M., Nicot, C., and Waks, M.,** Solubilization and insertion into reverse micelles of the major myelin transmembrane proteolipid, *FEBS Lett.,* 172, 343, 1984.

Chapter 8

ENZYME ACTIVITY IN REVERSE MICELLES: ROLES OF DIFFUSION AND EXTENT OF HYDRATION

Sérgio T. Ferreira and Enrico Gratton

TABLE OF CONTENTS

I. INTRODUCTION

The kinetic characterization of catalysis by enzymes entrapped in reverse micelles of amphiphiles in organic solvents has received a great deal of attention.[1-3] Reverse micelles can be formed by phospholipids or detergents and are usually capable of taking up relatively large amounts of water in the internal cavity, forming water pools dispersed in the organic phase.[3,4] Several hydrophilic, water-soluble enzymes retain enzymatic activity in reverse micelles, which presents interesting questions regarding their mechanisms of catalysis and the possible role of solvent in their function.

One of the best-studied reverse micellar systems is that of detergent Aerosol OT (AOT)[1] in organic solvents. Phase-equilibria are well characterized for this system,[5-7] and several physicochemical techniques have been employed for the study of the hydration of detergent polar heads in the interior of the micelle as well as of the structure of water in the water pool.[8-12] These studies indicated that up to a ratio of concentrations of water to AOT (Wo = $[H_2O]/[AOT]$) of 8 to 10, water is preferentially bound to the detergent polar heads and the properties of this bound water may differ significantly from those of bulk water.[8-12] Upon increasing hydration, the observed properties of the water pool gradually approach those of bulk water so that at high hydration levels (Wo = 70 to 80), it is generally assumed that the micelle contains a water pool which may resemble a pure aqueous solution. An advantage of the use of AOT is that this system is very homogeneous with respect to size dispersity.[13-15] Thus, it is likely that average micelle dimensions calculated from known aggregation parameters will reflect to a close approximation the true average dimensions of the micellar population.

Despite the large amount of work on enzyme activity in reverse micelles, several questions pertaining to the mechanisms of catalysis are still debated. For example, studies on the dependence of catalysis on the extent of hydration have revealed a bell-shaped dependence[1-3,16] for most enzymes investigated. In other words, at very low molar ratios of water to surfactant the enzyme is usually inactive, and activity increases with increasing hydration up to a certain Wo level beyond which activity again decreases. Furthermore, in some cases the activities observed at the maximum of the hydration-dependent activity curve have been reported to be higher than the activities of the enzymes in solution.[1-3] This striking feature has led to the concept that the particular environment of the reverse micelles (e.g., differences in polarity or structure of water as compared to bulk water) may result in hyperactivity of the enzymes.[2,3,16] While the physical basis for this hyperactivity remains mostly obscure,[2,3] some models have been proposed to account for this unusual behavior. For example, enzyme activity has been correlated with changes in the size of the micelles as a function of hydration and/or surfactant concentration.[16]

In the present work, we show that some of the unusual properties displayed by enzymes in reverse micelles (e.g., the bell-shaped water dependence and hyperactivity) can be qualitatively explained on the basis of considerations of: (1) the compartmentalization of micellar systems, (2) the influence of diffusion of enzyme-containing and substrate-containing micelles on the observed reaction rates, and (3) the hydration-dependent activation of enzyme catalytic properties. A connection is proposed between hydration studies in reverse micelles and previous results on the effects of water on several dynamic properties of proteins including enzyme activity.[17-19] In the next sections, we will briefly outline our basic assumptions and calculate, from literature values of structural parameters of AOT reverse micelles, the expected rates of reaction between enzymes and substrates in reverse micelles as a function of hydration level.

II. BASIC ASSUMPTIONS

A few initial considerations should be made regarding the effective concentrations of

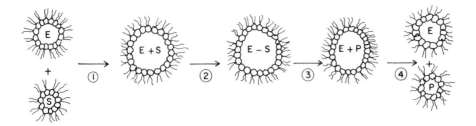

FIGURE 1.   Reaction scheme for enzyme (E) and substrate (S) to give the product (P).

reactive species in reverse micellar systems. As a simplification, we will disregard any significant partition of enzyme or substrate molecules into the main organic solvent phase; more general kinetic treatments which take into account the partition have been presented.[20-23] This should be a fair simplification for water-soluble proteins and substrates which possess negligible solubilities in the apolar solvents (such as octane or isooctane) which constitute the organic phase. In reverse micellar enzymology, measured kinetic parameters are usually expressed in terms of either overall concentrations or as "water-pool" concentrations of reactants.[3,4] For water-soluble, hydrophilic enzymes and/or substrates, the use of an overall concentration (i.e., the number of moles present divided by the total reaction volume) clearly presents a large underestimation of the effective concentrations inside the micellar cavities. To account for the nonsolubility of reactants in the organic phase, "water-pool" concentrations (i.e., the number of moles divided by the total volume of water in the assay) have been introduced;[3,5] however, a few words of caution should be said about water-pool concentrations as they have been previously defined. Typical AOT concentrations used in reverse micellar enzymology are in the 50- to 100-mM range, while overall enzyme and substrate concentrations range from 1 to 500 μM.[20,24-28] As a consequence, a large number of micelles will in fact not be occupied by either enzyme or substrate. The relative proportions of empty and occupied micelles can be calculated from the Poisson distribution[29] showing that under most experimental conditions only a fraction of the micelles will contain either enzyme or substrate (see below). Thus, the volume of water that should be used to calculate effective concentrations of reactants is not the total volume of water which was added to the assay, but the total volume inside each individual micelle which *actually contains* enzyme or substrate. Thus, we adopt here the use of *effective* or *local* concentrations, which can be obtained dividing the number of reactant molecules contained in a micelle (given by the Poisson distribution) by the volume of water in this micelle.[30] The effective concentrations thus calculated should provide a description of the micellar system which is closest to the true physical situation, and an immediate conclusion is that, due to the small volume of the water pool inside an individual micelle, the effective concentrations of enzymes and substrates can be extremely high — of the order of several millimolar — as compared to micromolar water-pool concentrations. Possible consequences of these high local concentrations on the reaction rates will be discussed later.

The primary condition for reaction between enzyme and substrate, when both are entrapped in reverse micelles, is collision between enzyme-containing and substrate-containing micelles. The basic reaction scheme we adopt is shown in Figure 1. The diffusion-controlled collision between enzyme- and substrate-containing micelles (step 1) gives rise to a transient larger micelle containing both enzyme and substrate. A complex is formed between enzyme and substrate (step 2), which will eventually lead to conversion of substrate into product (step 3). The larger micelle is again split (step 4), and the enzyme-containing micelle is ready for another cycle, while the product-containing micelle may diffuse away.

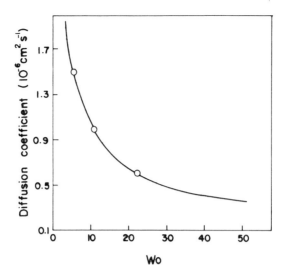

FIGURE 2.   Diffusion coefficient for enzyme-containing micelles as a function of water content (Wo = [H$_2$O]/[AOT]).

III. DIFFUSION-CONTROLLED COLLISION BETWEEN M$_E$ AND M$_S$

Structural characterization of reverse micelles of AOT (50 mM) in isooctane has been presented.[30] In addition to micellar parameters (aggregation number, radius, number of water molecules per micelle, and diffusion coefficient) as a function of Wo for empty micelles, structural data on reverse micelles containing lysozyme, ribonuclease, and liver alcohol dehydrogenase at various hydration levels are also available.[30] Therefore, we will use available structural data[30] for the calculation of the rates of collision between M$_S$ and M$_E$.

Diffusion coefficients for substrate-containing micelles (D$_S$) were calculated using the classical expression for a spherical particle:

$$D_S = kT/f \tag{1}$$

where k is Boltzmann's constant, T is the absolute temperature, and f is Stokes' frictional coefficient (f = 6$\pi\eta$r). The viscosity (η) of isooctane (at 298 K) is 0.4478 cP.[31] Internal radii for M$_S$ were taken from values of Bonner et al.[30] for empty micelles (which should be a fair approximation given the small size of most substrate molecules — generally a few angstroms in length — as compared to the radii of the micelles, which can be one to two orders of magnitude larger, depending on Wo). Total radii (r) were obtained by adding 12 Å (corresponding to the extended length of an AOT molecule)[31] to the internal radii.

Diffusion coefficients for enzyme-containing micelles (D$_E$) were obtained through an empirical fit of literature values[30] for micelles containing ribonuclease (MW $\cong$ 13,700) in 50 mM AOT in isooctane. Figure 2 shows diffusion coefficients as a function of Wo for ribonuclease-containing micelles; open circles are experimental values.[30] The solid line is an empirical fit of the data (r^2 = 0.998) to the equation:

$$D_E = (4.8946 \times 10^{-6})Wo^{-0.67235} \text{ cm}^2/\text{s} \tag{2}$$

While no further assumption was made regarding the dependence of D$_E$ on Wo, this empirical relationship was used to estimate D$_E$ at various hydration levels. Table 1 shows diffusion coefficients calculated according to Equations 1 and 2 for M$_S$ and M$_E$, respectively, at

TABLE 1

Diffusion Parameters for Enzyme-Containing (M_E) and Substrate-Containing (M_S) Micelles

Wo	r_E^a (Å)	r_S^a (Å)	D_E^b (10^{-6} cm²/s)	D_S^b (10^{-6} cm²/s)	k^c (10^{10} M^{-1} s^{-1})
3	31	20	2.34	2.43	1.84
5	33	24	1.66	2.03	1.59
10	37	30	1.04	1.62	1.35
15	44	40	0.79	1.22	1.28
20	49	46	0.65	1.06	1.23
30	65	64	0.50	0.76	1.23
40	81	80	0.41	0.61	1.24
50	97	97	0.35	0.50	1.25

[a]   Micellar radius, equal to the internal radius (from Ref. 30) plus 12 Å (the length of an extended AOT molecule).[32]

[b]   Diffusion coefficients calculated with Equations 1 and 2 for substrate-containing (D_s) and enzyme-containing (D_E) micelles, respectively.

[c]   Rate constant for collision between M_E and M_S, calculated according to Equation 3.

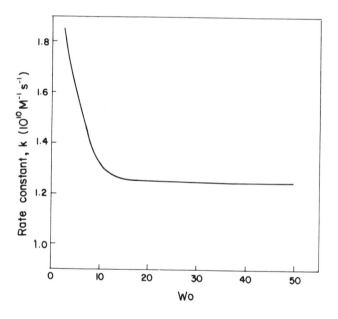

FIGURE 3.   Diffusion-controlled rate constant for collision between a micelle containing enzyme and one containing substrate as a function of Wo = [H$_2$O]/[AOT].

selected Wo values. The micelles grow considerably upon increasing Wo from 3 to 50 (Table 1), which results in a significant decrease in their diffusion coefficients. The rate constant (k) for diffusion-controlled collision between M_S and M_E (Table 1 and Figure 3) was then calculated from the expression:

$$k = 4\pi r_o(D_E + D_S)10^{-3}N_o \qquad (3)$$

where r_o is the effective collision radius equal to the sum of the radii of the individual particles, and N_o is Avogadro's number.

TABLE 2
Internal Areas and Aggregation Numbers
for Enzyme-Containing (M_E) and
Substrate-Containing (M_S) Micelles

Wo	$\mathring{A}_E^a$ ($\mathring{A}^2$)	$\mathring{A}_S^a$ ($\mathring{A}^2$)	f_A^b ($\mathring{A}^2$)	n_E^c	n_S^c
3	4,536	804	32.7	139	25
5	5,542	1.810	41.0	135	44
10	7.854	4,072	46.5	169	88
15	12,868	9,852	50.0	257	197
20	17,203	14,527	52.0	331	279
30	35,229	33,979	53.3	662	638
40	59,828	58,107	53.8	1,112	1,080
50	90,792	90,792	54.0	1,681	1,681

[a] Internal areas calculated from the values of internal radii from Reference 30.
[b] Area occupied by each AOT polar head (interpolated from data of Reference 11).
[c] Aggregation numbers calculated with Equation 4.

Figure 3 shows that as Wo is increased (from Wo = 3 to 10), there is an initial rapid decrease in k which then levels off at a constant value up to Wo = 50. The initial decrease in k reflects the decrease in diffusional mobility of the growing micelles. As the micelles grow, however, the effective collision radius (r_o) also increases and the increase in r_o counterbalances the decrease in diffusional mobility of the micelles above Wo = 10.

As seen above, increasing Wo results in larger micelles. This means that the aggregation number also increases with increasing hydration, and, hence (since total AOT concentration is kept constant), the overall concentration of micelles decreases. This will, of course, affect the reaction rate between M_E and M_S. Thus, we will now proceed to calculate the concentrations of M_E and M_S as a function of Wo.

The micellar aggregation number can be calculated dividing the internal area of the micelle by the average area occupied by each detergent polar head (f_A). Values for the area occupied by an individual AOT polar head as a function of Wo have been presented.[11] We will use the assumption[30] that f_A will not be modified much by the presence of a protein inside a micelle so that this packing parameter can be used to calculate aggregation numbers for both M_S and M_E. Aggregation numbers (n_o) were calculated as:

$$n_o = 4\pi r_i^2/f_A \tag{4}$$

where r_i is the internal radius of the micelle. Table 2 shows the internal areas of M_S and M_E, as well as the area parameter f_A (from Eicke and Kvita[11]), as a function of Wo. The aggregation number for a substrate-containing micelle increases by approximately 60-fold when Wo is increased from 3 to 50 (Table 2) which implies that the concentration of micelles will decrease by the same factor. On the other hand, a tenfold increase in n_o occurs for an enzyme-containing micelle.

The question, then, is to calculate, for the conditions of a typical enzymatic activity assay in reverse micelles, the concentrations of enzyme-containing and substrate-containing micelles as a function of Wo. For example, let us consider the measurements reported on the hydrolysis of *N*-glutaryl-L-phenylalanine *p*-nitroanilide by α-chymotrypsin in AOT reverse micelles.[27] In this case, 50 m*M* AOT in isooctane was used and overall enzyme and substrate concentrations were 1.2 and 50 to 400 μ*M*, respectively. The concentrations of

TABLE 3
Overall Micellar Concentrations[a] and Probabilities of Finding
Different Micellar Species[b] in AOT-Isooctane Reverse Micelles

Wo	[micelles]$_{ov}$ (μM)	P(0;μ)	P(1;μ)	P(2;μ)	P(3;μ)	P(4;μ)
3	2,000	0.975	0.024	<0.001	<0.001	<0.001
10	568	0.916	0.081	0.004	<0.001	<0.001
20	179	0.757	0.211	0.029	0.003	<0.001
25	114	0.645	0.283	0.062	0.009	0.001
30	78	0.527	0.338	0.108	0.023	0.004
40	46	0.337	0.367	0.199	0.072	0.020
50	30	0.189	0.315	0.262	0.146	0.061

Note: Probabilities are expressed as P $(x;\mu)$, where x is the number of substrate molecules per micelle and μ = [substrate]$_{ov}$/[micelle]$_{ov}$.

[a] Overall micellar concentration calculated for a 50-mM AOT solution in isooctane with aggregation numbers from Table 2.
[b] Probabilities calculated with Equation 5 as described in the text for [substrate]$_{ov}$ = 50 μM and u = [substrate]$_{ov}$/[micelles]$_{ov}$.

M_E and M_S are given by the Poisson distribution[29] which yields the probability of finding micelles containing x molecules of reactant, given a certain ratio (μ) of overall reactant and micelle concentrations:

$$P(x;\mu) = (\mu^x/x!)e^{-\mu} \qquad (5)$$

Table 3 lists overall micelle concentrations as a function of Wo for a 50-mM AOT solution in isooctane, as well as the probabilities — given by the Poisson distribution — of finding micelles containing x substrate molecules for an overall substrate concentration of 50 μM. At low hydration levels (e.g., Wo = 3) most of the micelles are actually empty, and there is only a 2.4% probability of finding a micelle containing one substrate molecule (Table 3). The probability of finding micelles containing two or more substrate molecules at Wo = 3 is negligible (<0.1%, Table 3). Increasing Wo to 50 results in a marked decrease in the overall concentration of micelles which significantly changes their occupancy. At Wo = 50 (Table 3), there is only an 18.9% chance of finding empty micelles and there are significant probabilities of finding micelles containing one, two, three, or even four substrate molecules.

Once the probability of finding a given species of micelle (containing x substrate molecules) is known, the overall concentration of this micellar species is given by the product of this probability by the total overall concentration of micelles. Figure 4A shows the concentration of each different micellar species (containing x substrate molecules, with $1 < x < 4$) as a function of Wo. For our purposes, all substrate-containing micelles (irrespective of the number of substrate molecules they contain) may potentially collide with an enzyme-containing micelle which may lead to catalysis. Therefore, the concentration of all substrate-containing micelles is of relevance (Figure 4B). It is apparent from Figure 4B that above Wo = 10 the concentration of substrate-containing micelles decreases rapidly as a function of Wo.

Due to the very low enzyme concentration employed (1.2 μM)[27] the occupancy of the micelles by enzyme is always very low. The Poisson distribution shows that, irrespective of Wo, the vast majority of the micelles does not contain enzyme and the concentration of micelles containing two or more enzyme molecules is negligible (data not shown). Thus, irrespective of Wo, the concentration of enzyme-containing micelles is, in this example, constant at 1.2 μM.

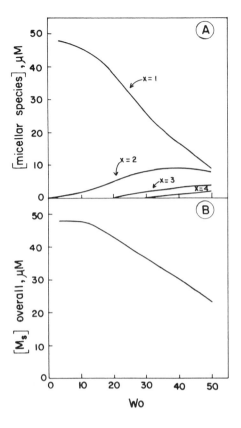

FIGURE 4.   (A) Micelles containing one, two, three, and four substrate molecules as a function of Wo = [H$_2$O]/[AOT]. (B) Overall concentration of micelles containing substrates. The substrate overall concentration of 50 μM in a solution of 50 mM AOT in isooctane.

Finally, the rate of collision between M$_E$ and M$_S$ at a given Wo value can be calculated from the product of the collision rate constant (k) times the concentrations of M$_E$ and M$_S$:

$$V = k[M_E][M_S] \tag{6}$$

The rate of collision between M$_E$ and M$_S$, calculated for overall enzyme and substrate concentrations (in a 50-mM AOT solution in isooctane) of 1.2 and 50 μM, respectively, is shown in Figure 5 as a function of Wo. Increasing Wo from 3 to 50 results in a marked decrease (approximately fivefold) in the rate of collision.

IV. HYDRATION-DEPENDENT ACTIVATION OF CATALYSIS

The discussion above took into account only the influence of micellar diffusion on the rates of reaction in reverse micelles. However, one important aspect that has yet to be considered is the role of hydration in the activation of catalysis. Several lines of evidence have led to a dynamic view of protein structure,[32-36] and it is now widely accepted that hydration plays a pivotal role in the modulation of structural fluctuations of the protein matrix which may be essential to biological function.[17-19,33] The effect of water is minimal on the protein structure, but it is essential to the dynamic properties of the protein.[19] In the

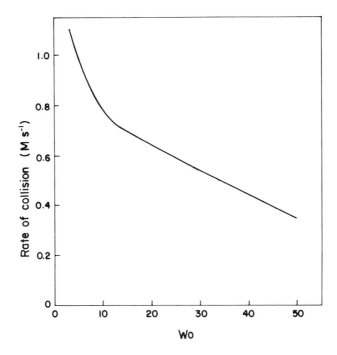

FIGURE 5. Rate of collision between a micelle containing at least one substrate molecule and a micelle containing an enzyme molecule as a function of $Wo = [H_2O]/[AOT]$.

absence of the hydration shell, the protein is essentially frozen and inactive,[18] and there is very little, if any, motion occurring in the time range of 1 ns and longer. As water is gradually added to the protein, nothing major occurs with respect to the protein dynamics (here including catalysis)[18] until about 0.2 g of water per gram of protein. Above this value, there is a sudden increase in internal dynamics and enzyme activity which gradually approach their solution values.[18,33] Since hydration appears to be paramount in the modulation of enzyme activity, a proper description of catalysis in reverse micelles should include these previous results.

The hydration-dependent activation of catalysis has been measured in hydrated lysozyme powders[18,33] as a function of grams of water added per gram of protein. To correlate these previous results with our results on micelle diffusion, we will express our results in terms of grams of water per gram of protein (h) inside the micelles instead of using the usual parameter Wo. The conversion from Wo into h can be made as:

$$h = (Wo_n/MW_{H_2O})/MW_{prot} \qquad (7)$$

where MW_{H_2O} is the molecular weight of water (18 g) and MW_{prot} is the molecular weight of the protein. In our case, for chymotrypsin (MW $\cong$ 25,000):

$$h = (7.2 \times 10^{-4})Wo_n \qquad (8)$$

Figure 6A (open circles) shows a plot of the data available on the hydration-dependent activation of lysozyme (from Refs. 18 and 33) as a function of h. The experimental data available cover a limited hydration range (<1 g of water per gram of protein) as compared to the hydration range which can be probed in reverse micelles (for comparison, see scale in Figure 6A, expressed in terms of Wo). To calculate a form of the function which describes

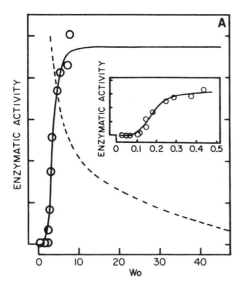
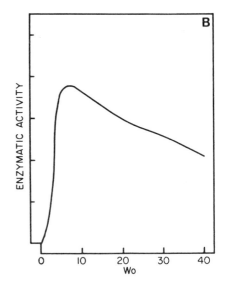

FIGURE 6.   (A) Hydration-dependent activity of lysozyme from References 18 and 34. h = (mg H$_2$O)/(mg proteins). (Inset) Fit of the hydration-dependent data using the Hill equation (Equation 9; see text). Dashed line is from Figure 5. (B) Activity profile obtained using the rate of encounter between enzyme and substrate, modulated by the hydration-dependent enzyme activity obtained from panel A.

activity in reverse micelles to bring into account that enzyme activity is hydration dependent, the data reported in Figure 6A were fit using an analytical expression. The purpose here is not to prove the validity of this particular equation but to describe the enzyme-activity curve using an analytical function for computational purposes. Hydration data were fit in terms of a cooperative model of hydration-dependent activation of catalysis which accounts for cooperative binding of water molecules to the protein; the inset in Figure 6A shows a fit of hydration data with the Hill equation:

$$v = V_{max}h^n/(K_{0.5})^n + h^n \qquad (9)$$

where $K_{0.5}$ is the concentration of water which promotes half-maximal activation of catalysis and n is the Hill coefficient. This fit yielded $V_{max} = 0.94$ s^{-1}, $K_{0.5} = 0.35$ g of water per gram of protein, and a Hill coefficient of 4.95.

For comparison, Figure 6A (dashed line) also shows the profile for diffusion-controlled rate of collision between M_S and M_E, as previously shown in Figure 5.

Figure 6B shows the profile resulting from the product of the hydration-dependent activation of catalysis times the rate of collision. Figure 6B shows that the reaction measured under such conditions will display an approximately bell-shaped dependence on Wo. In the low-hydration regime (Wo < 10), the activation of catalysis by increasing hydration is sufficiently sharp (open circles and solid line, Figure 6A) to overcome the decrease in diffusional mobility of M_E and M_S. Above Wo = 10, however, the effect of hydration on the activation of catalysis is complete (Figure 6A, solid line) and the decrease in diffusion is large enough to produce a decrease in the product. Thus, our results show that a bell-shaped dependence of enzymatic reaction on Wo may be obtained by considering simultaneously a monotonic profile of activation of catalysis upon hydration and changes in the encounter rate of M_E and M_S.

V. LOCAL CONCENTRATION OF REACTANTS AND HYPERACTIVITY

Reaction in reverse micellar systems necessarily requires collision between M_E and M_S. After one such collision, substrate and enzyme are confined to the small internal volume of this newly formed micelle, which means that their local concentrations may be extremely high. It is illustrative to consider, for example, the case of the hydrolysis of N-glutaryl-L-phenylalanine p-nitroanilide by chymotrypsin[27] which has been analyzed in the previous sections. At Wo = 9, a hydration level at which hyperactivity has been reported,[27] M_E has an inner radius of about 25 Å (as interpolated from the values of Table 1). The local or effective concentration of protein in this micelle is given by:

$$[\text{Protein}]_{\text{eff}} = 3/4\pi N_o 10^3 r_i^3 \qquad (10)$$

This calculation yields a local concentration of chymotrypsin of 25 mM. The same calculation performed for a substrate-containing micelle (and 50 μM overall substrate concentration) yields a local substrate concentration (in a micelle containing a single substrate molecule) of 68 mM. Assuming that fusion of M_E and M_S results in a new micelle with a volume equal to the sum of the volumes of M_E and M_S, final concentrations in this micelle are 18 mM for both enzyme and substrate. Naturally, higher overall substrate concentrations in the assay will result in higher occupancy of the micelles leading to even higher local concentrations of substrate. These conditions clearly go against the usual Michaelis-Menten assumptions used in enzyme kinetics. Namely, an important assumption in Michaelis-Menten analysis is that the concentration of free substrate is very nearly equal to the total substrate concentration, i.e., that enzyme concentration is much smaller than substrate concentration. In addition, usually steady-state conditions are assumed, which implies that the concentration of the enzyme-substrate complex (ES) is constant in time (i.e., d[ES]/dt = 0). Finally, it is generally accepted that none of the product reverts to substrate, which stems from the fact that, at initial stages, the concentration of product is always very low. Clearly, none of these three basic requirements for Michaelis-Menten analysis are met in a micellar situation such as described above. For one, enzyme and substrate concentrations in a potentially reactive micelle are both extremely high, and formation of a single ES complex inside a micelle which originally contained a single substrate molecule results in the total consumption of substrate. As a consequence, no steady-state approximation can be made for reaction inside an individual micelle, since the concentration of ES rises abruptly from zero to several millimolar upon complex formation and then falls abruptly back to zero when substrate is converted into product. Finally, diffusion of the product is dependent on the splitting of the micelle or on the collision and exchange of contents with another micelle so that, at least at an early stage after conversion of substrate into product, concentrations of product can be of the same order of enzyme concentrations. Thus, it may be that the observation of hyperactivity arises from the fact that enzymatic reactions confined to the water pool inside a reverse micelle are subject to extreme conditions which deviate significantly from Michaelis-Menten assumptions. Extremely high local concentrations of enzyme and substrate, as well as the relative immobilization of reactants and geometric hindrances imposed by the micellar cavity, may result in a significant increase in the effectiveness of intramicellar collisions between enzyme and substrate, giving rise to increased rates of conversion of substrate into product. While the considerations presented in this work may qualitatively explain some of the unusual features of enzyme behavior in reverse micelles, it is clearly impossible to rule out specific effects of the micellar environments on enzyme structure and function. Further investigation of enzymatic reactions carried out in micellar environments or with enzymes and/or substrates immobilized in a solid matrix may give more insight into these phenomena.

In conclusion, our analysis shows that a bell-shaped curve of enzyme activity in reverse micelles as a function of Wo is a natural consequence of two contrasting effects: (1) At low hydration, the high rate of encounters gives a limited catalysis due to the incomplete hydration of enzyme, and (2) At high hydration, the rate of encounters decreases, but now the enzyme is fully active. The contribution of the two effects gives an apparent hyperactivity as compared to the solution value, i.e., in the limit of very high Wo.

ABBREVIATIONS

AOT   Aerosol OT, *bis*-(2-ethylhexyl) sulfosuccinate, sodium salt
D_E Diffusion coefficient of enzyme-containing micelles
D_S Diffusion coefficient of substrate-containing micelles
f_A Area occupied by each AOT polar head
h   Grams of water per gram of protein
M_E Enzyme-containing micelle
M_S Substrate-containing micelle
n   Hill coefficient
n_o Micellar aggregation number
Wo   Ratio of the concentration of water to the concentration of AOT

ACKNOWLEDGMENTS

This work was supported by grants from Fundagào de Amparo à Pesquisa do Estado do Rio de Janeiro (FAPERJ), Conselho Nacional de Desenvolvimento Cientifico e Tecnológico (CNPq), IBM-Brazil (Ferreira), and National Institutes of Health (NIH) grant RR 03155 (Gratton).

REFERENCES

1. **Luisi, P. L.**, Enzymes as guest molecules in reverse micelles, *Angew. Chem.*, 24, 439, 1985.
2. **Martinek, K., Levashov, A. V., Klyachko, N., Khmelnitskii, Y. L., and Berezin, I. V.**, Micellar enzymology. Catalytic activity of peroxidase in colloidal aqueous solution in an organic solvent, *Eur. J. Biochem.*, 155, 453, 1986.
3. **Luisi, P. L., Giomini, M., Pileni, M. P., and Robinson, B. H.**, Reverse micelles as host for proteins and small molecules, *Biochim. Biophys. Acta*, 947, 209, 1988.
4. **Luisi, P. L. and Magid, L. J.**, Solubilization of enzymes and nucleic acids in hydrocarbon micellar solutions, *CRC Crit. Rev. Biochem.*, 20, 409, 1986.
5. **Kon-no, K. and Kitahara, A. J.**, Secondary solubilization of electrolytes by Di-(2-ethyl hexyl) sodium sulfosuccinate in cyclohexane solutions, *Colloid Interface Sci.*, 41, 47, 1972.
6. **Ekwall, P.**, Composition, properties, and structures of liquid crystalline phases in systems of amphiphilic compounds, in *Adv. Liq. Cryst.*, Vol. 1, Brown, G. H., Ed., Academic Press, New York, 1975, 1.
7. **Maitra, A., Vasta, G., and Eicke, H.-F.**, Head group interactions in membrane-cholesterol association: studies of Aerosol OT model membrane systems by ^{13}C chemical shifts, relaxation and nuclear Overhouser enhancement effects, *J. Colloid Interface Sci.*, 93, 383, 1983.
8. **Higuchi, W. I. and Misra, J.**, Physical degradation of emulsion via the molecular diffusion route and possible prevention, *Pharm. Sci.*, 51, 455, 1962.
9. **Morel, J. P., Morel-Desrosiers, N., and Lhermet, C.**, Heat capacities and volumes in the reverse micellar phase of the OAT-water-decane systems, *J. Chim. Phys.*, 81, 109, 1984.
10. **Maitra, A. N.**, Determination of size parameters in water-Aerosol OT-oil reverse micelles from their nuclear magnetic resonance data, *J. Phys. Chem.*, 88, 5122, 1984.
11. **Eicke, H.-F. and Kvita, P.**, Reverse micelles and aqueous microphases, in *Reverse Micelles*, Luisi, P. L. and Straub, B. E., Eds., Plenum Press, New York, 1984, 21.

12. **Magid, L. J. and Martin, C. A.,** Aggregation of sulfosuccinate surfactants in water, in *Reverse Micelles,* Luisi, P. L. and Straub, B. E., Eds., Plenum Press, New York, 1984, 121.

13. **Gulari, E., Bedwell, B., and Alkhafaji, S.,** Quasi-elastic light scattering investigation of microemulsions, *J. Colloid Interface Sci.,* 77, 202, 1980.

14. **Robinson, B. H., Toprakcioglu, C., Dore, J. C., and Chieux, P.,** Small-angle neutron-scattering studies of microemulsions stabilized by aerosol-OT. I. Solvent and concentration variations, *J. Chem. Soc. Faraday Trans. 1,* 80, 13, 1984.

15. **Toprakcioglu, C., Dore, J. C., Robinson, B. H., Howe, A., and Chieux, P.,** Small-angle neutron-scattering studies of microemulsions stabilized by aerosol-OT. II. Critical scattering and phase stability, *J. Chem. Soc. Faraday Trans. 1,* 80, 413, 1984.

16. **Kabanov, A. V., Levashov, A. V., Klyachko, N. L., Namyotkin, S. N., Pshezhetsky, A. V., and Martinek, K.,** *J. Mol. Biol.,* 20, 275, 1987 (Russ.).

17. **Yang, P. H. and Rupley, J. A.,** Protein-water interactions. Heat capacity of the lysozyme-water system, *Biochemistry,* 18, 2654, 1979.

18. **Careri, G., Gratton, E., Yang, P.-H., and Rupley, J. A.,** Correlation of IR spectroscopic, heat capacity, diamagnetic susceptibility and enzymatic measurements on lysozyme powder, *Nature,* 284, 572, 1980.

19. **Rupley, J. A., Gratton, E., and Careri, G.,** Water and globular proteins, *Trends Biochem. Sci.,* 8, 18, 1983.

20. **Martinek, K., Levashov, A. V., Klyachko, N. L., Pantin, V. I., and Berezin, I. V.,** The principle of enzyme stabilization. VI. Catalysis of water-soluble enzymes trapped into reversed micelles of surfactants in organic solvents, *Biochim. Biophys. Acta,* 657, 277, 1981.

21. **Hilhorst, R., Sprujit, R., Laane, C., and Veeger, C.,** Rules for the regulation of enzymic activity in reverse micelles as illustrated by the conversion of apolar steroids by 20/3-hydroxysteroid dehydrogenase, *Eur. J. Biochem.,* 144, 459, 1984.

22. **Fletcher, P. D. I., Robinson, B. H., Freedman, R. B., and Oldfield, C.,** Activity of lipase in water-in-oil microemulsions, *J. Chem. Soc. Faraday Trans. 1,* 81, 2667, 1985.

23. **Martinek, K.,** Micellar enzymology, *Eur. J. Biochem.,* 155, 453, 1986.

24. **Douzou, P., Keh, E., and Balny, C.,** Cryoenzymology in aqueous media: micellar solubilized water clusters, *Proc. Natl. Acad. Sci. U.S.A.,* 76, 681, 1979.

25. **Menger, F. M. and Yamada, K.,** Enzyme catalysis in water pools, *J. Am. Chem. Soc.,* 101, 6731, 1979.

26. **Wolf, R. and Luisi, P. L.,** Micellar solubilization of enzymes in hydrocarbon solvents, enzymatic activity and spectroscopic properties of ribonuclease in *n*-octane, *Biochem. Biophys. Res. Commun.,* 89, 209, 1979.

27. **Barbaric, S. and Luisi, P. L.,** Micellar solubilization of biopolymers in organic solvents. V. Activity and conformation of α-chymotrypsin in isooctane-AOT reverse micelles, *J. Am. Chem. Soc.,* 103, 4239, 1981.

28. **Grandi, C., Smith, R. E., and Luisi, P. L.,** Micellar solubilization of biopolymers in organic solvents. Activity and conformation of lysozyme in isooctane reverse micelle, *J. Biol. Chem.,* 256, 837, 1981.

29. **Bevington, P. R.,** *Data Reduction and Error Analysis for the Physical Sciences,* McGraw-Hill, New York, 1969, 53.

30. **Bonner, F. J., Wolf, R., and Luisi, P. L.,** Micellar solubilization of biopolymers in organic solvents. I. A structural model for protein-containing reverse micelles, *J. Solid-Phase Biochem.,* 5, 255, 1980.

31. **Zulauf, M. and Eicke, H. F.,** Inverted micelles and microemulsions in the ternary system water/aerosol-OT/isooctane as studied by photon-correlation spectroscopy, *J. Phys. Chem.,* 83, 480, 1979.

32. **Careri, G., Fasella, P., and Gratton, E.,** Statistical time events in enzymes: a physical assessment, *CRC Crit. Rev. Biochem.,* 3, 141, 1975.

33. **Frauenfelder, H. and Gratton, E.,** Protein dynamics and hydration, *Methods Enzymol.,* 127, 11, 1985.

34. **Ansari, A., Berendzen, J., Bowne, S. F., Frauenfelder, H., Iben, I. E. T., Sauke, T. B., Shyamsunder, E., and Young, R. O.,** Protein states and proteinquakes, *Proc. Natl. Acad. Sci. U.S.A.,* 82, 5000, 1985.

35. **Karplus, M. and McCammon, J. A.,** Dynamics of proteins: elements and function, *Annu. Rev. Biochem.,* 53, 263, 1983.

36. **Karplus, M., Bruenger, A. T., Elber, R., and Kuriyan, J.,** Molecular dynamics: applications to proteins, *Cold Spring Harbor Symp. Quant. Biol.,* 52, 381, 1987.

Chapter 9

REVERSED MICELLES: RECENT ADVANCES AND FUTURE PROSPECTS

Janos H. Fendler

TABLE OF CONTENTS

I. INTRODUCTION

Structures of reversed micelles and interactions and reactions therein continue to receive increasing attention. Surfactant-solubilized water molecules in reversed micelles or in water-in-oil (w/o) microemulsions can be considered as polar droplets in the sea of nonpolar solvent. As such, they provide nanometer-sized reactors with highly specific properties. Much recent progress has been made in preparing, characterizing, and exploiting enzymes,[1-3] catalysts,[4] and semiconductors[5,6] in reversed micelles.

Relatively few surfactant systems have been utilized in these fundamental investigations. Thus, Aerosol-OT (1,2-*bis*-2-ethylhexylsulfonate), PEDGE (pentaethylene glycol dodecyl ether), dodecylammonium propionate, and lecithin have been used most often. There is no *a priori* reason to assume that reversed micelles prepared from other surfactants would have properties identical to those formed from Aerosol-OT. Indeed, many of the reversed micellar systems utilized in a variety of industrial applications behave differently.

Recent work from our laboratory will be highlighted in this chapter. Emphasis will be placed on solubilizate reorganization in calcium alkylarylsulfonate reversed micelles,[7] on the observation of intimate details during the sulfate-ion-triggered fusion of dioctadecyldimethylammonium bromide (DODAB) vesicles,[8] and on the formation of ultrathin metal-island particulate films by the transfer of monolayers of reversed-micellar-entrapped colloidal particles to solid support.[9]

II. NOT ALL REVERSED MICELLES ARE EQUAL

Reversed micelles, freshly prepared from 2.0×10^{-3} M calcium alkylarylsulfonate surfactant (**1**)* and 4.0×10^{-4} M TbCl$_3$ containing 2.2×10^{-2} M water ([H$_2$O]/[**1**] = Wo = 11.1), showed no luminescence upon excitation at 300 nm (the wavelength at which $\epsilon_1 = 4200$ M^{-1} cm^{-1} and $\epsilon_{TbCl_3} = 0.12$ M^{-1} cm^{-1}). Allowing this solution to stand at room temperature for several hours resulted, however, in the appearance of a weak luminescence which grew in intensity with time and became identifiable with that corresponding to TbCl$_3$ emission.

Similar behavior has been observed for different concentrations of TbCl$_3$ added to reversed micellar **1**. These data are shown graphically in Figure 1. A typical time dependence of luminescence development is shown in the insert in Figure 1.

Addition of hydroxyphenylacetic acid to TbCl$_3$ in 1:1 molar ratio increased the amount of luminescence and its rate of appearance (Figure 2). Hydroxyphenylacetic acid is known to complex with Tb^{3+} and, hence, to increase its luminescence. Incorporation of TbCl$_3$ in methanolic, rather than in aqueous solutions, into reversed micelles of **1** led to a much faster development of luminescence. Thus, for example, weak luminescence could be observed 2 h subsequent to introducing 2×10^{-4} M TbCl$_3$ in methanol into 2×10^{-3} M **1** heptane. Within 5 h of its preparation, the luminescence intensity of this solution corresponded to that observed 40 h subsequent to the introduction of identical concentrations of aqueous TbCl$_3$ into 2.0×10^{-3} M **1** in heptane. Precipitation (30 to 40 h subsequent to their preparation) precluded, however, a detailed examination of energy transfer in reversed micelle-solubilized methanolic solutions.

Since fluorescence lifetimes of **1** in heptane ($\tau_1 = 1.03 \pm 0.02$ ns, $\tau_2 = 5.12 \pm 0.06$ ns, $\tau_3 = 12.3 \pm 0.11$ ns, and $\chi^2 = 1.17$) did not change upon the addition of 4×10^{-3} M TbCl$_3$ ($\tau_1 = 1.02 \pm 0.02$ ns, $\tau_2 = 5.01 \pm 0.06$ ns, $\tau_3 = 12.1 \pm 0.12$ ns, $\chi^2 = 1.21$), the observed luminescence (L$_{Tb}$) has been attributed to energy transfer from the triplet domain of the aromatic moieties of the surfactant to (^{7}F)Tb(III).

---

*   Texaco sulfonate, a proprietary branched chain alkylaryl (60 to 70% monoaryl and 30 to 40% diaryl) surfactant containing one-half mole of Ca(OH)$_2$ per surfactant.

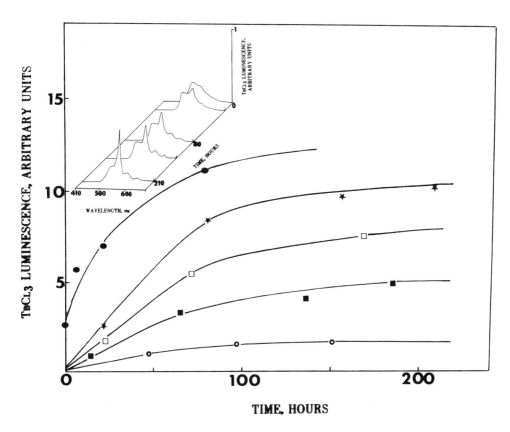

FIGURE 1. Time-dependent development of TbCl$_3$ luminescence in 2.0 × 10^{-3} M **1** in heptane in the presence of 1.0 × 10^{-4} M (○), 2.0 × 10^{-4} M (■), 3.0 × 10^{-4} M (□), and 4.0 × 10^{-4} M (★) TbCl$_3$ and in the presence of 4.0 × 10^{-4} M TbCl$_3$ and 4.0 × 10^{-4} M hydroxyphenylacetic acid (●). In all samples, [H$_2$O]/[**1**] = Wo = 11.1 λ_{ex} = 300 nm. Emission spectra were taken front face with a 490 nm cutoff filter in the emission side to eliminate fluorescence from **1**. Slits: 0.5 nm, 0.5 nm (excitation), and 0.2 nm (emission). Time-dependent development of emission spectra in 2.0 × 10^{-3} M **1** in the presence of 4.0 × 10^{-4} M TbCl$_3$ (i.e., ★ in the figure) is shown in the insert.

Photoexcitation of the aromatic moieties (Ar) is described in the following equations:

$$^1Ar + h\nu \xrightarrow{k_1} {}^1Ar^* \tag{1}$$

$$^1Ar^* \xrightarrow{k_2} {}^1Ar + h\nu^F_{Ar} \tag{2}$$

$$^1Ar^* \xrightarrow{k_3} {}^1Ar \tag{3}$$

$$^1Ar^* + {}^1Ar \xrightarrow{k_4} \text{quenching} \tag{4}$$

$$^1Ar^* + O_2 \xrightarrow{k_5} \text{quenching} \tag{5}$$

$$^1Ar^* \xrightarrow{k_6} {}^3Ar^* \tag{6}$$

$$^3Ar^* \xrightarrow{k_7} {}^1Ar + h\nu^P_{Ar} \tag{7}$$

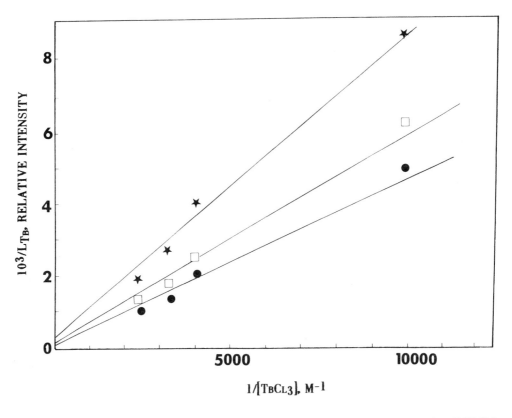

FIGURE 2.   Double-reciprocal plots of luminescence intensities ($/L_{Tb}$) against acceptor concentrations ($1/[TbCl_3]$) at 50 h (★), 100 h (□), and 200 h (●) subsequent to the introduction of $TbCl_3$ into $2.0 \times 10^{-3}\ M$ **1** in heptane.

$$^3Ar^* \xrightarrow{k_8} {}^1Ar \tag{8}$$

$$^3Ar^* + O_2 \xrightarrow{k_9} \text{quenching} \tag{9}$$

$$^3Ar^* + {}^3Ar^* \xrightarrow{k_{10}} \text{annihilation} \tag{10}$$

The singlet excited Ar (formed in Equation 1) can return to the ground state by fluorescence emission (Equation 2), by radiationless decay (Equation 3), by concentration (Equation 4), or by oxygen quenching (Equation 5). Alternatively, it can cross intersystem to form triplet Ar (Equation 6). The triplet Ar, in turn, may decay by phosphorescence emission (Equation 7) or by radiationless transition (Equation 8). It can also be quenched by oxygen (Equation 9) or be returned to the ground state by triplet-triplet annihilation (Equation 10).

If terbium chloride, Tb(III), is present along with Ar, additional photophysical processes (Equations 11 to 16) need to be considered:

$$(^7F)Tb(III) + h\nu \xrightarrow{k_{11}} (^5D)Tb(III) \tag{11}$$

$$^1Ar^* + (^7F)Tb(III) \xrightarrow{k_{12}} {}^1Ar + (^5D)Tb(III) \tag{12}$$

$$^3Ar^* + (^7F)Tb(III) \xrightarrow{k_{13}} {}^1Ar + (^5D)Tb(III) \tag{13}$$

$$(^5D)Tb(III) \xrightarrow{k_{14}} (^7F)Tb(III) + h\nu L_{Tb} \tag{14}$$

$$(^5D)Tb(III) \xrightarrow{k_{15}} (^7F)Tb(III) \tag{15}$$

$$(^5D)Tb(III) + O_2 \xrightarrow{k_{16}} quenching \tag{16}$$

Terbium chloride may be excited directly (Equation 11) or, alternatively, it can act as an energy acceptor from the singlet (Equation 12) or the triplet (Equation 13) manifold of Ar. The excess energy of the terbium cation may be dissipated by returning to the ground state via luminescence (Equation 14), radiationless decay (Equation 15), or oxygen quenching (Equation 16).

Irradiation at 300 nm and at relatively low intensities ensures energy deposition predominantly in Ar since the molar absorptivity of Ar is far in excess of that of TbCl$_3$. Furthermore, the intensity of TbCl$_3$ emission is some 100-fold less than that observed subsequent to sufficient incubation in reversed micelles, rendering Equation 13 to be the main photophysical pathway for excitation of TbCl$_3$ and, thus, allowing Equation 11 to be neglected. Additionally, energy transfer from the singlet manifold of Ar to TbCl$_3$ is spin-forbidden (Equation 12). Sufficient degassing obviates the possibility of oxygen quenching (Equations 5, 9, and 16 can be neglected). Taking advantage of this simplification, the observed luminescence intensity of TbCl$_3$ in reversed micelles, L$_{Tb}$, can be described by:

$$L_{Tb} = k_{14}[(^5D)Tb(III)] = \frac{k_{13}k_6 I_N[(^5D)Tb(III)]}{(k_{14} + k_{15})(k_2 + k_3 + k_6)(k_7 + k_8 + k_{13})[(^5D)Tb(III)]} \tag{17}$$

where I$_N$ is the fluorescence intensity of Ar in the absence of TbCl$_3$, governed by Equation 1. Equation 17 is analogous to that utilized in the triplet-triplet energy-transfer mechanisms of organic molecules and it can be rearranged to:

$$\frac{1}{L_{Tb}} = \frac{(1 + k_6 I_N \tau_N^S)k_{13}\tau_N^T}{k_6 k_{13} k_{14} I_N \tau_{(^5D)Tb(III)}\tau_N^S \tau_N^T} + \left(\frac{1 + k_6 I_N \tau_N^S}{k_{14} k_{13} I_N \tau_{(^5D)Tb(III)}\tau_N^S \tau_N^T}\right)\frac{1}{[TbCl_3]} \tag{18}$$

where $\tau_{(^5D)Tb(III)}$, τ_N^T, and τ_N^S are luminescence lifetimes of terbium chloride, triplet lifetimes of Ar, and singlet lifetimes of Ar. Plots of the left-hand side of Equation 18 against the reciprocal TbCl$_3$ concentration gave good straight lines for reversed micellar **1** at different times subsequent to the introduction of TbCl$_3$ (Figure 2). The ratio of the intercept to the slope of each lines, $k_{13}\tau_N^T/k_6$, is referred to as the triplet sensitization constant, K_S^{T}.[11] K_S^T values of 438, 167, and 100 M^{-1} have been obtained from plots shown in Figure 2 for the triplet sensitization constants for reversed micellar solutions of **1** in heptane 50, 100, and 200 h subsequent to the introduction of TbCl$_3$. Assuming that the rate constant for triplet energy transfer remains constant during the long-term development of terbium luminescence, these values indicate a decrease of the triplet lifetime (τ_N^T) with increasing incubation time. This is in accord, of course, with increased efficiency of the energy transfer with increasing luminescence intensity of TbCl$_3$. Triplet sensitization constants obtained in the present reversed micellar solution are appreciably larger than that found for the naphthalene-to-TbCl$_3$ energy transfer in aqueous micellar sodium dodecyl sulfate ($K_S^T = 40\ M^{-1}$).[11] This is in accord with the high degree of fluidity of the aqueous micellar systems which then facilitates the donor-acceptor encounter and, hence, increases the energy-transfer efficiencies which manifests in a smaller K_S^T value.

Requirements of the observed energy transfer are twofold. First, the energy level of the donor has to be similar to that required for raising an acceptor molecule to its excited state.[12,13] This requirement is clearly met. The triplet energy of naphthalene, the most likely donor in **1**, is on the order of 21,250 to 21,300 cm^{-1} in hydrocarbon solvents, and it is relatively independent of solvent changes.[14] The $^7F_6 \rightarrow {}^5D_4$ Tb(III) transition with emission at 490 nm (20,370 cm^{-1})[10] appropriately matches the donor for the required optimum overlap for energy transfer. The second requirement for energy transfer is that the terbium ions can encounter the donors in their triplet state. This encounter requires an apparent solubilization reorganization which occurs on a markedly slow scale.

Infrared spectroscopic data (Figure 3) clearly indicate that TbCl$_3$ most profoundly affects the aryl sulfonate moieties of the surfactant, which forms the polar core of reversed micellar **1**. Introduction of aqueous TbCl$_3$ results in the enlargement of the polar core of the reversed micelle, where the terbium ions are initially effectively shielded from the surfactant headgroups by tightly bound water molecules. The water-to-surfactant molar ratios under the experimental conditions were kept at 11.1 ([H$_2$O]/[**1**] = Wo = 11.1). This relatively small number of water molecules is tied up by strong hydration forces, both to the surfactant headgroups and to terbium ions. Indeed, mobility of the aromatic moiety of the surfactant headgroups is drastically restricted, as indicated by the observed anisotropy on the nano-second time scale. It is important to recognize that no such time-dependent anisotropies have been observed for AOT reversed micelles.[1-4] The slow appearance of TbCl$_3$ lumines-cence represents the reorganization of the hydration shell and, hence, the gradual decrease of the average distance between the donors and acceptors. Stating it differently, with in-creasing incubation time, complexation of the terbium ions to the surfactant headgroups becomes more important, as manifested by the progressive increase of the intensity of the asymmetric S($=$O)$_2$ band (Figure 3).

III. ROLE OF REVERSED MICELLES IN FUSION

Fusion is vitally important in physiological processes such as endocytosis, exocytosis, and fertilization.[1] It converts two cells, or their surfactant vesicle analogs, into one and thereby removes the barrier between the contents of the two merging entities. Fusion of membranes and their models (liposomes and surfactant vesicles, for example) have been extensively investigated by several different approaches.[15,16]

Addition of equimolar sodium sulfate to vesicles, prepared from 4.0×10^{-4} M diocta-decyldimethylammonium bromide (DODAB) in their gel state at room temperature, slowly increased the mean hydrodynamic diameters, D$_H$, of the aggregates from 80 to 740 nm over several days. Fluorescence resonance energy transfer efficiency in DODAB vesicles prepared from 4.0×10^{-4} M DODAB, 0.8 mol% of N-(7-nitrobenz-2-oxa-1,3-diazol-4-oyl)phosphatidylethanolamine (donor, **D**) and 0.8 mol% of N-(lissamine rhodamine B sul-fonyl)phosphatidylethanolamine (acceptor, **A**) was also found to decrease slowly upon di-lution by equal volumes and equal concentrations of DODAB vesicles and Na$_2$SO$_4$ (Figure 4). Thus, the slow increase of D$_H$ following Na$_2$SO$_4$ addition corresponds to fusion rather than to nonproductive adhesion of the DODAB vesicles.

Greater insight was obtained upon using DPH as a fluorescence probe sensitive to its microenvironment. Changes in the apparent relaxation time, τ_R values, of DPH in DODAB vesicles undergoing fusion at 20.0°C also occur on a long timescale (Figure 5). A possible mechanism of one cycle of vesicle fusion is illustrated in Figure 6. One can see, in the simplified overall picture, that the process starts with two small vesicles and ends with a large one. From start to finish, the curvature of the membrane and, accordingly, the average headgroup area of the amphiphile are decreased, while the packing is increased.[17] The τ_R of DPH in membranes reflects the rate of rotational diffusion and the orientation order.[18] Increasing the relative amount of amphiphiles in a given area hinders the motion of DPH

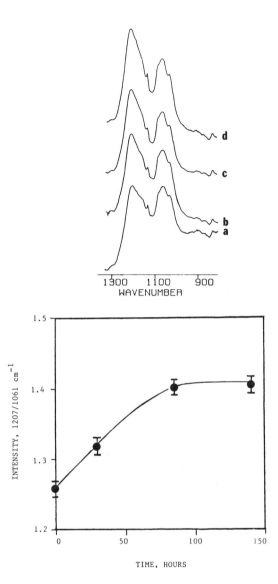

FIGURE 3.   Top: FTIR spectra of $2.0 \times 10^{-3} M$ **1** in heptane, $[H_2O]/[\mathbf{1}] = W_o = 11.1$, immediately (a), 18 h (b), 24 h (c), and 72 h (d) subsequent to the $TbCl_3$. Bottom: Plot of the intensity of the 1207 cm^{-1} peak relative to that of the 1061 cm^{-1} peak as a function of $TbCl_3$ incubation time.

and increases τ_R. The probability of quenching DPH emission by water molecules also decreases with tighter surfactant packing and, as a result, τ_F also increases. Similar considerations lead us to conclude that the observed (not shown) steady-state anisotropy, $\bar{r}$, and residual anisotropy, r_∞, follow the same trend as τ_R, while the "average" angle of distribution, $\langle\Theta\rangle$, changes in the opposite direction.

A closer inspection of the data in Figure 5 shows unexpectedly large fluctuations of the τ_R values of DPH, recorded subsequent to the addition of Na_2SO_4 to DODAB vesicles. Expanding these fluctuations revealed previously unrecognized behavior. In the 19- to 26- and 45- to 50-h time windows, for example, τ_F, τ_R, $\bar{r}$, r_∞, and $\langle\Theta\rangle$ underwent sets of systematic decreases and increases (see curves in Figure 7). It is important to note that these parameters were calculated using several different and independent approaches. Such systematic

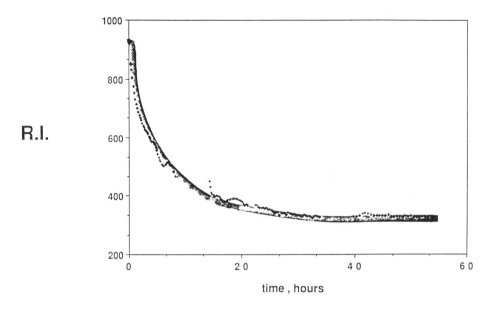

FIGURE 4.   Relation **A-D** emission intensity, measured at $I_{588\,nm} - I_{544\,nm}$, in solutions which contained 4.0×10^{-4} *M* DODAB, 4.0×10^{-4} *M* Na_2SO_4, 3.2×10^{-6} *M* **A**, and 3.2×10^{-6} *M* **D** at $20.0°C$ as a function of time. $\lambda_{EXC} = 470$ nm. Points are experimental data.

fluctuations of the five experimental parameters were reproduced in several Na_2SO_4-induced fusions of DODAB vesicles. In contrast, no systematic fluctuations were observed in two separate measurements of the fluorescence parameters of DPH in DODAB vesicles in the absence of added Na_2SO_4; 24 separate measurements of each sample over a 6-h period resulted in $\tau_F = 9.23 \pm 0.01$ ns, $\tau_R = 3.04 \pm 0.1$ ns, and in random variations of successively determined τ_F and τ_R.

Combined experimental and theoretical analysis led to the prediction that spherical vesicles necessarily flatten against each other (Step iii in Figure 6) prior to adhesion which destabilizes the bilayer.[19] Vesicle flattening is likely to be caused by the formation of a *trans* complex between the fusogenic agent (SO_4^{2-}, for example) and the headgroups of lipids in the two apposed bilayers.[20] Flattening events decrease the curvature and dehydrate the headgroups of the adjacent bilayers[21] with resultant increases in the "average" τ_F, τ_R, $\bar{r}$, and r_∞ values of DPH and corresponding decreases of $\langle \Theta \rangle$. Subsequent bilayer tension and deformation (Steps iv and v in Figure 6) result in the destabilization of vesicles and the likely formation of lipidic intramembranous particles.[22] Exposure of DPH to water molecules in Steps iv to v (Figure 6) increases dramatically since the lipidic particles formed (reversed micelles and related structures) represent equilibrium states which may give rise to fusion by the intermingling of the two aqueous compartments of both vesicles or, alternatively, which may reform the antecedent structures. The overall consequence of lipidic structure formation is increased surfactant hydration and, hence, observable decreases in the "average" τ_F, τ_R, $\bar{r}$, and r_∞ values of DPH, but increases in $\langle \Theta \rangle$. Finally, merging of adjacent bilayers through an ellipsoid-like shape to enlarged spherical structures (Steps vi to viii in Figure 6) changes the bilayer curvature and the relative amount of surfactant molecules per unit area in a way which, once again, increases the "average" τ_F, τ_R, $\bar{r}$, and r_∞ values of DPH, but decreases $\langle \Theta \rangle$. The whole process (Steps i to viii in Figure 6) is then repeated until fusion terminates. Indeed, the overall increase of D_H for DODAB vesicles from 80 nm to 740 nm indicates the occurrence of several complex fusion cycles.

Individual data points (in Figures 4, 5, and 7) cannot be related, of course, to particular events depicted for one fusion cycle in Figure 6. The observed decrease in the **A-D** emission

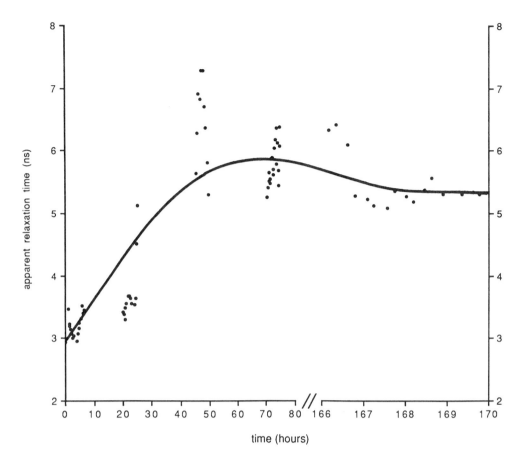

FIGURE 5. Apparent relaxation times (τ_R values) of DPH in DODAB vesicles ($\lambda_{EXC} = 300$ m and $\lambda_{EM} = 450$ nm) as a function of time after injection of Na_2SO_4 (the first point at time $= 0$ represents a τ_R value prior to the injection of Na_2SO_4) at 20.0°C. The fluorescence decay curves, $I_{\parallel}(t)$, were deconvoluted and fitted with a two-exponential function, $I_{\parallel}(t) = A\,e^{-t/\tau_1} + B\,e^{-t/\tau_2}$, where $\tau_1 > \tau_2$. The values of τ_R and τ_F (fluorescence lifetime) shown in Figure 4 were calculated as follows: $\tau_R = (\tau_2^{-1} - \tau_1^{-1})^{-1}$ and $\tau_F = \tau_1$. The goodness of the fittings was within the range $1.0 < \chi^2 < 1.5$. Determination of a data point took three to six minutes. All of the data points in this Figure (as well as in Figure 4) have been passed through a smoothing process which did not change the trends.

intensity requires approximately 30 h (Figure 4) and corresponds to monitoring fusion events until very few vesicles remain unlabled by **A** and **D**. Fusion continues further, of course (as indicated by the progressive increase in D_H values), but cannot be seen by the decrease in the **A-D** emission intensity. Using the environmentally sensitive DPH fluorescence probes has allowed, however, monitoring of fusion on a longer timescale and the capture of subtle changes (Figure 7) which have indicated additional consecutive and concurrent events in the large ensemble of vesicles undergoing sulfate ion-induced slow fusion (Figure 6). The long timescale of observable fusion is likely to be the consequence of the relatively low concentration of sulfate ions used and the high degree of stability in DODAB vesicles, which are present in their gel state under the experimental conditions used here.

IV. TRANSFER OF REVERSED MICELLE-ENTRAPPED COLLOIDAL PARTICLES TO SOLID SUPPORT

Molecular organization of ultrasmall colloidal particles is an intensively active current area of research.[23-25] The assembly of individual molecules to small clusters, size-quantized

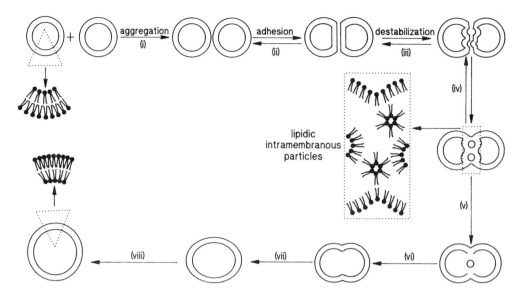

FIGURE 6.   A grossly oversimplified schematic diagram of one fusion cycle.

particles, and larger crystallites having solid-state properties is inherently interesting and provides an entry to novel materials which have desirable chemical, mechanical, electrical, and electro-optical properties. Molecular beam epitaxy has been used primarily to construct desired structures by atomic, layer-by-layer deposition in ultrahigh vacuum.[25] An alternative "wet" colloid chemical approach was launched in our laboratories some years ago. Surfactant vesicles, bilayer lipid membranes, monolayers, and Langmuir-Blodgett films have been used as templates in the *in situ* generation of nanosized semiconductor and magnetic particles.[23] Silver particles, *in situ* formed in the aqueous pools of reversed micelles, have been transferred as a "monolayer" of metal-island particulate films onto solid substrates.

Irradiation of Aerosol-OT reversed micelle-entrapped silver ions resulted in silver particle formation which manifested in the development of a broad absorbance with a maximum at around 450 nm.[26]

Silver particle-containing reversed micelles were transferred to freshly cleaved highly oriented pyrolytic graphite (HOPG, Union Carbide Corporation). The transfer was effected by vertically pulling the HOPG substrate from the silver particle-containing reversed-micellar solutions at a slow speed (1 mm/min). Slow deposition was found to be critical for good deposition of one layer of reversed micelles. STM images of reversed micelle-coated HOPG substrates were taken subsequent to drying in air for 1 h.

STM images revealed the presence of silver islands on atomically smooth HOPG (Figure 8). Concentrations of reversed micelles affected the inter-island separation distance. In undiluted (1.0×10^{-2} *M* AOT) samples, the substrate was fairly completely covered by interconnected silver islands (not shown). Tenfold dilution of the silver particle-containing reversed micelles by heptane (prior to transfer) appeared to be optimal for assessing inter-island distances (Figure 8). At further dilutions, it became increasingly difficult to find images. Most significantly, heights of silver islands were consistently the same (4.0 ± 0.5 nm; see z-x plots in Figure 8), thus indicating the effective transfer of only one layer of reversed micelles on HOPG. The proposed mode of transfer is illustrated in Figure 9.

V. FUTURE PROSPECTS

Three recent research projects have been described here. Development of microemulsion-based gels (organogels) is also significant.[27] These investigations well illustrate the diversity,

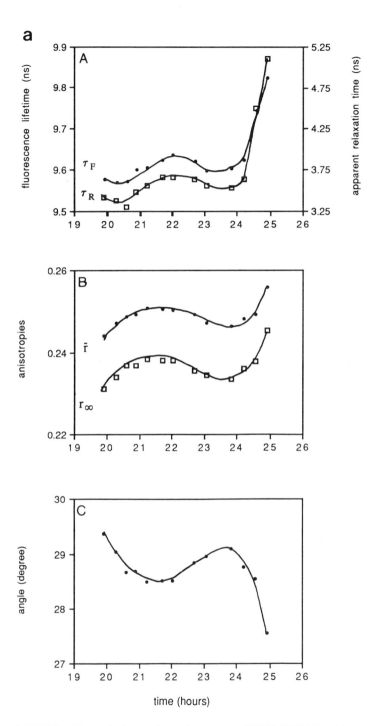

FIGURE 7. Changes in the experimental parameters of DPH in DODAB vesicles in the (a) 19- to 26-h and (b) 45- to 50-h windows at 20.0°C. (A) Fluorescence lifetime (τ_F values) and apparent relaxation time (τ_R values). (B) Steady-state anisotropy ($\bar{r}$) calculated from $\bar{r} = [\Sigma_0^\infty GI_\|(t) - \Sigma_0^\infty I_\perp(t)]/[\Sigma_0^\infty GI_\|(t) + 2 \Sigma_0^\infty I_\perp(t)]$ (G = normalization factor) and residual anisotropy determined by plotting r(t) = $[GI_\|(t) - I_\perp(t)]/ [GI_\|(t) + 2 I_\perp(t)]$ and measuring the anisotropy value at the tail of r(t), according to r(t) = $r_\infty + (r_0 - r_\infty) e^{-t/\tau_R}$. (C) Average angle of distribution, $\langle \Theta \rangle$, of the distribution of DPH in DODAB vesicular membranes calculated from $r_\infty/r_0 = {}^3/_2 \langle \cos^2\Theta \rangle - {}^1/_2$. The anisotropy at time equals zero (r_0) was determined to be 0.362 for DPH.

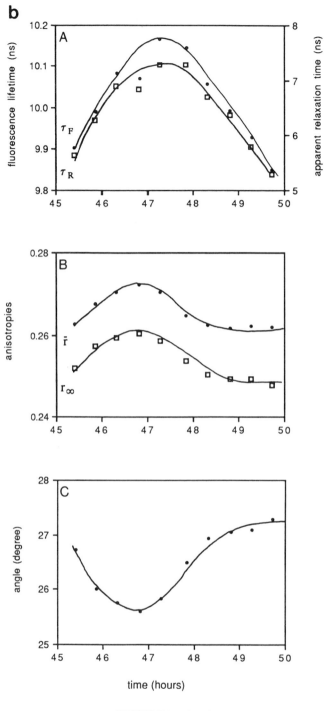

FIGURE 7 (continued)

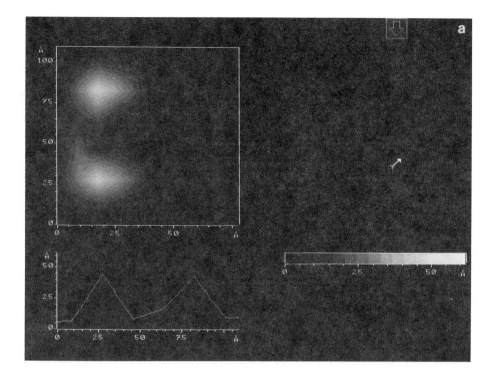

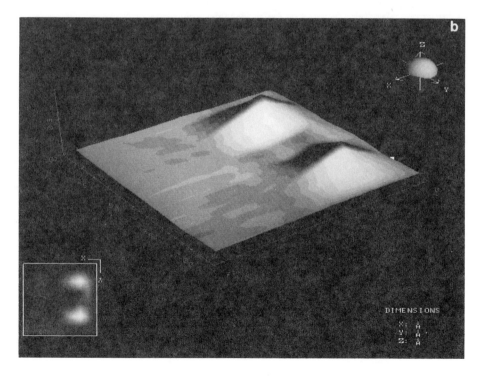

FIGURE 8. STM images of silver islands on HOPG. Samples were prepared by the slow vertical transfer of heptane-diluted (tenfold), irradiated (60 min), $1.0 \times 10^{-2}\ M$ AOT and $7.3 \times 10^{-3}\ M$ Ag$^+$ in heptane containing $1.0 \times 10^{-1}\ M$ H$_2$O, w = 10 (a and b), and $2.0 \times 10^{-1}\ M$ H$_2$O, w = 20 (c and d). Inserts in the three images give bird's eye views of the same areas. In the two-dimensional x-y projections (a and c), the dark and white contours correspond to the lowest and highest heights for each image. The heights of each image and their separations can be best ascertained from the z-x plots.

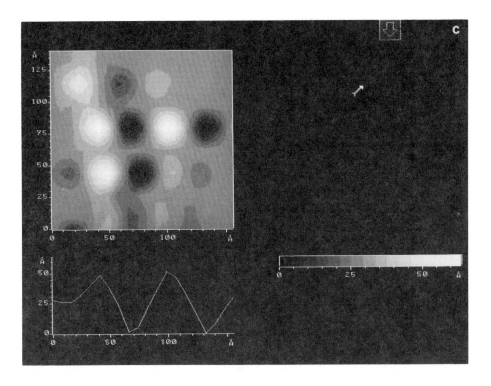

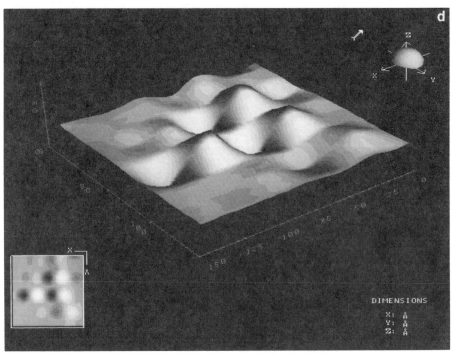

FIGURE 8 (continued)

FIGURE 9.   Schematic diagram (*not drawn to scale*)
of the proposed mode of transfer of reversed micelle-
entrapped silver particles to a solid support.

progress, and potential of reversed micelles. The subject has reached maturity to an extent of undesirable specialization. This is most evident in four subfields involving reversed micelles: industrial approach to lubrication, enzyme incorporation and related biotechnology, generation of size-quantized particles and their utilization in solid-state electronic devices, and separation and other analytical applications. Exploitation of the full potential of any of these subfields requires much improved cross-fertilization of ideas. If this happens, future progress will only be limited by our imagination.

ACKNOWLEDGMENT

Support of this work by a grant from the National Science Foundation (U.S.) is gratefully acknowledged.

REFERENCES

1. **Fendler, J. H.,** *Membrane Mimetic Chemistry,* John Wiley & Sons, New York, 1982.
2. **Luisi, P. L. and Magid, L. J.,** *CRC Crit. Rev. Biochem.,* 20, 409, 1986.
3. **Martinek, K., Levashov, A. V., Khmelnitski, Yu. L., Klyachko, N. L., and Berezin, I. V.,** *Eur. J. Biochem.,* 155, 453, 1986.
4. **B.Nagy, J., Deroune, E. J., Gourgue, A., Lufimpadio, N., Ravet, I., and Verfaillie, J. P.,** Proc. 6th Int. Symp. Surfactants in Solution — Modern Aspects, New Delhi, India, August 18, 1986.
5. **Steigerwald, M. L. and Brus, L. E.,** *Acc. Chem. Res.,* 23, 183, 1990.
6. **Henglein, A.,** *Top. Curr. Chem.,* 143, 113, 1988.
7. **B.Nagy, J., Yuan, Y., Jao, T.-C., and Fendler, J. H.,** *J. Phys. Chem.,* 94, 863, 1990.
8. **Yogev, D., Guillaume, B. C. R., and Fendler, J. H.,** *Langmuir,* 7, 623, 1991.
9. **Dolan, C., Yuan, Y., Jao, T.-C., and Fendler, J. H.,** *Chem. Mater.,* 3, 215, 1991.
10. **Heller, A. and Wasserman, E.,** *J. Chem. Phys.,* 43, 949, 1965.
11. **Escabi-Perez, J. R., Nome, F., and Fendler, J. H.,** *J. Am. Chem. Soc.,* 99, 7749, 1977.
12. **Kropp, J. L. and Windsor, M. W.,** *J. Chem. Phys.,* 42, 1599, 1965.
13. **Birks, U. B.,** *Photophysics of Aromatic Molecules,* John Wiley & Sons, New York, 1970.

14. **Wilkinson, F.,** in *Fluorescence Theory, Instrumentation and Practice,* Gilbault, G. G., Ed., Marcel Dekker, New York, 1967.
15. **Sowers, A. E. Ed.,** *Cell Fusion,* Plenum Press, New York, 1987.
16. **Ohki, S., Doyle, D., Flanagan, D., Hui, S. W., and Mayhew, E.,** *Molecular Mechanisms of Membrane Fusion,* Plenum Press, New York, 1987; **Papahadjopoulos, D., Nir, S., and Düzgünes, N.,** *J. Bioenerg. Biomembr.,* 22, 157, 1990.
17. **Israelachvilli, J. M., Marcelja, S., and Horn, R. G.,** *Q. Rev. Biophys.,* 13, 2, 1980.
18. **Kinosita, K., Ikegami, A., and Kawato, S.,** *Biophys. J.,* 37, 461, 1982; **Kinosita, K., Kawato, S., and Ikegami, A.,** *Biophys. J.,* 20, 289, 1977; **Lentz, B. R.,** *Chem. Phys. Lipids,* 50, 171, 1989.
19. **Rand, R. P. and Parsegian, V. A.,** *Annu. Rev. Physiol.,* 48, 201, 1986.
20. **Rupert, L. A. M., Engberts, J. B. F. M., and Hoekstra, D.,** *J. Am. Chem. Soc.,* 108, 3920, 1986.
21. **Wilshut, J. and Hoekstra, D.,** *Trends Biochem. Sci.,* 9, 479, 1984.
22. **Verkleij, A. R.,** *Biochim. Biophys. Acta,* 779, 43, 1984.
23. **Fendler, J. H.,** *Chem. Rev.,* 87, 877, 1987.
24. **Henglein, A.,** *Chem. Rev.,* 89, 1861, 1989.
25. **Ploog, K.,** *Angew. Chem. Int. Ed. Engl.,* 27, 593, 1988.
26. **Henglein, A.,** *J. Phys. Chem.,* 83, 2209, 1979.
27. **Luisi, P. L., Scartazzini, R., Haering, G., and Schurtenberger, P.,** *Colloid Polym. Sci.,* 268, 356, 1990.

Chapter 10

BIOLOGICAL ELECTRON TRANSFER IN REVERSE MICELLAR SYSTEMS: FROM ENZYMES TO CELLS

Edgardo Escamilla, Martha Contreras, Laura Escobar, and Guadalupe Ayala

TABLE OF CONTENTS

I. INTRODUCTION

Enzymology in colloidal solutions of water in organic apolar solvents stabilized with amphilic compounds has become an attractive research area. The physicochemical properties of the encased aqueous solutions differ from those of bulk water in several respects and depend on the degree of hydration of the micelles.[1,2] An inverted micelle aggregate represents a water droplet of about 20 to 200 Å in diameter surrounded by a monolayer of surfactant molecules. The polar or charged heads, with their counterions, are localized in the interior of the aggregate, while the hydrophobic tails are directed towards the bulk, continuous organic phase. Water, hydrophilic solutes, and water-miscible solvents are readily solubilized in the micellar core, the so-called "water pool".

In these systems, it is possible to generate different and well-defined microenvironments which affect the properties of a sequestered enzyme. Both micellar diameter and the apparent kinetic properties of an encased enzyme strongly depend on the water content of the micelle. The term "reverse micelle" is better applied to small aggregates occurring at low water contents. Using the Wo parameter, defined as the molar ratio of water to surfactant (i.e., [H$_2$O]/[surfactant]), the upper limit for reverse micelles is at Wo values lower than 15. At higher Wo values, it is more appropriate to use the term "water-in-oil microemulsions" or simply "microemulsions" to describe ternary systems (H$_2$O/surfactant/oil) containing higher water fraction volumes. Although the physicochemical properties of the water core of true reverse micelles compared to water-in-oil microemulsions differ greatly, it is operationally acceptable to use the term "reverse micelle" in referring to both types of system.

Reverse micellar solutions in organic solvents are spontaneously formed upon addition of surfactants and suitable amounts of water. For this purpose a variety of anionic, cationic, zwitterionic, and nonionic surfactants have been used.[2-6] The most studied have been the anionic AOT (sodium *bis* [2-ethylhexyl] sulfosuccinate) the cationic CTAB (hexadecyltriethylammonium bromide), and the nonionic derivatives of Tween, Span, and polyoxyethylene alcohols. In addition, an upsurge of interest in reverse micellar systems based on natural and synthetic phospholipids has been seen in recent years[9] (see also Chapters 2, 9, and 11), probably because phospholipids are natural constituents of biological membranes.

Several organic solvents have been used as dispersion media; *n*-octane, isooctane, benzene, and heptane are the most common choices. The addition of a cosurfactant such as benzyl alcohol, cholesterol, or hexanol is required in some cases.[7-9] AOT-based systems are more attractive since no cosurfactant is required.[2]

Up to now, several dozen enzymes have been studied in reverse micellar solutions. Those include hydrophilic,[10-12] surface-active,[11,12] and membrane enzymes.[11] In many of the cases reported, an enzyme, under optimal assay conditions, exhibits only a fraction of the activity displayed in bulk water.[4,13-15] More often, the activity in the organic system is quantitatively close to that shown in bulk water systems;[7,14] however, in some few and well-recognized cases, the enzyme under study may reach an activity higher than in bulk water.[16-18] This latter phenomenon has been described as "superactivity".[11,19,20]

Three methods have been used for transfer and solubilization of protein into reverse micellar solutions.[4,10,11,21] In all cases, the procedure is started with a solution of the surfactant (and cosurfactant in some cases) in an organic solvent. Thereafter, the protein can be incorporated in three different ways: from small volumes of an aqueous concentrated solution (microinjection method); as a two-phase system containing similar volumes of an aqueous protein and reverse micelle medium (two-phase method); and by adding the dry protein to the micellar system already containing a fixed amount of water (solid-state method).

Solubilization of the protein is finally achieved by gentle or vigorous stirring. In the cases of microinjection and solid-state methods, the end point of solubilization is usually reached within seconds to minutes, while in the two-phase transfer method, it could take hours before solubilization equilibrium is attained.[10,21,22] This method can be modified by sonication (in a water bath sonicator) for a few minutes to speed the solubilization process.[23] However, care must be taken to avoid protein denaturation caused by overtime sonication.

A risk involved in the use of microinjection and solid-state transfer methods is oversaturation with water or with protein, respectively. Water in excess (cloudy preparations) can be eliminated by blowing nitrogen on the micellar preparation until a clear solution is obtained.[23] Nondissolved protein is eliminated by low-speed centrifugation. In the two-phase transfer method, the upper organic phase (containing the transferred protein) is recovered with a Pasteur pipette. The technical advantages and disadvantages associated with each of the above protein transfer methods have been amply discussed.[2,11]

Different types of enzymes have been studied in micellar systems; however, hydrolases (specially proteases) and oxidoreductases are by far the most intensively explored groups. This might be due to the commercial availability of these enzymes in their pure state and also to the simple and reliable spectroscopic methods available for the assay of this group of enzymes. On the other hand, it is also notable that workers have focused their attention mainly on hydrophilic enzymes,[10-12] compared to the number of studies so far performed with surface-active[11,12] and membrane enzymes.[11] This interest has occurred in spite of the fact that reverse micelle systems have been considered as a mimetic model of biological membranes and of its aqueous environment.

Protein-containing colloidal solutions of water in organic solvents are optically transparent; thus, they can be studied by conventional spectroscopic techniques. However, the increasing interest in micellar enzymology is demanding a wider range of quantitative techniques for the recording of enzymic activity in apolar solvents. In this work, the application of the Clark oxygen electrode to reactions evolving or consuming oxygen in these microheterogeneous media is described. The amperometric technique for titration of the oxygen contained in different organic media is first described and validated. Afterward, the suitability of the oxygen electrode is shown in the kinetic analysis of the catalase reaction. A second section describes the use of this method to assay the enzymic activity of individual respiratory enzyme complexes, and finally, the respiratory activity of whole cell preparations in organic solvent systems is described.

II. POLAROGRAPHIC TITRATION OF OXYGEN IN REVERSE MICELLE SYSTEMS

A. THE REACTION CELL

Polarography of oxygen in apolar organic media is a recent technique[24,25] with potential applications in micellar enzymology. As originally described,[24] the technique uses a standard Clark oxygen electrode and an oxygraph. However, keeping in mind that oxygen titration is performed in apolar organic solvents, an all-glass reaction cell (Figure 1) must be used. In addition, the reaction chamber is provided with a water jacket to control the temperature and a lateral hole for insertion of the electrode, which is tightly fitted with the same O-ring that holds the membrane. To avoid back-transfer of oxygen from the atmosphere to the sample solution, a glass stopper provided with a capillary bore tube (1 to 2 cm long) is used. Through this conduit, additions can be made during the reaction.

B. CALIBRATION AND STABILITY OF THE ELECTRODE

Since oxygen molecules are nonpolar, their solubility in apolar organic solvents is several times larger than in water[26] and can be determined by either of the two following methods.[24]

Chemical titration of O_2 with an anaerobic solution of $Na_2S_2O_4$ (sodium dithionite) — The solution (i.e., 50 mM $Na_2S_2O_4$) is prepared in anaerobic cold aqueous medium contained in a bottle (neck filled) sealed with a rubber cap. Aliquots are withdrawn through

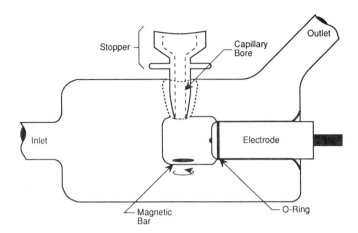

FIGURE 1.   A model reaction cell for O_2-polarography in organic solvents. A Clark oxygen electrode is inserted into an all-glass chamber. This chamber is surrounded by a water jacket to maintain a constant temperature. The electrode is tightly fitted with the same O-ring that holds a Teflon membrane. The glass stopper has a capillary bore for additions during the assay.

the cap with a microsyringe and added to the chamber containing an air-saturated buffer (Figure 2A) or a micellar organic medium (Figure 2B). Changes in the electrode reading brought about by successive additions of $Na_2S_2O_4$ solution (i.e., 3 μl for the aqueous solution and 15 μl for the organic medium) are recorded and converted to micromoles of oxygen assuming that 1 mol of $Na_2S_2O_4$ will consume 1.0 mol of O_2.

Enzymic titration of O_2 produced by decomposition of H_2O_2 by bovine liver catalase[27] — Aqueous solutions of catalase (Figure 2A) or reverse micelle preparations of catalase (Figure 2B) are rendered anaerobic with a N_2 current. Subsequently, the reaction is started by adding substrate aliquots (about 10 μl) containing known concentrations of H_2O_2 (0.2 to 2.0 μmol). O_2 concentration in aqueous and organic systems is calculated assuming that 1.0 mol of H_2O_2 will produce 0.5 mol of O_2.[27]

In our experience, titration of oxygen by either of the two methods described above gives similar results. Air-saturated (at 30°C and 582 mm partial gas pressure) 0.1 M K-phosphate buffer (Figure 2A) contains 0.2 μmol O_2 ml^{-1}, while an air-saturated 0.2 M AOT-toluene solution contains 1.4 μmol O_2 ml^{-1} (Figure 2B); thus, the latter mixture has seven times more dissolved O_2 than typical low-ionic-strength aqueous solutions.

The response of the oxygen electrode in AOT-toluene medium is proportional to the amount of dithionite consumed as indicated by successive additions of dithionite (Figure 2). A similar linear response can be obtained with the catalase-H_2O_2 calibration method (not shown). The O_2 concentration value (i.e., 1.4 μmol O_2 ml^{-1}) found for the toluene-AOT systems must be considered as the concentration in the overall volume (i.e., water pool plus hydrocarbon phase) and not the microenvironment concentration in the polar core of the AOT micelles. Indeed, a series of graded fluorescent probes of pyrene have been used to determine the relative oxygen content as a function of the apparent distance from the polar core of calcium alkylbenzene sulfonate reverse micelles in *n*-heptane; the conclusion reached is that oxygen forms a concentration gradient, lower in the aqueous and more polar micro-phase and several times larger in the bulk apolar solvent.[26] In fact, Wong et al.[28] have used fluorescence techniques to explore the nature of the aqueous core of reverse micelles and found that the oxygen solubility in water-swollen micelles of AOT in heptane was close to that obtained in pure water. Thus, those O_2-linked reactions taking place in the micelle core occur under O_2 tensions rather typical of water solutions, while the surrounding and continuous organic phase resembles a high-capacity O_2 buffer.

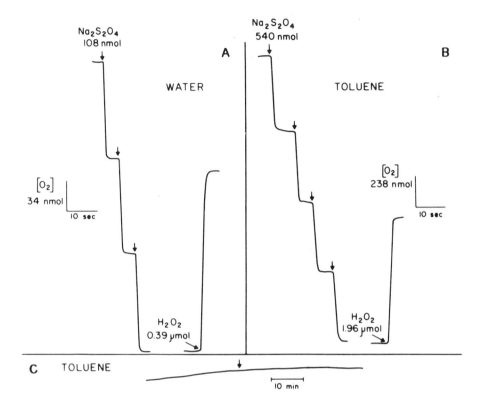

FIGURE 2.   Polarographic determination of O_2 concentration in air-saturated 0.1 M potassium phosphate buffer, pH 7.0 (**A**), and in 1.6 M H_2O/0.2 M AOT/toluene (**B**). O_2 titration was performed with an anaerobic solution of 50 mM $Na_2S_2O_4$. The changes in the electrode reading caused by successive additions of $Na_2S_2O_4$ solutions were compared and converted to micromoles of O_2, assuming that the aqueous system contains 0.2 μmol of O_2 per milliliter. The O_2 evolved by catalase from H_2O_2 was calculated from the O_2 concentration change brought about by the addition of known quantities of H_2O_2. Aqueous and organic reaction contained 18 and 7.0 μg of catalase, respectively. Assays were performed at 30°C. Part **C** shows a long-term recording of the electrode in 1.6 M H_2O/0.2 M AOT/toluene medium (O_2 depleted by bubbling of N_2). After the arrow, the oxygraph chamber was closed by the insertion of a conical plastic device in the capillary bore. Concentrations always refer to the overall volume. (From Escobar, L., Salvador, C., Contreras, M., and Escamilla, J. E., *Anal. Biochem.*, 184, 139, 1990. With permission.)

The Clark oxygen electrode has good stability in long-term recordings of organic solvent media. Figure 2C shows a 60-min recording in 0.2 M AOT toluene medium taken after depletion of O_2 by bubbling with N_2. The first part of the trace, free of noise, increases steadily due to the slow back-transfer of O_2 from the atmosphere through the capillary bore of the stopper. The value of "the oxygen leak" rate was calculated to be 0.07 μmol O_2 ml^{-1} h^{-1}. However, the O_2 leak stopped after the insertion of a plastic conical device in the capillary bore. Thus, long-term O_2 recording (hours) in organic solvents is stable, and the diffusion rate of O_2 through the capillary bore is orders of magnitude lower than typical rates of enzyme-catalyzed reactions.

C. SOLVENT APOLAR INDEX AND O_2 SOLUBILITY

Besides toluene, several other organic solvents are currently used in the study of enzymes in reverse micelles. Thus, it was important to know if the reliability of the Clark oxygen electrode could be extended to other organic solvents with different polarities. To this purpose, O_2 was titrated in hydrated AOT-solutions in benzene, toluene, cyclohexane, *n*-hexane, or isooctane (noted as B, T, CH, H, and IO, respectively, in Figure 3A). The

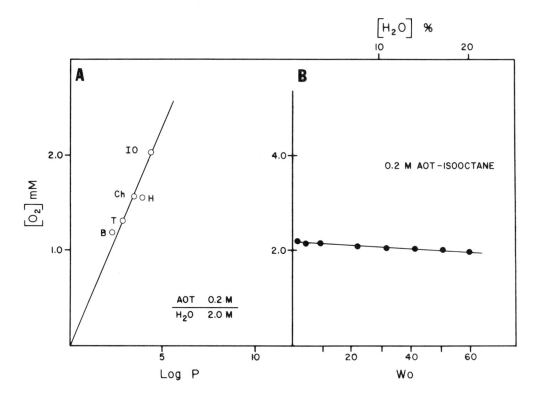

FIGURE 3.   (**A**) Polarographic determination of oxygen concentration in air-saturated 2 M H$_2$O/0.2 M AOT in different organic solvents with an increasing apolar index (log P): B, benzene; T, toluene; CH, cyclohexane; H, hexane; and IO, isooctane. The titration of O$_2$ was carried out through oxidation of known amounts of dithionite as described in Figure 2. (**B**) Effect of water volume fraction (from 3.6 to 20%, v/v) on the concentration of oxygen dissolved in 0.2 M AOT-isooctane medium. Variation of water concentration at constant surfactant concentration is also expressed as Wo (i.e., from 2 to 60). Assays were carried out at 30°C.

oxygen concentration values found were plotted against the corresponding log P values (solvent polarity index).[29] The plot obtained (Figure 3A) shows a linear correlation between O$_2$ dissolving capacity and the apolar index (log P) of the solvents tested.

D. WATER FRACTION VOLUME AND O$_2$ SOLUBILITY

One the most frequently studied phenomenon in reverse micelle enzymology is the dependence of the catalytic activity of the solubilized enzyme on the size of the water pool (Wo parameter). For most of the cases reported, this dependence shows a bell-shaped activity pattern.[4,6,9-11,17,19,30,31] Variations in Wo of reverse micelles can be imposed by either of two ways: (1) the water fraction volume is maintained constant while the concentration of surfactant is varied, or (2) surfactant concentration is kept constant while increasing the water fraction volume.

Since the O$_2$ solubility in the hydrocarbon phase is several times higher than in the water droplet,[26] it was judged important to know to what extent the size of the water fraction volume affects the overall concentration of O$_2$ when Wo values are achieved by increases in the amount of water at a constant concentration of surfactant.

Thus, "empty" reverse micelles were prepared with 0.2 M AOT in isooctane to which increasing volumes of water were added to give Wo values between 2 and 60. Thereafter, the O$_2$ concentration, at each Wo value, was determined (Figure 3B) with the dithionite calibration method, It should be noted that the oxygen concentration decreases steadily with the increment of the water fraction volume in the reverse micelle system. In going from

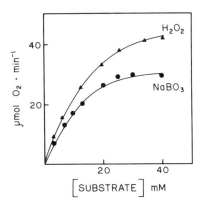

FIGURE 4.   Effect of the substrate concentration on the initial catalytic rate
of liver catalase trapped in reverse micelles formed with 0.2 *M* AOT in toluene,
Wo = 10. Catalase activity was titrated with the indicated concentrations of
NaBO$_3$ (●) or H$_2$O$_2$ (▲). The reaction was started by the addition of 5 μl (0.6 μg
of protein in 0.1 *M* K-phosphate, pH 7.0) of the micelle preparation of catalase;
the initial rates were calculated within the first 10 s of reaction time. Assays
were performed at 30°C. (From Escobar, L., Salvador, C., Contreras, M., and
Escamilla, J. E., *Anal. Biochem.*, 184, 139, 1990. With permission.)

Wo = 2.0 (3.6% H$_2$O v/v) to Wo = 60 (20% H$_2$O v/v), the overall concentration of oxygen
dropped from 2.2 m*M* to 1.95 m*M*, respectively. Thus, differences in O$_2$ concentration as
a function of Wo are small and can be neglected for most experimental purposes.

III. POLAROGRAPHIC CHARACTERIZATION OF THE CATALASE REACTION IN REVERSE MICELLES

The reliability of the oxygen electrode in enzyme-catalyzed reactions producing or
evolving oxygen in organic solvents was tested by comparing several kinetic parameters of
the catalase reaction in aqueous buffer and in AOT reverse micelles. Toluene or isooctane
was used as a hydrocarbon phase, and either H$_2$O$_2$ or NaBO$_3$ was used as a substrate.

A. CATALASE CATALYTIC CONSTANTS FOR SUBSTRATE
The enzyme response to substrate concentration followed simple hyperbolic kinetics
with both NaBO$_3$ and H$_2$O$_2$ as substrates in the AOT-toluene system (Figure 4) as well as
in the aqueous system (not shown).

The Michaelis constant (K$_m$) of water-soluble substrates in reverse micelle systems[5] can
be referred to the overall volume (K$_{mov}$) or to the water pool alone (K$_{mwp}$). K$_{mov}$ values for
H$_2$O$_2$ and NaBO$_3$ were 17 and 15 m*M*, respectively (Table 1), and were three times lower
than the corresponding K$_m$ values obtained in the aqueous medium (Table 1). Alternatively,
considering the water pseudophase volume (36 μl H$_2$O ml^{-1} at Wo = 10 and 0.2 *M* AOT)
in the toluene system, the K$_{mwp}$ values become 27-fold higher than those calculated from
overall volume concentrations and about ninefold higher when compared to K$_m$ values
obtained in aqueous medium (Table 1). The above results are in agreement with several
reports on reverse micelle systems[19,32,33] where the particular enzyme tested showed an
apparent lower affinity for its substrates when compared to the assay in water. This behavior
has been explained as a consequence of the substrate partition between the micellar water
pseudophase, the micellar interphase, and the bulk organic solvent.[6,10,15,34,35]

Indeed, the chemical structure of H$_2$O$_2$ with a symmetrical distribution of the oxygens
and hydrogens suggests that a significant part partitions in the organic apolar phase.

TABLE 1
Comparison of Catalytic Constants of Catalase in Aqueous and Organic Media[a]

Medium	K_{mov} (mM)		K_p[b] ($\times 10^6$ min^{-1})	
	NaBO$_3$[c]	H$_2$O$_2$	NaBO$_3$	H$_2$O$_2$
100 mM K-phosphate, pH 7.0	45	54	3.2	3.9
0.2 M AOT/toluene, pH 7.0, Wo = 10	15	17	1.2	1.8
0.2 M AOT/isooctane, pH 7.0, Wo = 60	40	55	3.0	3.2

[a] Catalytic activity constants in AOT-toluene medium at Wo = 10 and in AOT-isooctane medium at Wo = 60 were calculated by the method of Lineweaver-Burk, assuming substrate concentrations in the overall volume (i.e., water plus hydrocarbon phases).
[b] K_p catalytic center activity.
[c] Substrate solutions were entrapped in the corresponding reverse micelle preparation and the reaction was started by the addition of the corresponding reverse micelle preparation of catalase.

Accordingly, Takahashi et al.[36] have reported the use of benzene H$_2$O$_2$-saturated solutions for the peroxidase assay in surfactant-reverse micelle systems.

The catalytic center activity (K_p) of catalase was calculated from the Lineweaver plots for substrate. At Wo = 10.0 in the AOT-toluene system, the K_p values (Table 1) for NaBO$_3$ and H$_2$O$_2$ were 1.2 and 1.8 $\times 10^6$ min^{-1}, respectively. On the other hand, in the case of AOT-isooctane at Wo = 60, K_p values of 3.0 and 3.2 $\times 10^6$ min^{-1} were obtained for the same substrates. This difference, as will be seen later, depends on the hydration degree of the preparation and not on the type of solvent used. It is also seen (Table 1) that the maximal K_p values obtained at Wo = 60 (i.e., 3.2 $\times 10^6$ min^{-1}) represent about 82% of the K_p values calculated for the aqueous medium. On the assumption that all molecules of catalase remain active after transfer to the organic solvent, it seems that the catalytic performance of catalase in AOT reverse micelles at optimal Wo values is close to that observed in water.

B. VARIATION OF THE MICELLE HYDRATION DEGREE

Enzyme activity in reverse micelle systems is greatly affected by the hydration degree of the system[2-4,11,19] and, according to Khmelnitski et al.,[31] the hydration value where maximal activity is observed occurs when the hydrated core of the micelle reaches a diameter that closely corresponds to the diameter of the protein molecule encased. Thus, the effect of Wo at constant water fraction volume was explored for catalase in AOT-toluene and AOT-isooctane systems (Figure 5). With the former system, the value of K_p increased and reached a plateau at Wo values of 15 to 20. This behavior seems to be due to the fact that toluene, as an aromatic solvent, produces micellar systems with limited capacity of hydration.[37] Consequently, the system becomes saturated with water at about Wo = 20. On the other hand, isooctane-based systems have a higher water uptake capacity[38] and assays can be performed at Wo values up to 70. Under these circumstances, the activity of catalase follows a nearly bell-shaped curve having its maximal K_p at Wo = 60. At Wo = 60, the AOT-micelle will have a water pool diameter higher than 260 Å,[39] a dimension that seems to be larger than that required to contain a catalase tetramer (being an ellipsoid of 46 $\times$ 92 $\times$ 146 Å[40]). Thus, our result does not seem to agree with Khmelnitski et al.[31] and Klyachko (cited in Reference 31), who reported that the optimal Wo value is about 28 for catalase in AOT micelles. A possible explanation for this disagreement is that the sonication step used in our transfer protocol could be too harsh and hence would cause oligomeric rearrangements with higher aggregation numbers. This possibility must be further tested.

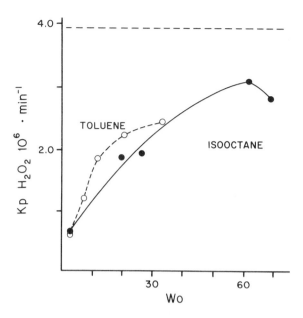

FIGURE 5. Effect of the H_2O/AOT molar ratio (Wo) on the catalytic center activity (K_p) of the liver catalase in H_2O_2. Catalase reverse micelles of 0.2 M AOT in toluene (○) and in isooctane (●) were prepared at the indicated Wo values. Reverse micelles containing the substrate (26 μmol of H_2O_2) were added to 1.7 ml of 0.2 M AOT micellar medium. The reaction was started by addition of 5 μl (1 μg of protein) of catalase.

C. EFFECT OF SURFACTANT CONCENTRATION AT FIXED Wo

When the surfactant concentration is raised at a fixed Wo, the number of the micelles is also increased without altering their size and other properties. Hence, entrapped enzymes should maintain their activity level.[31] This prediction has been demonstrated for a number of enzymes.[19,41,42] However, there are several documented cases[11,17,31] where the particular enzymes are progressively inhibited by increasing surfactant concentration, the inhibition being independent of the surfactant used.[31]

Figure 6 shows that catalase activity in an AOT-isooctane system at a constant value of Wo = 30 is progressively inhibited by raising the concentration of AOT from 0.02 to 0.25 M. Therefore, catalase seems to follow the kinetic behavior previously described for peroxidase, acid phosphatase, lactase, and prostaglandin synthetase.[31] Khmelnitski et al.[31] have proposed that the kinetic behavior of this group of enzymes can be explained as a result of the interaction of the micellar membrane with anchoring groups present in the enzymes mentioned above. This interaction would account for the dependence of catalytic activity on the surfactant concentration. Moreover, the same authors have suggested that this dependence can be used to test the ability of an enzyme to interact with micellar and biological membranes. In support of this idea, they showed that native α-chymotripsin, which is insensitive to the surfactant concentration, becomes deeply dependent on surfactant concentration after its covalent modification with stearoyl residues,[11,43] thus making it more hydrophobic.

D. THERMOSTABILITY OF CATALASE IN REVERSE MICELLES

Water is essential to enzyme catalysis since water enhances conformational flexibility of the enzyme molecule which in turn promotes catalysis[44] (see Chapters 1 and 2). However, only a very small part of the total water that surrounds an enzyme (water solutions) seems

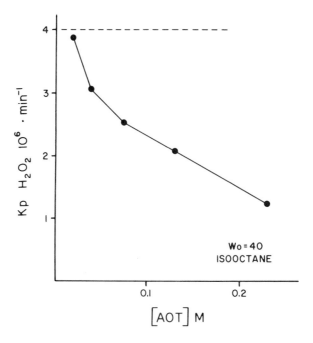

FIGURE 6. Regulation of catalase activity by variation of the surfactant concentration at a constant Wo. Assays were carried out with 5 μl (1 μg) of catalase in AOT-isooctane reverse micelles. Catalase was solubilized in reverse micelles containing the indicated AOT concentrations at a constant Wo value of 40. The reaction was started by addition of H_2O_2 (15 mM final concentration). The dashed line represents catalase activity obtained in 50 mM K-phosphate buffer, pH 7.0.

to be instrumental for catalysis. As a matter of fact, this relevant problem has been conveniently tackled by studying enzyme behavior in anhydrous organic solvents.[45] Results obtained with several enzymes[44-47] have revealed that the minimal quantity of water that is required in enzyme catalysis varies from a few tens of mole-equivalent bound to the enzyme to several hundreds of molecules.[44] Accordingly, this hydration range will be sufficient to create either a few water clusters bounded to charged groups in the enzyme or a complete monolayer of water surrounding the enzyme molecule.[44]

Thermal denaturation of proteins requires ample conformational mobility,[45] and free water has an active role in this process. Supporting this view, enzymes[48,49] suspended in anhydrous organic solvents are more thermostable (i.e., several hours at 100°C) than enzymes in water solutions. Thermostability in organic solvents decreases when water content of the solvent increases.[45]

Thermostability of enzymes has also been studied in reverse micelle systems.[4,13] Ayala et al.[13] reported that at 70°C, ATPase and cytochrome c-oxidase had half-lives of 11 h and 100 s, respectively, in an asolectin-toluene system containing 1.3% of water; however, thermostability of cytochrome c-oxidase could be increased more than 100 times by decreasing the water concentration to 0.3%. Under these circumstances, ATPase had a half-life of 96 h.

Along this line, Figure 7 shows that the thermostability of catalase in AOT reverse micelles in toluene at 60°C decreases as the water content of the micelle preparation increases (after incubation at 60°C, catalase assays were performed at 30°C). Water-saturated preparations (noted as emulsion in Figure 7) showed a time course of inactivation at 60°C similar to that obtained in the aqueous system (not shown). The effect of the Wo parameter on

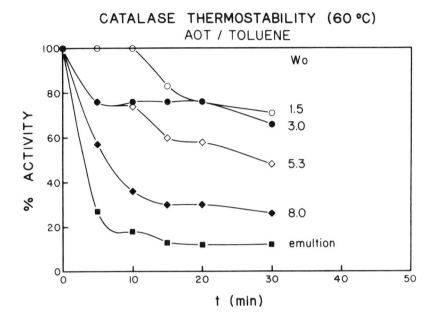

FIGURE 7. Time course for the inactivation of catalase trapped in AOT-toluene reverse micelles at 60°C. Catalase (1 μg) in AOT-isooctane reverse micelles were prepared at the Wo values indicated in the figure and incubated at 60°C. Aliquots were withdrawn at the indicated times to measure remaining activity at 30°C as described in Figure 4. The catalytic activity registered at time zero for the different Wo values assayed were ($K_p \times 10^6$ min^{-1}): 0.6, 0.7, 1.0, 1.5, and 2.3 for the Wo values of 1.5, 3.0, 5.3, and 8.0 for the emulsion, respectively.

catalytic performance (Figure 5) as well as on thermal inactivation (Figure 7) cannot be considered as the result of merely specific interactions of water molecules with the encased enzyme. Indeed, enzyme catalytic response to water content (i.e., Wo parameter) in reverse micelle systems does not seem to be due to the direct sensing of water molecules by the enzyme, as occurs with enzymes in anhydrous organic solvents; rather, the size of the micellar cavity containing the enzyme seems to be the factor involved in the behavior of the enzyme at various Wo values.[31] In support of this view, Khmelnitski et al.[31] have reported that the optimal Wo value decreases when water is partially substituted by a water-miscible solvent (i.e., glycerol) in the micelle cavity. The same workers found a good correlation between the diameter of the micelle cavity and the diameter of the trapped enzyme, hence producing the maximal activity response; i.e., whenever the micelle cavity reaches a critical size, geometrically fitted to the enzyme molecule enclosed, maximal activity is expressed. Moreover, under optimal conditions of solvation by water-glycerol mixtures, it was found that the higher the glycerol fraction volume, the higher the activity. This result has been interpreted[31] in terms of the growing microviscosity imposed by the glycerol concentration on the micelle inner fluid and to the resulting increase in the rigidity of the surfactant shell. Thus, the enzyme is progressively restricted in its conformation mobility until the more active conformation is stabilized.[31]

Another interesting point to consider is the effect of temperature on catalysis at low-water-content water. In this respect, Klibanov[45] has claimed that lipase remains catalytically active in anhydrous organic solvents at 100°C, indicating that the proper active conformation is maintained even at such extreme temperature. On the other hand, Sánchez Ferrer et al.[8] reported that polyphenoloxidase in brij 96-cyclohexane reverse micelles exhibited a broad optimal temperature range between 25 and 40°C, as in water; however, above 45°C the activity decreased to almost zero. This behavior was explained as a consequence of a

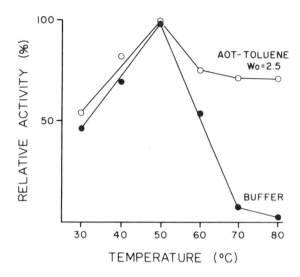

FIGURE 8.   Effect of temperature on the catalytic performance
of catalase in aqueous and AOT-toluene reverse micelle mixtures.
1.75 μg of catalase in 1 ml of 50 mM K-phosphate buffer, pH
7.0, and 3.5 μg of catalase in reverse micelles in 1 ml of 0.2 M
AOT were preincubated for 5 min at the indicated temperatures;
the activity assays were performed at the temperature used in the
preincubation. Reaction was started by the addition of 5 μl of 30%
H_2O_2 to the aqueous and organic assays. After 30 s of reaction
time, the assays were stopped with 2 ml of 2% H_2SO_4. Titration
of the remaining H_2O_2 was performed with 1 × 10^{-3} N KMnO$_4$.[50]

temperature-induced distortion of the micellar system, probably causing percolation of the
aqueous phase. This experiment was carried out at Wo = 10 (the optimum) and a water-
volume fraction of 0.6%.

Considering that hydration (bound water) accelerates enzymatic reactions and that water
in excess brings about the conditions required for thermal inactivation, we decided to test
if catalase in AOT micelles in toluene, at low hydration values (i.e., Wo = 2.5), carried
out catalysis at high temperatures. To this purpose, catalase reverse micelle preparations
were pre-equilibrated for 5 min at the indicated temperatures (Figure 8); thereafter, the
catalytic assay was run at the temperature used in pre-equilibration.

As the Clark oxygen electrode is not reliable at high temperatures, we measured catalase
activity by titrating H_2O_2 with KMnO$_4$[50] (see Figure 8 caption). Catalase activity increased
linearly up to 50°C; this was to the point of maximal activity for both aqueous and AOT-
toluene systems. At temperatures higher than 50°C, the activity in water decreased, and at
70°C it was almost zero. On the other hand, catalase activity in AOT micelles at Wo =
2.5 remained fairly active up to 80°C; at this temperature, the activity was about 75% of
the maximum obtained at the optimal temperature (i.e., 50°C). It is stressed that at Wo =
2.5 and 50°C, catalase activity was only 30 to 35% of that obtained at Wo 20 (not shown).
However, at Wo = 20 the catalytic response to temperature closely corresponds to that
obtained in the all-water medium (Figure 8).

At Wo = 2.5, all water molecules are committed in the solvation of charged groups;
thus, free water required for thermal inactivation would be not available, i.e., the conformational fluctuation of catalase would be severely restricted. This condition is responsible
for both the lower catalytic performance observed and for the catalytic stability occurring
at such high temperature. Further studies should explore higher levels of hydration in search
for a condition in which activity at high temperatures is maximal considering the thermal
stability of the enzyme.

IV. RESPIRATION IN APOLAR ORGANIC SOLVENTS

One of the most interesting applications of the O_2-polarographic technique to enzyme kinetics in organic solvents is the study of the respiratory chain activities using O_2 as terminal electron acceptor.

Reverse micelle systems have been considered as a model in the study of the physical properties of biological membranes.[4] Moreover, the existence of lipid particles resembling reverse micelles in biological membranes has been suggested.[4] Thus, the study of the respiratory complexes in reverse micelle systems is particularly relevant to the understanding of the functional interactions occurring between molecular components of biological membranes.

A. MITOCHONDRIAL RESPIRATORY ENZYME COMPLEXES IN REVERSE MICELLE SYSTEMS

Early studies[51] with cytochrome c in reverse micelles suggested the possibility that redox reactions involving physiological components of the respiratory chain could be achieved in such systems. In the first studies, it was shown that oxidized and reduced cytochrome c in reverse micelles conserved their spectroscopic properties. It was also realized that the normal spectroscopic pattern of cytochrome c in reverse micelles was altered at low hydration degrees, suggesting that the heme c chromophore could be used as a spectroscopic probe of protein conformation in reverse micelles.

Further studies[52] with horse cytochrome c and purified cytochrome c-oxidase in reverse micelles of asolectin in toluene demonstrated for the first time the electron transfer between two redox components of the respiratory chain. In these experiments, the reduction of cytochrome c-oxidase by cytochrome c (reduced with ascorbate) was followed spectroscopically.

The above results prompted us to ascertain if mitochondrial or bacterial membrane particles, once transferred into a reverse micelle system, could carry out electron transfer across the full span of the respiratory chain (i.e., from NADH to O_2). To this purpose, mitochondrial membrane particles were dispersed by sonication in an asolectin-toluene reverse micelle system. The water content of the preparation was adjusted to Wo = 13, giving an optically clear preparation suitable for spectrophotometric recording.

After recording the air-oxidized SMP minus air-oxidized SMP base line (trace **a** in Figure 9A), NADH plus KCN was added to the sample cuvette and reduction of the cytochromes was recorded after incubation for 30 min at 30°C (trace **b** in Figure 9A). The reduced forms of cytochrome c (555 nm) and aa_3 (607 nm) become clearly apparent due to KCN inhibition of cytochrome aa_3 activity. Unexpectedly, reduced cytochrome b (562 nm) was not accumulated even after 45 min of incubation. This result suggested that electron transfer activity across cytochrome b could be the rate-limiting step, probably due to a preferential partition of Coenzyme Q_{10} to the toluene phase. Indeed, natural ubiquinones and menaquinones are highly hydrophobic molecules. In the experiments, it could be considered that reduced cytochrome c undergoes autooxidation in reverse micelle systems.[52] Thus, the low transport activity across cytochrome b sector and the spontaneous oxidation of cytochrome c could account for the low reduction levels of cytochromes b and c in the presence of cyanide. In fact, the addition of menadione (functional analogue of physiological quinones) to the preparation already containing NADH plus KCN produced a dramatic increase in the reduction levels of cytochromes b and c; the reduction levels of cytochrome aa_3 also increased. Between 60 and 75% of the total cytochrome concentration (as calculated by dithionite reduction) were reduced by the combination of NADH, KCN, and menadione. The finding that reduced cytochrome aa_3 accumulated only in the presence of KCN suggested that electron transport was taking place in the full span of the respiratory chain (i.e., from NADH to oxygen).

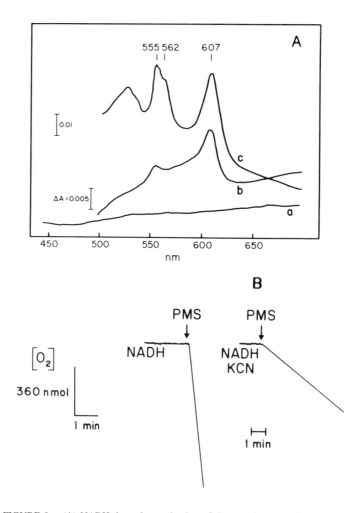

FIGURE 9. (**A**) NADH-dependent reduction of the cytochromes of mitochondrial membrane particles from bovine heart (SMP) solubilized in a phospholipid-
toluene reverse micelle medium. SMP (1 mg of protein) was solubilized in 1 ml
of a reverse micelle system with 1% asolectin and 0.3% water in toluene. (a)
First, the air-oxidized SMP minus the air-oxidized SMP baseline was recorded.
(b) 0.3 mM NADH and 0.5 mM KCN were added to the sample cuvette and
incubated for 30 min at room temperature. (c) As preparation (b), but in the
presence of 0.1 mM menadione (dissolved in toluene). (**B**) Polarographic recording
of mitochondrial cytochrome oxidase activity. SMP (1 mg) solubilized in asolectin-
isooctane reverse micelles (Wo = 19) were incubated with 5.6 mM NADH and
4 mM PMS at 30°C. In the right trace, cytochrome oxidase activity was inhibited
with 0.5 mM KCN. Cytochrome oxidase specific activity was 0.5 μmol O_2 mg^{-1}
min^{-1}.

In the presence of NADH, menadione, and KCN, the electron transport rate was calculated to be about 0.4 nmol of reduced cytochrome aa_3 per minute per milligram of protein.
This slow rate precludes the utilization of the polarographic technique to measure the NADH-
oxidase activity in reverse micelle preparations containing mitochondrial membrane particles.
However, cytochrome oxidase remains fairly active in the micelle system and its catalytic
activity could be measured by direct O_2 uptake measurements upon the addition of NADH
(or ascorbate) plus phenazine methosulfate (PMS) (Figure 9B). The activity of cytochrome
oxidase in reverse micelle systems was close to that expressed in water (not shown) and
sensitive to KCN.

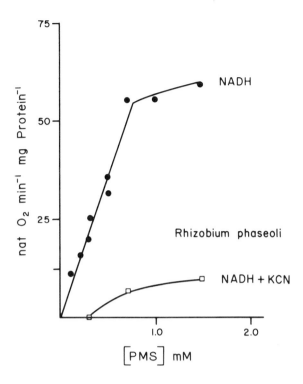

FIGURE 10.   Cytochrome oxidase activity of *Rhizobium phaseoli* membrane particles solubilized in 1% asolectin-isooctane vs. the effect of the PMS concentration. The O_2 uptake reaction was started by addition of 10 μl of 400 mM NADH aqueous solution to 2 ml of reverse micelle preparations containing 2 mg ml^{-1} of membrane protein and the indicated PMS concentrations. Water was adjusted to 2% final concentration. After recording of activity at each PMS concentration (●), the reaction was inhibited by addition of 10 μl of 100 mM KCN (□). Control and KCN-inhibited O_2 uptake were calculated and plotted as η atoms oxygen per minute per milligram of protein against PMS concentration.

B. BACTERIAL RESPIRATORY COMPLEXES IN REVERSE MICELLES

In addition to mitochondrial preparations, bacterial membranes also exhibit electron transfer activity when transferred to a reverse micelle system. Membrane particles of *Bacillus cereus* and *Rhizobium phaseoli* were active in asolectin-isooctane medium. Figure 10 shows the catalytic response of *R. phaseoli* membranes to the PMS concentration; here, NADH (or ascorbate) was used as electron donor to PMS. The dependence of activity on PMS concentration was hyperbolic with a K_{mapp} = 0.5 mM.

The high K_{mapp} value obtained can be explained as follows. PMS is very soluble in water, but its reduced form, $PMSH_2$, is not.[53] Hence, $PMSH_2$ will partition to the bulk organic phase, lowering the effective $PMSH_2$ concentration in the aqueous pseudophase. In agreement with this conclusion, it was found that the spontaneous oxidation of $PMSH_2$ was negligible at PMS concentrations lower than 1.0 mM. This was recorded after inhibition of cytochrome oxidase with KCN (Figure 10). As the oxidation of $PMSH_2$ involves the release of two protons, $PMSH_2$ in the organic phase is stable.

Figure 11 shows some of the coupled reactions that can be applied to the study of the mitochondrial or bacterial respiratory chains in reverse micelle systems. In all cases, O_2 is the final electron acceptor.

NADH-dependent reactions — The physiological pathway for the oxidation of NADH

I. NADH-DEPENDENT REACTIONS

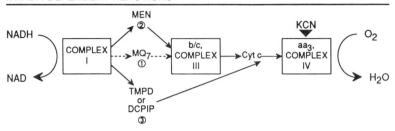

II. SUCCINATE-DEPENDENT REACTIONS

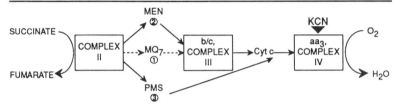

III. CYTOCHROME OXIDASE

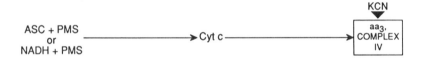

FIGURE 11.   Electron transfer coupled reactions in mitochondrial and bacterial respiratory systems dispersed in asolectin-organic solvent preparations. NADH and succinate act as electron donors to Complex I and Complex II, respectively. Electron carriers to Complex III are either natural quinones (MQ$_7$ in *B. cereus* and CoQ$_{10}$ in mitochondria) or synthetic analogs (i.e., menadione). Alternatively, electrons can be removed from Complex I or Complex II and be donated directly to the cytochrome *c*-Complex IV sector by using artificial electron carriers such as DCPIP or TMPD (NADH-liked reactions) or PMS (succinate-liked reactions). The activity of cytochrome oxidase can be measured with PMS as electron carrier and NADH or ascorbate as bulk reductants.

(route 1) can be accelerated by addition of menadione (route 2) or other suitable quinones. Alternatively, electrons from NADH-dehydrogenase (Complex I) can be delivered directly to cytochrome aa_3 (Complex IV) by including N,N,N', N' tetramethyl, *p*-phenylenediamine (TMPD) or 2,6-dichlorophenol-indophenol (DCPIP) as electron carriers. With both carriers, Complex III is bypassed, and in both cases the reaction is strongly inhibited by KCN. The strong spontaneous reaction between NADH and PMS[53] precludes utilization of PMS as electron acceptor of NADH-dehydrogenase.

Succinate-dependent reactions — As with NADH, the physiological pathway for the oxidation of succinate (route 1) can be accelerated by the menadione (route 2). Alternatively, electrons from succinic dehydrogenase (Complex II) can be delivered directly to cytochrome *c* and to cytochrome aa_3 by including PMS in the assay. PMS is a good electron acceptor during succinate oxidation by the dehydrogenase, and no spontaneous reaction between succinate and PMS takes place. In our experience, NADH and succinate-dependent activities in reverse micelles systems are much better preserved in preparations made with bacterial membrane particles than with bovine heart mitochondrial particles, probably because respiratory complexes of bacteria are smaller and simpler than those from mitochondria. We have also found that, similarly to the native membranes of *B. cereus*[54,55] in all water media, the rate-limiting step in the oxidation of NADH or succinate is not the activity of cytochrome oxidase, but the activity of NADH and succinic dehydrogenases. In fact, the highest respiratory rates are obtained when ascorbate or NADH together with PMS are used as electron donors (*vide infra*).

Cytochrome oxidase — Figures 9B and 10 show that cytochrome oxidase activity can be measured with ascorbate or NADH as bulk reductants and PMS as electron carrier to cytochrome oxidase. $PMSH_2$ is able to donate electrons to cytochrome c[53] or directly to cytochrome aa_3, thus the assay system can also be applied to pure preparations of cytochrome aa_3 oxidase free of cytochrome c. Membrane fluidity is a requisite for redox reactions of the respiratory chain. Respiratory enzyme complexes find their partners by lateral displacements in the membrane plane. Redox reactions in reverse micelle systems will take place in a three-dimensional pathway.

Figure 12 shows our hypothetical view of reverse micelles systems containing all the molecular components of the *B. cereus* respiratory chain.[54,55] Enzymic Complexes I to IV are randomly distributed in the reverse micelle solution. Hydrophilic domains in each protein complex are patched by hydrated phospholipid shells, while hydrophobic areas are directly exposed to the organic apolar solvent.[52,56]

Water-soluble substrates (S) and inhibitors (I) are trapped within reverse micelles. Thus, random collisions will provide the pathway for the exchange of substrates and products with the corresponding Complex I and Complex II. Electrons find their way to cytochrome b/c_1 (Complex III) using the hydrophobic menaquinone-7 (MQ)[57,58] as carrier.

Electron transfer to the alternative cytochrome oxidases aa_3[59] and o is mediated by water-soluble cytochrome c entrapped in reverse micelles.[52] Final reaction with molecular O_2 produces H_2O which is solubilized into a micelle compartment.

C. RESPIRATION OF WHOLE BACTERIAL CELLS IN ORGANIC SOLVENTS

Suspended or immobilized bacterial dead cells in organic solvents have been used to carry out catalytic processes with potential commercial applications.[43] Whole cell preparations, compared to pure enzymes, have shown to be more stable microreactors in organic solvents.[43] In 1985, Häring et al.[60] first reported the transfer of bacterial cells to a reverse micelle system composed of Tween 85 and 3 to 4% water (V:V) in isopropylpalmitate as a solvent. They prepared organic micellar systems containing up to 10^8 cells per milliliter. This number of cells yielded turbid suspensions in water; however, the same amount in the reverse micelle systems gave transparent solutions. This finding has been repeated in several laboratories using yeast or bacterial cells that have been transferred to a variety of surfactant-organic solvents systems (see Chapter 11).

Viability has been tested in several cases and values reported (Chapter 11) range from a few hours to a few days (the viability reported for *B. cereus* spores[56] is not considered since these cells are cryptobiotic forms and are naturally resistant to pure organic solvents).

Microbial cells dispersed in reverse micelle solutions are able to display metabolic activities using endogenous or added substrates. In this respect, the application of the Clark oxygen electrode to measure the respiratory response of *S. cerevisiae* to ethanol, in a reverse micellar solution of Tween 88-isopropylpalmitate, has been reported (Chapter 11).

We have recently applied the Clark oxygen electrode to measure respiration, associated to the germinative behavior of the *B. cereus* spore, in a system composed of asolectin in isooctane (unpublished results). Under these conditions the germination of the spore (in asolectin-isooctane) was specifically induced (adenosine plus alanine) with a yield better than 80% and a survival that is higher than 90%. The sequence of morphological changes and metabolic events registered (i.e., respiratory activation) followed the same ordered pattern that has been described for germination in water.[61]

ACKNOWLEDGMENTS

This work was supported by grants IN202191/DGAPA (Universidad Nacional Autónoma de México, México) and D111-903667 (Consejo Nacional de Ciencia y Tecnología, México). We thank Jennifer Cooper for her help during preparation of this manuscript.

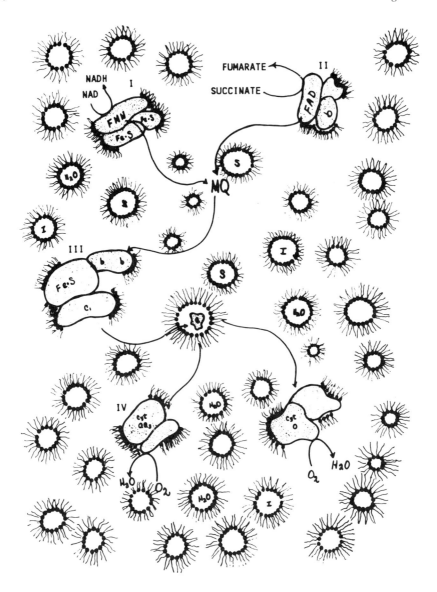

FIGURE 12. Hypothetical model of the respiratory chain complexes from *Rhizobium phaseoli* in a 0.1% asolectin-reverse micelle system in isooctane. NADH:ubiquinone oxidoreductase (Complex I), succinate:ubiquinone oxidoreductase (Complex II), ubiquinol:cytochrome *c* oxidoreductase (Complex III), and the terminal oxidases ferrocytochrome *c*:oxygen oxidoreductase (Complex IV), and cytochrome *o* are covered by hydrated patches of surfactant monolayers at their hydrophilic regions, while the hydrophobic domains are in direct contact with the organic solvent. Together with menaquinone (MQ) and cytochrome *c* (c), the respiratory complexes move in three dimensions, lowering the probability of an effective electron transfer between the complexes. Cytochrome *c* (c), hydrophilic substrates (S), and inhibitors (I) are dissolved in the aqueous pool of the reverse micelles.

REFERENCES

1. **Wong, M., Tomas, J. K., and Nowak, T.,** Structure and state of H₂O in reversed micelles, *J. Am. Chem. Soc.,* 14, 4730, 1977.
2. **Luisi, P. L., Giomini, M., Pileni, M. P., and Robinson, B. H.,** Reverse micelles as hosts for proteins and small molecules, *Biochem. Biophys. Acta,* 947, 209, 1988.
3. **Luisi, P. L. and Magid, L. J.,** Solubilization of enzymes and nucleic acids in hydrocarbon micellar solutions, *CRC Crit. Rev. Biochem.,* 20, 409, 1986.
4. **Martinek, K., Levashov, A. V., Klyachko, N., Khmelnitski, Y. L., and Berezin, I. V.,** Micellar enzymology, *Eur. J. Biochem.,* 155, 453, 1986.
5. **Luisi, P. L. and Steinamann-Hofmann, B.,** Activity and conformation of enzymes in reverse micellar solutions, in *Methods in Enzymology,* Vol. 136, Moabach, K., Ed., Academic Press, Orlando, 1987, chap. 19.
6. **Bru, R., Sánchez-Ferrer, A., and García-Carmona, F.,** A theoretical study on the expression of enzymic activity in reverse micelles, *Biochem. J.,* 259, 355, 1989.
7. **Hilhorst, R., Laane, C., and Veeger, C.,** Enzymatic conversion of apolar compounds in organic media using an NADH-regenerating systems and dihydrogen as reductant, *FEBS Lett.,* 159(1,2), 225, 1983.
8. **Sánchez-Ferrer, A., Bru, R., and García-Carmona, F.,** Kinetic properties of polyphenoloxidase in organic solvents. A study in Brij 96-cyclohexane reverse micelles, *FEBS Lett.,* 233, 363, 1988.
9. **Walde, P., Giuliani, A. M., Boicelli, A., and Luisi, P. L.,** Phospholipid-based reverse micelles, *Chem. Phys. Lipids,* 53, 265, 1990.
10. **Khmelnitski, Y. L., Levashov, A. V., Klyachko, N. L., and Martinek, L.,** A microheterogenous medium for chemical (enzymic) reactions based on a colloidal solution of water in an organic solvent, *Russ. Chem. Rev.,* 53(4), 319, 1984.
11. **Levashov, A. V., Khmelnitski, Y. L., Klyachko, N. L., and Martinek, K.,** Reversed micellar enzymology, in *Surfactants in Solution,* Vol. 2, Mittal, K. L. and Lindman, B., Eds., Plenum Press, New York, 1984, 1069.
12. **Kavanov, A. V., Nametkin, S. N., and Levashov, A. V.,** The principal difference in regulation of the catalytic activity of water-soluble and membrane forms of enzymes in reversed micelles: γ-glutamyltransferase and aminopeptidase, *FEBS Lett.,* 267, 236, 1990.
13. **Ayala, G., Gómez-Puyou, M. T., Gómez-Puyou, A., and Darszon, A.,** Thermostability of membrane enzymes in organic solvents, *FEBS Lett.,* 203(1), 41, 1986.
14. **Hilhorst, R., Spruijt, R., Laane, C., and Veeger, C.,** Rules for the regulation of enzyme activity in reversed micelles as illustrated by the conversion of apolar steroids by 20β-hydroxysteroid dehydrogenase, *Eur. J. Biochem.,* 144, 459, 1984.
15. **Verhaert, R. M., Tyrakowska, B., Hilhorst, R., Shaafsma, T. J., and Veeger, C.,** Enzyme kinetics in reversed micelles. Behaviour of enoate reductase, *Eur. J. Biochem.,* 187, 73, 1990.
16. **Menger, F. M. and Yamada, K.,** Enzyme catalysis in water pools, *J. Am. Chem. Soc.,* 111(22), 6731, 1979.
17. **Klyachko, N. L., Levashov, A. V., and Martinek, K.,** Catalysis by enzymes incorporated in reverse micelles of surface-active substances in organic solvents. Peroxidase in the aerosol AOT-water-octane system, *Mol. Biol.,* 18(4), 830, 1984.
18. **Larsson, K. M., Adlercreutz, P., and Mattiasson, B.,** Activity and stability of horse-liver alcohol dehydrogenase in sodium dioctylsulfosuccinate/cyclohexane reverse micelles, *Eur. J. Biochem.,* 166, 151, 1987.
19. **Martinek, K., Levashov, A. V., Klyachko, V. I., Pantinin, V. I., and Berezin, I. V.,** The principles of enzyme stabilization. VI. Catalysis by water-soluble enzymes entrapped into reversed micelles of surfactants in organic solvents, *Biochem. Biophys. Acta,* 657, 277, 1981.
20. **Martinek, K., Berezin, I. V., Khmelnitski, V. L., Klyachko, N. L., and Levashov, A. V.,** Enzymes entrapped into reversed micelles of surfactants in organic solvents: key trend in applied enzymology, *Biocatalysis,* 1, 9, 1987.
21. **Luisi, P. L.,** Enzymes hosted in reverse micelles in hydrocarbon solution, *Angew. Chem. Int. Ed. Engl.,* 24, 439, 1985.
22. **Lesser, E. M., Genshuan, W., Luisi, P. L., and Maestro, M.,** Application of reverse micelles for the extraction of proteins, *Biochem. Biophys. Res. Commun.,* 135(2), 629, 1986.
23. **Ayala, G., Nacimiento, A., Gómez-Puyou, A., and Darszon, A.,** Extraction of mitochondrial membrane proteins into organic solvents in a functional state, *Biochem. Biophys. Acta,* 810, 115, 1985.
24. **Escobar, L., Salvador, C., Contreras, M., and Escamilla, J. E.,** On the application of the Clark oxygen electrode to the study of enzyme kinetics in apolar solvents: the catalase reaction, *Anal. Biochem.,* 184, 139, 1990.
25. **Hochkoeppler, A. and Palmieri, S.,** Polarographic measurement of oxygen uptake using lipoxygenase in reverse micelles, *Biotechnol. Bioeng.,* 36, 672, 1990.

26. **Jao, T.-C. and Kreuz, L. K.,** Solubility of oxygen in inverted micelles of calcium alkylbenzene sulfonates, *J. Colloid Interface Sci.,* 102(1), 308, 1984.

27. **Goldstein, D. B.,** A method for assay of catalase with the oxygen cathode, *Anal. Biochem.,* 24, 431, 1968.

28. **Wong, M., Thomas, J. K., and Gratzel, M.,** Fluorescence probing of inverted micelles. The state of solubilized water clusters in alkane/diisooctylsulfosuccinate (aerosol OT) solution, *J. Am. Chem. Soc.,* 98, 2391, 1976.

29. **Inove, A. and Horikoshi, K.,** A Pseudomonas thrives in high concentrations of toluene, *Nature,* 338(16), 264, 1989.

30. **Martinek, K., Klyachko, N. L., Kabanov, A. V., Khmelnitski, Y. L., and Levashov, A. V.,** Micellar enzymology: its relation to membranology, *Biochem. Biophys. Acta,* 981, 161, 1989.

31. **Khmelnitski, Y. L., Kabanov, A. V., Klyachko, N. L., Levashov, A. V., and Martinek, K.,** Enzymatic catalysis in reverse micelles, in *Structure and Reactivity in Reverse Micelles,* 10th ed., Pileni, M. P., Ed., Elsevier/North-Holland, Amsterdam, 1989, 230.

32. **Grandi, C., Smith, R. E., and Luisi, P. L.,** Micellar solubilization of biopolymers in organic solvents. Activity and conformation of lysozyme in isooctane reverse micelles, *J. Biol. Chem.,* 256(2), 837, 1981.

33. **Barbaric, S. and Luisi, P. L.,** Micellar solubilization of biopolymers in organic solvents. V. Activity and conformation of α-chymotrypsin in isooctane-AOT reverse micelles, *J. Am. Chem. Soc.,* 103(14), 4239, 1981.

34. **Martinek, K., Levashov, A. V., Khmelnitski, N. L., Klyachko, N. L., and Berezin, I. V.,** Colloidal solution of water in organic solvents: a microheterogenous medium for enzymatic reactions, *Science,* 218(26), 889, 1982.

35. **Hilhorst, R.,** Applications of enzyme containing reversed micelles, in *Structure and Reactivity in Reverse Micelles,* Pileni, M. P., Ed., Elsevier/North-Holland, Amsterdam, 1989, 323.

36. **Takahashi, K., Nishimura, H., Yoshimoto, T., Saito, Y., and Inada, Y.,** A chemical modification to make horseradish peroxidase soluble and active in benzene, *Biochem. Biophys. Res. Commun.,* 121(1), 261, 1984.

37. **Frank, S. G. and Zografi, G.,** Solubilization of water by dialkyl sodium sulfosuccinates in hydrocarbon solutions, *J. Colloid Interface Sci.,* 29, 27, 1969.

38. **Kon-No, K. and Kitahara, A.,** Solubility behavior of water in nonaqueous solutions of oil-soluble surfactants: effect of molecular structure of surfactants and solvents, *J. Colloid Interface Sci.,* 37(2), 469, 1971.

39. **Zulauf, M. and Eicke, H.-F.,** Inverted micelles and microemulsions in the ternary system H_2O/aerosol-AT/isooctane as studied by photon correlation spectroscopy, *J. Phys. Chem.,* 83(4), 480, 1979.

40. **Longley, W.,** The crystal structure of bovine liver catalase: a combined study by X-ray diffraction and electron microscopy, *J. Mol. Biol.,* 30, 323, 1967.

41. **Fletcher, P. D. I., Rees, G. D., Robinson, B. H., and Freedman, R. B.,** Kinetic properties of α-chymotrypsin in water-in-oil microemulsion systems, *Biochim. Biophys. Acta,* 832, 204, 1985.

42. **Klyachko, N. L., Levashov, A. V., Pshezhetzky, A. V., Bogdanora, N. G., Berizin, I. V., and Martinek, K.,** Catalysis by enzymes entrapped into hydrated surfactant aggregates having lamellar or cylindrical (hexagonal) or ball-shaped (cubic) structure in organic solvent, *Eur. J. Biochem.,* 161, 149, 1986.

43. **Aldercreutz, P. and Mattiasson, B.,** Aspects of biocatalysis stability in organic solvents, *Biocatalysis,* 1, 99, 1987.

44. **Zaks, A. and Klibanov, A. M.,** The effect of water on enzyme action in organic media, *J. Biol. Chem.,* 263(17), 8017, 1988.

45. **Klibanov, A. M.,** Enzymatic catalysis in anhydrous organic solvents, *Trends Biochem. Sci.,* 14(4), 141, 1989.

46. **Zaks, A. and Klibanov, A. M.,** Enzymatic catalysis in nonaqueous solvents, *J. Biol. Chem.,* 263, 3194, 1988.

47. **Zaks, A. and Klibanov, A. M.,** Enzymes-catalyzed processes in organic solvents, *Proc. Natl. Acad. Sci. U.S.A.,* 82, 3192, 1985.

48. **Zaks, A. and Klibanov, A. M.,** Enzymatic catalysis in organic media at 100°C, *Science,* 224, 1249, 1988.

49. **Russell, A. J. and Klibanov, A. M.,** Inhibitor induced enzyme activation in organic solvents, *J. Biol. Chem.,* 263(24), 11624, 1988.

50. **Chance, B. and Maehly, A. C.,** Assay of catalases and peroxidases, in *Methods in Enzymology,* Vol. 2, Colowick, S. P. and Klapan, N. O., Eds., Academic Press, New York, 1955, 764.

51. **Douzou, P.,** Cryoenzymology in aqueous media, *Adv. Enzymol.,* 51, 1, 1980.

52. **Escamilla, J. E., Ayala, G., Gómez-Puyou, M., Gómez-Puyou, A., Millán, L., and Darszon, A.,** Catalytic activity of cytochrome oxidase and cytochrome c in apolar solvents containing phospholipids and low amounts of water, *Arch. Biochem. Biophys.,* 272(2), 332, 1989.

53. **Jones, K. M.,** Artificial substrates and biochemical reagents, in *Data for Biochemical Research,* 2nd ed., Dawson, R. M., Elliott, D. C., Elliott, W. H., and Jones, K. M., Eds., Carendon Press, Oxford, 1969, chap. 18.

54. **Escamilla, J. E. and Benito, M. C.,** Respiratory system of vegetative and sporulating *Bacillus cereus, J. Bacteriol.,* 160(1), 473, 1984.

55. **Escamilla, J. E., Ramírez, R., Del Arenal, I. P., Zarzoza, G., and Linares, V.,** Expression of cytochrome oxidases in *Bacillus cereus:* effects of oxygen tension and carbon source, *J. Gen. Microbiol.,* 133, 3549, 1987.

56. **Darszon, A., Escamilla, J. E., Gómez-Puyou, A., and Gómez-Puyou, M.,** Transfer of spores, bacteria and yeast into toluene containing phospholipids and low amounts of water: preservation of the bacterial respiratory chain, *Biochem. Biophys. Res. Commun.,* 151(3), 1074, 1988.

57. **Escamilla, J. E., Ramírez, R., Del Arenal, P., and Aranda, A.,** Respiratory systems of the *Bacillus cereus* mother cell and forespore, *J. Bacteriol.,* 167(2), 544, 1986.

58. **Escamilla, J. E., Barquera, B., Ramírez, R., Garcia-Horsman, A., and Del Arenal, P.,** Role of menaquinone in inactivation and activation of the *Bacillus cereus* forespore respiratory system, *J. Bacteriol.,* 170(12), 5908, 1988.

59. **García-Horsman, J. A., Barquera, B., Gónzalez-Halphen, D., and Escamilla, J. E.,** Purification and characterization of two-subunit cytochrome aa_3 from *Bacillus cereus, Mol. Microbiol.,* 5(1), 197, 1991.

60. **Häring, G., Luisi, P. L., and Meussdoerffer, F.,** Solubilization of bacterial cell in organic solvents via reverse micelles, *Biochem. Biophys. Res. Commun.,* 127(3), 911, 1985.

61. **Setlow, P.,** Germination and outgrowth, in *The Bacterial Spore,* Vol. 2, Hurst, A. and Gould, G. W., Eds., Academic Press, London, 1983, chap. 6.

Chapter 11

SOLUBILIZATION OF MICROORGANISMS IN ORGANIC SOLVENTS BY REVERSE MICELLES

Nestor Pfammatter, Maria Famiglietti, Alejandro Hochköppler, and Pier Luigi Luisi

TABLE OF CONTENTS

I. INTRODUCTION

One of the most surprising findings about the solubilization of biopolymers in reverse micelles is that there seems to be no restriction to the molecular weight of the guest molecules. Thus, proteins with a molecular weight up to a half-million Daltons can be solubilized, as well as plasmids with molecular weight of the order of a few million Daltons.[1-5]

From this point of view, the observation that also bacterial cells can be solubilized in organic solvents with the help of reverse micelles[6] was perhaps not a surprise. It is, however, clear that the mechanism of solubilization must be different. For example, for proteins which have dimensions comparable to reverse micelles (a radius of the order of 50 to 100 Å), one can envisage that they are hosted in the water pool.[1-5] In the case of cells which have a radius of the order of microns, such a host-guest mechanism is more problematic as reverse micelles or water pools with such dimensions have never been observed. For this reason, in the case of cells we prefer the term "solubilization *by* reverse micelles" to the term "solubilization *in* reverse micelles". Also, the kinetics and thermodynamics of the solubilization are probably very different.

All these structural and mechanistic aspects of the solubilization process are still unexplored, and thus this first review on the solubilization of cells and large organelles by reverse micelles will be rather devoted to the phenomenology of solubilization and to the first activity (viability) data.

We will consider, in the following section, some conceptual problems and some nomenclature clarifications, then we will report chronologically the solubilization experiments described in literature. A section on viability data will follow, and the last section is devoted to applications.

II. SOME GENERALITIES OF THE SOLUBILIZATION PROCESS

At the beginning, we need to clarify the terms "solubilization" and "solution". First, one has clearly to discriminate between solutions and suspensions. There is nothing special in adding proteins or microorganisms to an organic solvent and thus forming a suspension. By using reverse micellar solutions, or water-in-oil (w/o) microemulsions, it is, however, possible to solubilize microorganisms in organic solvents and obtain a solution, i.e., a clear, thermodynamically stable liquid without significant light scattering in the near-UV region. Of course, the dividing line between solutions and suspensions is experimentally thin, as it depends on the concentration of the guest organelles and other operational parameters (temperature, water and surfactant concentration, aging, etc.). In this paper, we will discuss mostly "solutions" as they represent a much more interesting case from the scientific point of view, although occasionally suspensions may be advantageous for biotechnological purposes.

The lack of turbidity and light scattering of the micellar solutions containing cells permits one to use the expressions "solution" and "solubilization". The question, however, as to why there is no precipitation or aggregation is not so trivial. Actually, from the consideration of density, one would expect a faster sedimentation of the cells in the micellar solution than in water. Perhaps one factor contributing to the stability of the solution is the tendency of the cells in the micellar solutions to exist as single units and not to aggregate. Also, the higher viscosity of the organic solvent may inhibit aggregation. Finally, the lack of scattering can be due to a favorable optical matching with the refracting index of the solvent.

Control experiments are obviously important: in absence of surfactant, there is no cell solubilization in the organic solvent; also, when no water is added, solubilization does not take place in the organic solution of the surfactant. In other words, the initial presence of both surfactant and water — this means the presence of reverse micelles or w/o microemulsions — is necessary to realize a cell solubilization.

Still, in the field of definitions it is important to clarify that the concentration (e.g., of cells) in the microemulsion system can be expressed as "overall" concentration — relative to the total volume of solution (water to organic solvent) — or as "local" concentration — relative to the water microphase (water pool). The latter is obviously numerically larger by the factor P_w, which is the volume ratio of solvent to water.

Occasionally, the microorganisms solubilized in the microemulsion phase can be re-extracted into an aqueous milieu; this process is usually referred to as "back-transfer", an expression originally coined to describe the behavior of proteins under similar conditions.

The cell activity after solubilization is generally expressed by the term "viability", which is a measure of the capability of cells to remain alive. Cell viability can be monitored by the classic microbiological technique of plating out aliquots of the micellar solution after dilution with physiological solution on petri dishes. In other words, the viability of cells in the micellar solution is generally measured making reference to standard aqueous conditions by bringing them back into the normal aqueous medium.

Recently, a method for testing the activity of cells directly in the organic medium has been proposed. This polarographic method is based on the consumption of oxygen measured by a Clark electrode dipped directly in the micellar solution. In this way, the respiration rate of the cells is observed.[7]

In addition, other criteria can be used to test the cell functionality: the activity of a particular enzyme system, a chemical reaction catalyzed by the entire cell, or the photosynthetic activity. Examples of all these possibilities are already present in the literature, and we will review most of them.

III. THE PHENOMENOLOGY OF SOLUBILIZATION

The first experiments were carried out in Zurich with *Escherichia coli* K 12, harboring the recombinant plasmid PEMBL-8, and *Acinetobacter calcoaceticus,* using Tween 85 (polyoxyethylen sorbitan trioleate) as surfactant (about 10%) in isopropylpalmitate (IPP) as solvent[6] and 3 to 4% (v/v) water solution containing LB nutrient medium. Later, also *Corynebacterium equi* IFO 3730 could be solubilized in the same system.[9]

An organic micellar system containing up to 10^8 cells per milliliter (overall concentration, i.e., referring to the total organic solution) could be prepared, and a first interesting observation (already mentioned above) was that the micellar solution was clear, whereas the same cell concentration in water afforded a turbid suspension. Thus, in this case, cells in reverse micellar solutions do not tend to aggregate so much as in water and do not tend to form large particles which scatter light. With time, the trend to aggregation was noticed also in the organic milieu, but no turbidity and no precipitation were observed for several days.

The presence of cells in the micellar solution could be directly observed by light microscope: after adding ethidium bromide to the bacterial suspension prior to solubilization, under UV light, a red fluorescence was detected in the organic solution at the position of the bacteria. With this technique, it was also observed that the mobility of bacteria in the micellar solution was reduced compared to water. In Table 1, conditions of these first and some following experiments are summarized.

It is also interesting to notice that in this first work, the solubilization was not successful by using the system AOT/isooctane/water, the system commonly used for the solubilization of proteins: there was solubilization, but the cells treated with this ionic surfactant died within hours. The reason for this quick death was recently illustrated by Dekker et al. in Wageningen.[8] They reported that reverse micelle systems with ionic surfactants (cetyltrimethylammonium bromide [CTAB]) can be used for the disruption of cells to extract their intracellular enzymes.

In a second paper from the same group in Zurich,[9] first transmission electron micrographs

TABLE 1
Survey of the Solubilization Experiments of Microorganisms in W/O Microemulsions

Microorganism	Surfactant System	Wo[a]	Viability[b]	Ref.
A. calcoaceticus	Tw 85/IPP	28	65	6
E. coli			1	
C. equi				
A. calcoaceticus	Tw 85/IPP or IPP/hexadecane	28	90	7
B. cereus (spores)	Phospholipid/toluene	—	>0.1	11
			(100)	
S. cerevisiae			>0.1	
Mitochondria (soybean)	AOT/isooctane	>10	n.d.	12
S. cerevisiae	Tw 85/IPP	14	40 90 (5th gen.)[c]	13
	Aso/IPP		60 90 (2nd gen.)	
	Aso/Hd			
S. cerevisiae	Aso/crude oil	14	95	Figure 16
	Crude oil		90	
P. alcaligenes	Aso/crude oil	14	15	
B. subtilis			100	
Arthrobacter			40	
Cyanobacteria	TW 85/Sp80/IPP or hexade-cane	14	n.d.	Figure 5

[a]   Wo = [H₂O]/[Surfactant].
[b]   Viability in percent after 24 h, 100% is referred to the viability after solubilization.
[c]   For the meaning of generation, see text.

were presented. It was, however, not possible to say whether the solubilization in the organic milieu had brought structural changes of the cells.

The microemulsion system described by Balasubramanian and Kumar (i.e., Triton X-100/hexanol/cyclohexane) was found to solubilize cells, but the cells also died immediately after solubilization.[10]

The group of Darszon was able to show the transfer of bacteria (*Bacillus cereus*), spores from the same bacilli, and baker's yeast into a ternary system composed of toluene, phospholipids (essentially asolectins), and a low amount of water.[11] The solubilization technique was actually slightly different from that used in Zurich to solubilize proteins or microorganisms in reverse micellar solutions: after sonicating the cells in the ternary system, the emulsion was clarified by blowing nitrogen on the surface. A large decrease of turbidity of the system could be achieved with phospholipid concentration as low as 2 mg/ml toluene, but maximal transparency required ten times more surfactant. With yeast, a stable emulsion was obtained adding 70 μl of an aqueous suspension containing 10^9 cells per milliliter to 2 ml of toluene solution. The authors could show that the electron chain of yeast and bacteria preserved its functional integrity in the ternary system. However, only 1 out of 10^4 yeast cells was viable after 24 h. In contrast, the viability of spores was not affected up to 30 d.

In the same year, a paper from the Zurich group appeared relating to the solubilization of mitochondria (extracted from soybean nodules) in AOT/isooctane w/o microemulsions.[12]

The argument offered for this work was that mitochondria, although much smaller than entire cells, generally still display a large array of biotechnologically interesting reactions, and since they are structurally and chemically simpler, they may offer the advantage of better regulation. On the other hand, mitochondria are mechanically less stable than entire cells and it was not obvious whether the solubilization would have been feasible without denaturation.

It was found (see also Table 1) that mitochondria could not be solubilized at Wo — the molar ratio of water to AOT (Wo = [H₂O]/[AOT]) — less than 10 (Figure 1A). At the

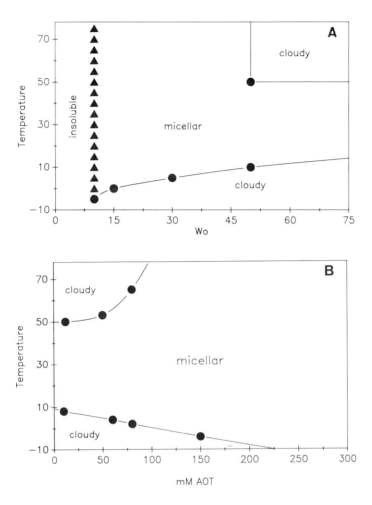

FIGURE 1. The stability of mitochondria-containing AOT micelles: (A) Against temperature and water content (Wo). The AOT concentration is constant (100 mM). The microemulsion contained 97 μg of mitochondrial protein per milliliter of micellar solution. (B) Against surfactant concentration. The water content is constant (Wo = 16.6). Same amount of mitochondria as in (A).

same time, it was observed (Figure 1B) that an increase of the concentration of surfactant extends the range of the clear micellar solution containing the organelles. Figure 1 illustrates that there is a rather large range of temperature, surfactant concentration, and water content which permits the solubilization of mitochondria in a clear micellar solution.

Preliminary light-scattering experiments have shown that the dimensions of the organelles in the micellar solutions are not significantly altered with respect to water.

The same information could also be obtained directly by electron microscopy. The dimensions of the particles back-transferred from AOT micelles to water are not significantly altered with respect to those of the native mitochondria (Figure 2), although, as apparent from Figures 2A and 2B, besides the intact organelles, some debris are also observed in the micellar solution.

Considerable effort has been devoted to the solubilization of yeast cells, and understandably so, since they are important both from the biotechnological and from the pure biological point of view. These cells are larger than the bacteria considered in the first works, and initially it was not obvious whether the solubilization would be successful.

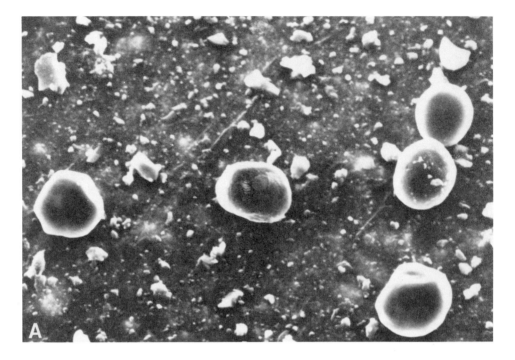

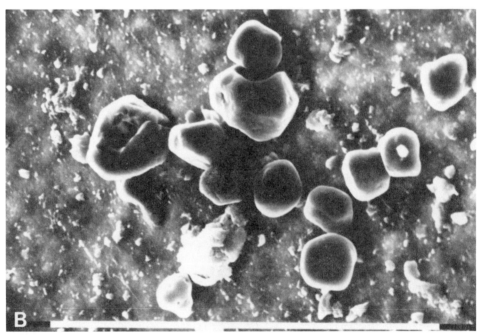

FIGURE 2. Scanning electron micrographs of mitochondria after back-transfer in water from AOT micellar solution (A and B) and of mitochondria in water (C and D). Original magnifications: A, × 13,661; B, × 10,635; C, × 30,171; and D, × 9,829. The white bars in B and D represent 10 μm.

The results obtained show again that the dimensions of the biomaterial are not a great hindrance to the solubilization and that solubilization was achieved in a first report already with different solvent/surfactant systems (Table 1). In particular, Tween/IPP, asolectin/IPP, and asolectin/hexadecane were tested in this first experiment with *Saccharomyces cerevisiae*.[13] The final concentration was around 10^6 cells per milliliter overall concentration, with

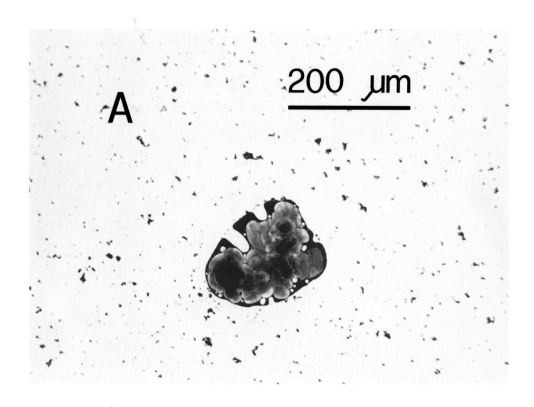

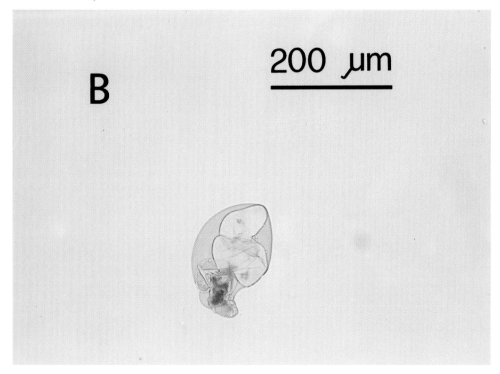

PLATE 1. Light micrographs of suspension-cultured carrot cells solubilized in water-in-oil microemulsion. The water-soluble stain Evans blue was added to the aqueous cell suspension before the cells were solubilized. (A) Water/AOT/isooctane system; (B) water/asolectin/hexadecane system.

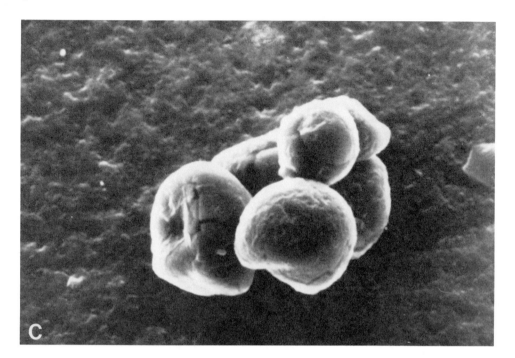

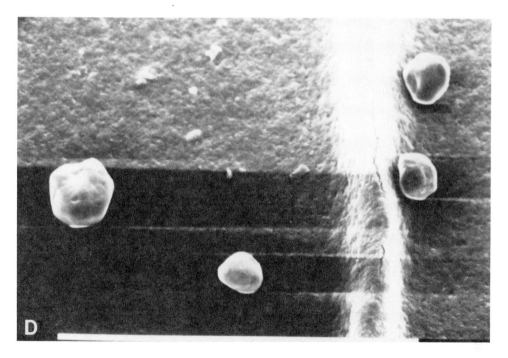

FIGURE 2 (continued).

Wo values around 14 (about 1 M water, i.e., 2% v:v). A TEM-micrograph after freeze-fracturing of the yeast cells in the micellar solution is shown in Figure 3. However, yeast cells are large enough to be comfortably observed by optical microscopy, which is indeed another advantage of working with this type of microorganism.

In the case of the yeast cells it was possible to develop a polarographic method to determine the activity and concentration of cells directly in the organic milieu.[7] In this way,

FIGURE 3.   Transmission electron micrograph after freeze-fracturing: *S. cerevisiae* in Tween/IPP/YPD (10%/ 88.2%/1.8% (w/w), Wo = 14) after 21 h solubilization. Original magnification: × 18,942; bar represents 1 μm.

it is possible to follow the kinetics of solubilization and to follow the activity as a function of different parameters, e.g., addition of ethanol. Also, the question of the maximal amount of cells which could be brought into solution was addressed. In particular, the question was raised whether there is a linearity between cell concentration and activity even beyond the limit of solubility, the point at which a solution becomes a cloudy suspension.

As shown in Figure 4, the linearity is present only in the lower concentration range. The limits of a clear solution are in this case between 1 and 10 mg of cells per milliliter (overall concentration); however, it is also clear that it can be advantageous — e.g., for biotechnological work — to increase the cell concentration above the solubility limits.

This was indeed done in the group of Haag et al.:[14] in this case, the commercially available pressed lager baker's yeast was used without adding water (traces of water are probably present in the solvent and in the commercial yeast preparation) to the organic medium, which was IPP containing phospholipids. In this way, the authors could achieve interesting results in the preparative reduction of certain ketoesters, as will be described later in this review.

Until now, all attempts to solubilize cells or organelles in the organic media were carried out empirically. The first attempt to follow a more systematic pattern was accomplished in the case of Cyanobacteria. Recognizing that the matching between the structure of the cell surface and the hydrophobicity of the surfactant system may be an important factor to affect the solubilization, the influence of the hydrophilic/lipophilic balance (HLB) of the surfactant system on the extent of solubilization was studied.

Results in Figure 5 show that indeed the HLB (as determined by varying the relative proportions of the two cosurfactants, Tween and Span) has an important influence on the solubilization: it increases remarkably at HLB values above 8 or 9.

Instead, the nature of the solvent does not seem to be so critical, as IPP or hexadecane give the same extent of solubilization for a given HLB value of the surfactant system. The same experiments are now in progress in our group for *Pseudomonas putida*.

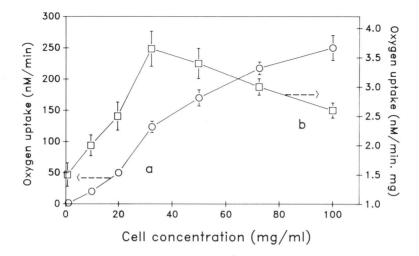

FIGURE 4. Dependence of the cell concentration on the oxygen uptake in 100 m*M* Tw 85/IPP, Wo = 14: (a) absolute O_2 consumption; (b) consumption normalized to the cell concentration (per milligram of fresh cells).

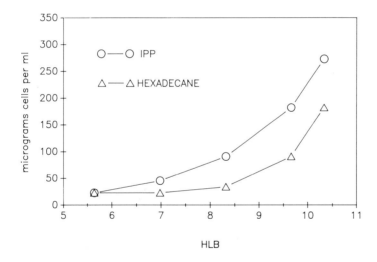

FIGURE 5. Influence of the HLB on the amount of cyanobacteria to get clear solutions. (Cells added with 15 μl growth medium in 2-ml micellar solution.)

Can organelles even larger than yeast cells be brought into organic solvents? Preliminary results obtained with plant cells seem to give a positive answer to this question. Plate 1* shows that carrot cells, with dimensions extending 50 μm, can be solubilized in the organic medium.

One report dealing in particular with the solubilization of *Rumex obtusifolius* cells in AOT/isooctane has been recently presented.[15]

IV. VIABILITY AND ACTIVITY OF MICROORGANISMS IN MICELLAR SOLUTIONS

Let us consider now in more detail the question of viability and/or activity of the solubilized cells.

* Plate 1 follows p. 246.

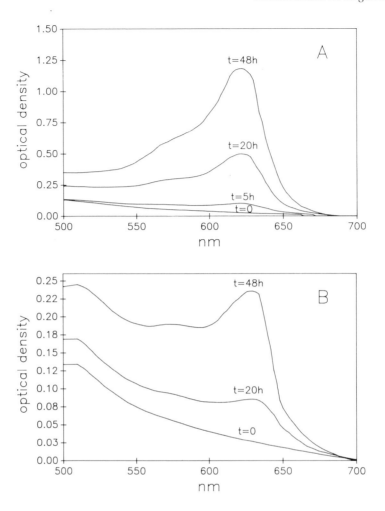

FIGURE 6.   Absorption spectra of a micellar solution of *E. coli* showing cleavage of X-gal by β-galactosidase at different times. The substrate is added to the aqueous cell suspension immediately prior to solubilization. (A) Cells were induced with isopropyl-D-thiogalactoside (IPTG) for 16 h prior to solubilization. (B) IPTG was added to the aqueous cell suspension immediately prior to solubilization.

In the first paper on cell solubilization in reverse micellar media,[6] the rate of survival of *A. calcoaceticus* was tested by plating out aliquots of the micellar solution after different time periods. In these first experiments, stability was good up to 5 h, and after 3 d 2% of the cells were still alive.

Although the *E. coli* strain was less stable than the other bacteria, this case was particularly interesting because the activity assay was based on an enzyme reaction; in fact, the strain produces β-galactosidase, and the measurement of this enzyme activity provides a very sensitive test of the cell activity. When isopropyl-D-thiogalactoside (IPTG) and the substrate X-gal were added to the cells immediately before the solubilization, a small but significant increase of β-galactosidase activity could be observed (Figure 6). This was taken to indicate that the cells in reverse micelles were capable of protein biosynthesis.

A comparison of the viability in different organic solvent systems is reported in Figure 7A. In all cases, the viability was limited to 60 h at the best, and no large difference in viability between mixtures of hexadecane and IPP was detected. However, it was shown that handling of the sample (e.g., standing vs. shaking) had a significant influence on viability. Also, the amount of water present in the system affects the viability to a certain

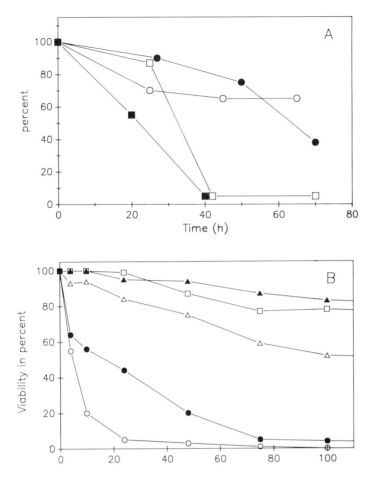

FIGURE 7.   Influence on the viability of cells solubilized by Tween 85 in micellar solutions. (A) *Acinetobacter calcoaceticus* in IPP (□, ○) and in a mixture of IPP/HD 7:2 (■, ●) as a function of the handling method: (●, ○) keeping them standing and (■, □) shaking. (B) Dependence of baker's yeast in IPP on water content. (Wo = [H₂O]/[surfactant]). (○) Wo = 7, (●) Wo = 11, (△) Wo = 14, (□) Wo = 20, and (▲) Wo = 25.

extent as Figure 7B displays. Yeast cells need systems with Wo values above 12 or 13 (more than about 1.5%) to stay alive for several days.

In the first Darszon experiments, as already mentioned, the viability was good in the case of spores, but rather modest in that of entire cells.[11]

The activity of mitochondria solubilized in the system AOT/isooctane/water[12] was measured with the polarographic technique, but not yet directly in the organic solution. Results in Figure 8 compare the time course of oxygen uptake in water and after back-transfer into water from the micellar solution. It is apparent that the respiration capability is very similar in the two cases, which indicates that the solubilization of organelles in the micellar system does not significantly destroy them: the structural integrity should remain satisfactory at least at the level of the internal membrane. It is also apparent by extrapolation of the curve in Figure 8 that the activity remains high for a relatively long time, at least 50% of the initial value after 1 d.

Is it possible to find conditions under which the viability of cells in the organic medium extend to periods of several days or several weeks? Of course, this question is important if one thinks of biotechnological applications.

Experiments with yeast cells first gave a positive answer to this question. For example,

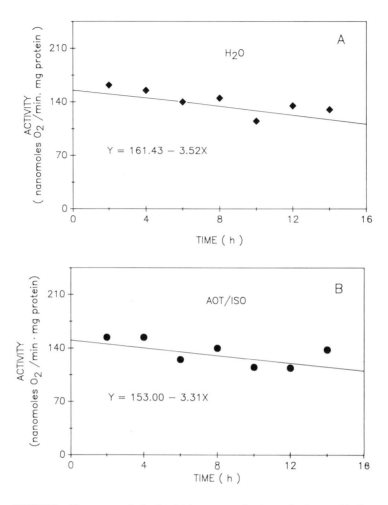

FIGURE 8.   Time course of mitochondrial oxygen uptake observed polarographically
with a Clark electrode: (A) in water and (B) after back-transfer into water from the
micellar solution.

Figure 9A shows that with the system asolectin/IPP, over 20% cell survival can be obtained
after 9 d. Note also that the viability decreases significantly when a minimal medium is
used instead of the complex YPD medium. This suggests that in the organic microemulsion
system, just as in water, cells are sensitive to the kind and amount of nutrient present in
the aqueous component.

In this first yeast paper,[13] the interesting observation was made that the viability can be
considerably improved by working with yeast cells obtained from parent cells which had
already experienced the micellar environment. One can actually define a "second generation", a "third generation", and so on, according to the number of times this operation has
been carried out with a given sample.

Figure 9B shows the results: in all three media tested, the viability of the second
generation was remarkably higher than that of the first generation (i.e., cells which had not
previously experienced the micellar medium).

As already mentioned, it was with yeast cells that the new polarographic method to
determine the activity directly in the organic medium was used.[7] Figure 10 shows how the
respiration of cells in the organic medium is affected by the addition of ethanol and how
the respiratory activity of cells, having decreased to one half after 1 d, can be brought to
the original level or even higher by addition of glucose.

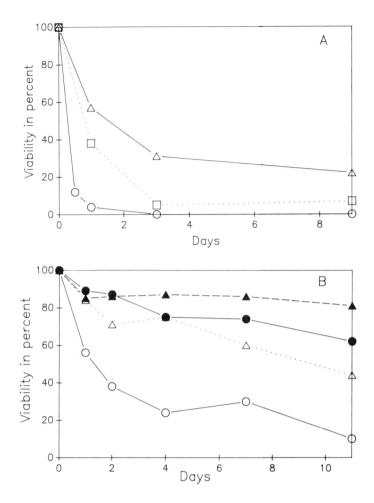

FIGURE 9.   Viability of *S. cerevisiae* solubilized in different w/o microemulsion systems (Tween 10% (78 m*M*), Asolectin 4% (52 m*M*); Wo = 14; room temperature). (A) Viability of the "first generation" in different media: (□) Tween/IPP/YPD; (△) Asolectin/IPP/YPD; and (○) Tween/IPP/minimal medium. (B) Viability of further "generations". (○) Tween/IPP (2nd gen.); (▲) Asolectin/IPP (2nd gen.); (△) Asolectin/hexadecane (2nd gen.); and (●) Tween/IPP (5th gen.)

In the case of enzymes applied to biotechnological reactions, it is known that their activity is generally increased by immobilizing them to an insoluble matrix. Does the same principle work in the case of cells? The answer appears to be positive, as judging from the experiments carried out by Fadnavis and co-workers[16] with yeast cells immobilized on calcium alginate beads. Figure 11 shows the viability of the immobilized yeast cells in an AOT/isooctane micellar suspension.

While the free cells are completely destroyed in a few hours, the immobilized cells are able to survive for several days. The data also appeared to indicate that cells in the micellar system are able to grow.

The authors also made the interesting observation that cell viability is rapidly lost if the positively charged surfactant hexadecyl-trimethyl-ammonium-bromide (CTAB) is used for the preparation of reverse micelles. The authors argue that the immobilized cells are protected in the case of AOT due to electrostatic repulsions between the charged groups of the surfactant and the alginate beads.

Polarographic methods have been used also to monitor the activity of plant cells. Plants have an obvious biotechnological relevance due to photosynthetic activity.

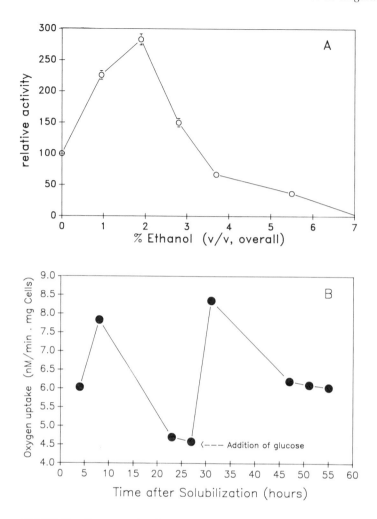

FIGURE 10.   Oxygen uptake of *S. cerevisiae* in 100 m*M* Tween/IPP containing 2.5 mg/ml yeast: (A) influence of ethanol concentration and (B) as a function of the solubilization time.

In the paper dealing with *R. obtusifolius*,[15] conditions giving a normal photosynthetic response as in water could not be found. However, a light-induced consumption of oxygen in the presence of ascorbate could be ascribed to a photosynthetic reaction. Figure 12A shows that the ascorbate-mediated oxygen uptake is a linear function of the amount of cells used per assay (expressed as micrograms of chlorophyll); Figure 12B reports the effect of water concentration in the microemulsion on the light-dependent oxygen uptake in the presence of ascorbate. The fact that the activity is higher at high Wo indicates that the water content in the water pools affects the cellular activity in the ascorbic acid reaction.

V. APPLICATIONS

The biotechnological relevance of cells in reverse micellar solutions is based on the possibility of extending microbiology to organic media. Thus, for example, the microbiological transformation of water-insoluble substances may become possible.

The first successful attempt has been described by Fadnavis and his group in Hyderabad.[17] In this experiment, yeast cells, cross-linked with glutaraldehyde (to minimize the protein extraction by reverse micelles), have been used to catalyze the enantioselective hydrolysis of amino acid derivatives, as shown in Scheme 1.

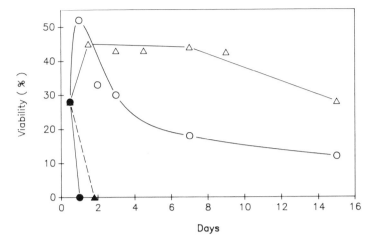

FIGURE 11.   Viability of yeast cells under different environmental conditions (room temperature, no nutrients added): (●) Free cells in reverse micelles of 100 mM AOT/isooctane, Wo = 20 with 50 mM glycine buffer (pH = 7). Immobilized cells suspended in (△) aqueous glycine buffer, (▲) reverse micelles of 100 mM CTAB in chloroform-isooctane (1:1 v/v), and (○) reverse micelles of AOT. (The cells lose more than 70% of their viability by the immobilization treatment; starting point is 28%).

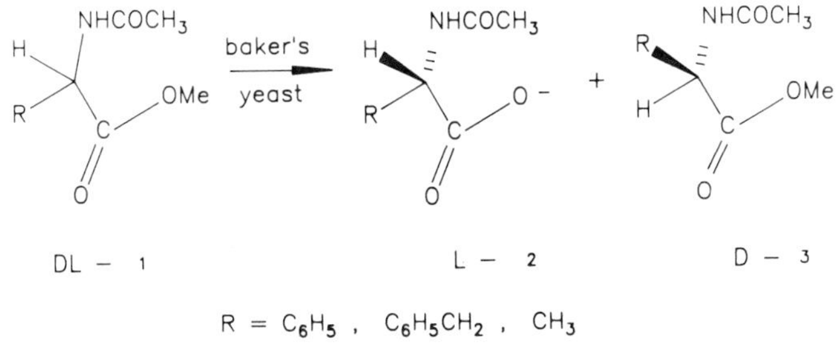

SCHEME 1.

First experiments with a suspension of yeast cells containing CTAB as surfactant in isooctane chloroform and 2% aqueous buffer were not satisfactory; instead, in chloroform isooctane (1:9 v:v) containing AOT as surfactant, things were considerably better: unreacted D-ester (3) and the L-acid (2) were present in two different media (the latter was extracted in water from the micellar solution) and enantioselectivity was as high as 98%.

In a following paper,[16] Fadnavis' group was able to utilize yeast cells for the synthesis of peptide bonds. This time the cells were immobilized on calcium alginate beads. Typical reactions which were successfully carried out by the immobilized yeast cells in the organic medium are shown below:

$$\text{Ac-Phe-OMe} \ + \ \text{H-Leu-NH}_2 \qquad \rightarrow \ \text{Ac-Phe-Leu-NH}_2 ..... \qquad (A)$$
$$\underset{\mathbf{1}}{\phantom{\text{Ac-Phe-OMe}}} \qquad \underset{\mathbf{2}}{\phantom{\text{H-Leu-NH}_2}} \qquad \qquad \underset{\mathbf{3}}{\phantom{\text{Ac-Phe-Leu-NH}_2}}$$

$$\text{Z-Ala-Phe-OMe} \ + \ \text{H-Leu-NH}_2 \qquad \rightarrow \ \text{Z-Ala-Phe-Leu-NH}_2 ..... \qquad (B)$$
$$\underset{\mathbf{4}}{\phantom{\text{Z-Ala-Phe-OMe}}} \qquad \qquad \qquad \underset{\mathbf{5}}{\phantom{\text{Z-Ala-Phe-Leu-NH}_2}}$$

$$\text{Z-Gly-Gly-Phe-OMe} \ + \ \text{H-Leu-NH}_2 \rightarrow \ \text{Z-Gly-Gly-Phe-Leu-NH}_2 ..... \qquad (C)$$
$$\underset{\mathbf{6}}{\phantom{\text{Z-Gly-Gly-Phe-OMe}}} \qquad \qquad \qquad \underset{\mathbf{7}}{\phantom{\text{Z-Gly-Gly-Phe-Leu-NH}_2}}$$

SCHEME 2.

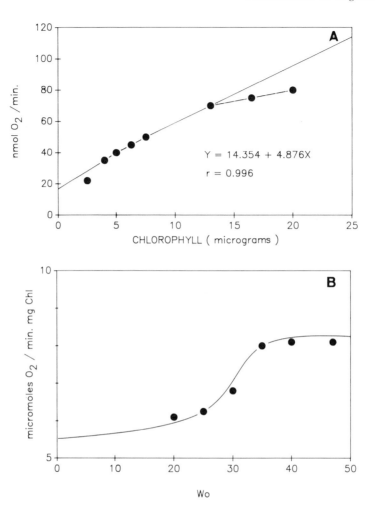

FIGURE 12.   Photosynthetic activity of *Rumex obtusifolius* cells with ascorbic acid (2 m*M*) as substrate. (A) Influence of the amount of cells used per assay (240 m*M* AOT, Wo = 46, pH = 7.5 (10 m*M* HEPES), 25°C). (B) Effect of the Wo; 8.8 μg of chlorophyll for each determination under the same conditions as in (A).

The water-insoluble substrates were dissolved in a micellar system consisting of AOT in isooctane/aqueous buffer (1 to 5% v:v). This system is analogous to peptide synthesis catalyzed by proteases solubilized in reverse micellar solution.[18,19]

Figure 13 taken from this work shows the effect of water content of the micellar medium on the yield of the peptide product for Reactions A, B, and C.

It could be shown that reaction yield for Reaction A is twice as large as in water. Yield decreases because as the amount of water increases, the hydrolysis of the ester becomes important. In this reaction, it is interesting to notice that in the presence of surfactant, lipophilic substrates are able to interact with the enzymes in the cell, i.e., they must overcome both the barriers of the gel matrix and the cell wall.

An elegant organic chemistry application of yeast cells suspended in organic solvent has been offered by Seebach and his group in Zurich.[14] The reaction studied was the reduction of β-keto-esters and ester hydrolysis to yield optically active products. The system used was IPP containing asolectins as surfactants. As already mentioned, in this case yeast cells were suspended in the organic medium without adding water; in this way, about 50 g of β-keto-

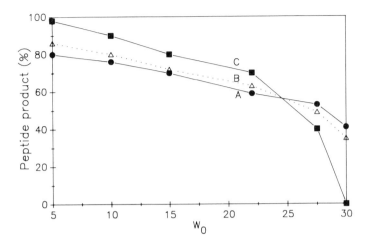

FIGURE 13. Dependence of the water content in the micellar medium on peptide bond formation of reactions A, B, and C. 100 mM AOT/isooctane, pH 10, [ester] = 1 mM, [H-Leu-NH$_2$HCl] = 2 mM (overall).

ester could be easily converted by 500 to 700 g of yeast in 1.5-l volume; and typically 22 g of ester were hydrolyzed by 220 g of yeast in a volume of 0.5 l, thus surpassing the reported procedures in aqueous media. The product alcohols (Scheme 3, 1 to 4) were thus obtained from the corresponding achiral or racemic ketones, and the acetate and octanol (Scheme 3, 5 and 6) were prepared by hydrolysis of the corresponding racemic acetate rac-5.

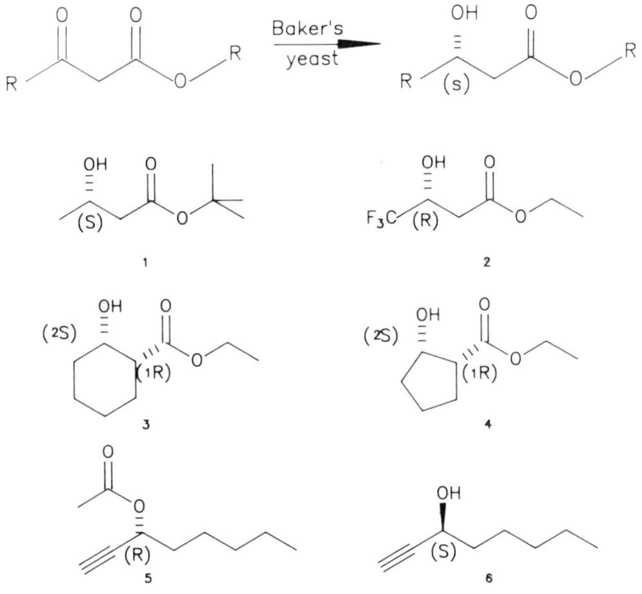

SCHEME 3.

Yields and enantio- and diastereoselectivities of these reactions were generally better than those obtained in water. The restriction is that the procedure is only applicable when the products are sufficiently volatile.

A quite different concept in the application of cells solubilized in organic solvents is the microbiological purification of the solvent itself from undesired compounds. The most

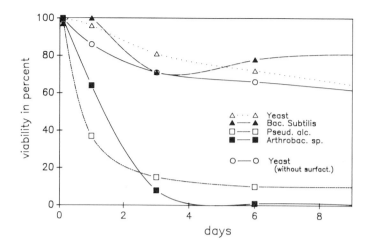

FIGURE 14.   Viability of cells in Asolectin/crude oil (50 mg/ml). Wo = 14.

obvious case is perhaps the degradation of sulfur-containing chemicals present in mineral oil and/or the products of petrol distillation. An attempt to do so using reverse micellar solution was presented by Lee and Yen.[20] They used a biphasic system consisting of water which contains sulfur-oxidizing bacteria (or enzymes) and an organic solution of the oil containing a surfactant. In this case, the cells are not in the micellar solution. Actually, the authors could show in this study that the use of cell-free enzyme extracts of *Thiobacillus ferroxidans* was better than that of the bacteria.

Experiments to desulfurize mineral oil by using microorganisms solubilized directly in the oil with the help of surfactant are in progress in our group. At the beginning of this research, one has to find conditions under which microorganisms are stable in the mineral oil or its corresponding distillation products. Figure 14 shows some typical viability results in this anaqueous medium.

Another possible application is based on the growth of solubilized cells at the expense of the organic solvent itself (as primary energy source). In this case, we would have a transformation of organic solvent (oil) into growth mass of the microorganisms (protein). First preliminary results of such a behavior are presented in Figure 15. A significant increase of the cell concentration after solubilization can be detected.[21]

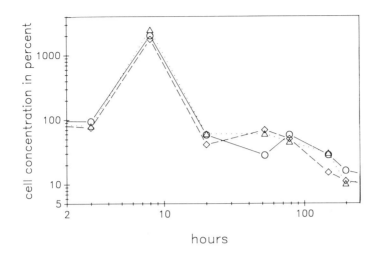

FIGURE 15. Cell growth at high cell concentrations: yeast cells in Tw 85/IPP, Wo = 14, prepared with nutrient medium (three independent trials).

REFERENCES

1. **Luisi, P. L.,** *Angew. Chem. Int. Ed. Engl.,* 24, 439, 1985.
2. **Martinek, K., Levashov, A. V., Klyachko, N., Khmelnitski, Y. L., and Berezin, I. V.,** *Eur. J. Biochem.,* 155, 453, 1986.
3. **Luisi, P. L. and Magid, L. J.,** *CRC Crit. Rev. Biochem.,* 20, 409, 1986.
4. **Luisi, P. L., Giomini, M., Pileni, M. P., and Robinson, B. H.,** *Biochim. Biophys. Acta,* 947, 209, 1988.
5. **Battistel, E., Imre, E. V., and Luisi, P. L.,** Solubilization and structural properties of nucleic acids in reverse micelles, in *Controlled Release of Drugs: Polymers and Aggregate Systems,* Rosoff, M., Ed., VCH Publishers, New York, 1989, 255.
6. **Haering, G., Luisi, P. L., and Meussdoerffer, F.,** *Biochem. Biophys. Res. Commun.,* 127, 911, 1985.
7. **Hochkoeppler, A., Pfammatter, N., and Luisi, P. L.,** *Chimia,* 43, 348, 1989.
8. **Dekker, M., Hilhorst, R., and Laane, G.,** *Anal. Biochem.,* 178, 217, 1989.
9. **Haering, G., Pessina, A., Meussdoerffer, F., Hochkoeppler, A., and Luisi, P. L.,** *Annu. N.Y. Acad. Sci.,* 506, 337, 1987.
10. **Balasubramanian, D. and Kumar, C.,** Water-in-oil microemulsions: structural features and application as biological models, in *Solution Behavior of Surfactants,* Vol. 2, Mittal, K. L. and Fendler, E. J., Eds., Plenum Press, New York, 1982, 1207.
11. **Darszon, A., Escamilla, E., Gómez-Puyou, A., and Tuena de Gómez-Puyou, M.,** *Biochem. Biophys. Res. Commun.,* 151, 1074, 1988.
12. **Hochkoeppler, A. and Luisi, P. L.,** *Biotechnol. Bioeng.,* 33, 1477, 1989.
13. **Pfammatter, N., Guadalupe, A. A., and Luisi, P. L.,** *Biochem. Biophys. Res. Commun.,* 161, 1244, 1989.
14. **Haag, T., Arslan, T., and Seebach, D.,** *Chimia,* 43, 351, 1989.
15. **Hochkoeppler, A. and Luisi, P. L.,** *Biotechnol. Bioeng.,* 37, 918, 1991.
16. **Fadnavis, N. W., Deshpande, A., Chauhan, S., and Bhalerao, U. T.,** *J. Chem. Soc. Chem. Commun.,* 21, 1548, 1990.
17. **Fadnavis, N. W., Prabhakar Reddy, N., and Bhalerao, U. T.,** *J. Org. Chem.,* 54, 3218, 1989.
18. **Lüthi, P. and Luisi, P. L.,** *J. Am. Chem. Soc.,* 106, 7285, 1984.
19. **Pessina, A., Lüthi, P., and Luisi, P. L.,** *Helv. Chim. Acta,* 71, 631, 1988.
20. **Le, K.-I. and Yen, T. F.,** *Am. Chem. Soc. Div. Fuel Chem. Prepr.,* 33, 573, 1988.
21. **Bohler, C.,** diploma work, Swiss Federal Institute of Technology, 1989.

INDEX

A

Acid phosphatase, 48, 177
Adenosine triphosphatase (ATPase), 50, 53, 54, 55, 58
  mitochondrial, 117, 121, 125
  soluble, 117
  of submitochondrial particles, 118
  of thermophilic bacteria, 125
Alcohol dehydrogenase, 59, 94, 106, 107, 145, 192
  in structural studies, 138, 140, 142
Alcohols, 38, see also specific types
Alkaline phosphatase, 48, 50, 58, 59, 94
Amino alcohols, 38, see also specific types
α-Amylase, 121, 136
Antigen-antibody interactions, 107
Apolar compounds, 6–7, 145, 231–235, see also specific types
Arabinose binding protein, 114
Asolectin-toluene, 231
Assays, 137, see also specific types
ATPase, see Adenosine triphosphatase
Azurin, 140

B

Bacteria, 52, 125, 136, 235, 248, see also Microorganisms; specific types
Bacterial respiratory complexes, 233–235
Bacteriorhodopsin, 54
Benzyl alcohol, 220
Bilayer lipid membranes, 212
Biological membranes, 37, 38, 43–44, 51, 231
Bioluminescence assays, 137
Bulk water, 69–72, 82

C

Calcium-adenosine triphosphatase, 117
Calcium alkylarylsulfonate reverse micelles, 204
Calorimetry, 43, 121
Cardiolipin, 43
Catalase, 222
  enzymic titration of oxygen and, 222–223
  polarographic characterization of reaction of, 225–230
  thermostability of, 227–230
Catalysis, 47–50, 115
  activation of, 196–198
  enzyme, see Enzyme catalysis
  hydration-dependent activation of, 196–198
  in reverse micelles, 119–121
  water, 114–115, 119–121
CD (circular dichroism), 137–138, 164
Ceramides, 38, see also specific types
Cholesterol, 37, 220

Cholesterol oxidase, 59
Chymotrypsin, 95, 96, 106, 121, 137, 151, 199
α-Chymotrypsin, 121, 177, 194
Circular dichroism (CD), 137–138, 164
Clathrate, 18
Colloidal particles, 211–212
Concentration, 143–146, 199–200
Crambin, 10, 18
Crystallography, 3, 18
Cyanobacteria, 248
Cytochrome a, 231, 232, 234
Cytochrome b, 231
Cytochrome c, 51, 118, 121, 138, 142, 231
Cytochrome c oxidase, 54, 60
Cytochrome oxidase, 51, 117, 121, 235

D

Denaturants, 125–127, see also specific types
Denaturation, 4, 125–127
1,2-Diacylglycerol, 50
2,6-Dichlorophenol-indophenol, 234
Dielectric relaxation spectroscopy, 115
Dielectric wall, 78–84
Differential scanning calorimetry, 43, 121
Diffraction, 2, 42, 43, 69, 86
  from protein crystals, 10–11
  from single crystals, 7–18
  technical considerations in, 8–10
Diffusion
  coefficients for, 192
  collisions and, 192–196
  of reverse micelles, 146, 167–168
Dilatometry, 43
Dipole-dipole correlation, 73, 81–84
Dipole movement of water, 115
Dipole orientation, 81
DMSO, 20

E

Electron microscopy, 42, 44
Electron paramagnetic resonance (EPR), 94, 105
Electron spin resonance (ESR), 43, 115
Electron transfer in reverse micelles, see under Reverse micelles
Electrophoresis, 53
Enantioselectivity, 107
Energy balance for proteins, 2
Enoate reductase, 147
Enthalpy, 43
Enzymatic conversions in reverse micelles, 134–135
Enzyme catalysis, 114–115, 119–121, 133–157
  activation of, 196–198
  at high temperatures, 123–125
  hydration-dependent activation of, 196–198